THE MOLLUSKS OF THE
ARID SOUTHWEST

THE MOLLUSKS OF
THE ARID SOUTHWEST

With an Arizona Check List

Joseph C. Bequaert

and

Walter B. Miller

THE UNIVERSITY OF ARIZONA PRESS

TUCSON   ARIZONA

About the Authors . . .

Joseph C. Bequaert turned to malacology as his main interest following his retirement as Curator of Insects at the Museum of Comparative Zoology of Harvard, after serving as assistant professor at the Harvard Medical School. An honorary visiting scholar at the University of Arizona from 1959 on, he pursued this interest he had begun first at the University of Houston following his retirement in 1956. Some fifty of his many publications are based on land and fresh-water mollusks. A native of Belgium, he received his degree of Doctor of Natural Sciences at the University of Ghent in 1908, was entomologist and botanist for the Belgian Colonial Office in the Congo, then came to the United States in 1916 and subsequently acquired citizenship. Prior to his affiliation with Harvard, he was Research Associate at the American Museum of Natural History. He travelled extensively in Africa and the Americas, studying the evolution and ecology of the tropical fauna and flora.

Walter B. Miller, whose particular fields of interest lie in the taxonomy, zoogeography, speciation, and evolution of the land snails of Western North America¬particularly the South-Western United States and adjacent Mexican states¬has described numerous new species from these areas. He began collecting and studying land snails as a boy in 1932, and accumulated a large, worldwide collection while serving in the U. S. Navy. He joned the Biological Sciences faculty at the University of Arizona in 1967, from which institution he received his doctoral degree, backgrounded by his master's degree from the California Institute of Technology and his B. S. from the U. S. Naval Academy.

The University of Arizona Press
www.uapress.arizona.edu

Printed in the United States of America
21   20   19   18   17   16   7   6   5   4   3   2

ISBN-13: 978-0-8165-0318-6 (paper)
ISBN-13: 978-0-8165-3516-3 (Century Collection paper)

L. C. No. 72-187825

∞ This paper meets the requirements of ANSI/NISO Z39.48-1992 (Permanence of Paper).

In Memory of
HENRY AUGUSTUS PILSBRY
1862-1957
Pioneer of Southwestern Malacology

CONTENTS

Introduction

When publication of a Check List of the mollusks of Arizona was first contemplated, following a suggestion by Professor Charles H. Lowe, it was agreed that it should not be a mere enumeration of names, which might appeal only to collectors and would hardly justify the effort. Instead, it should integrate and evaluate all available information of possible interest to students of variation, evolution, zoogeography, and ecology of the fauna of the arid Southwest. The Check List proper is therefore mainly the essential background for a detailed discussion of malacological geography. For purely pragmatic reasons, the taxonomic List has been restricted to the State of Arizona, the only area with which we feel sufficiently acquainted from personal experience. However, as more fully explained later, the discussion of zoogeography which introduces the Check List covers a much wider territory, namely the Southwestern Molluscan Province in the arid Southwest of the United States and northern Mexico.

The primary aim of a local list should be to record accurately the taxa (genera, subgenera, species, subspecies, etc) that actually live at present in the area and may therefore be accepted as part of its Recent biota. It is important also to dispose at the proper place of any synonyms or misidentifications that might otherwise result in fictitious additions to the List. Throughout this work the term "Recent" (capitalized) means extant or living at present, usage now customary with most paleontologists. This primary aim may seem a matter of course; yet it is sometimes the most delicate and most critical aspect of the undertaking. Published Arizona records, also to be considered for the Check List, do not always distinguish clearly between living or fresh (presumably Recent) material and dead (possibly fossil) specimens washed up in riparian drift of watercourses and lakes, or blown by wind in caves. For this reason locality records taken from museum labels are not necessarily valid evidence of present-day living occurrence in the area, particularly for species also known there as late Cenozoic fossils.

The selection of the species here accepted as Recent for Arizona is based primarily on our fieldwork during the past decade (1960 to 1972) at many stations and biotopes in most sections of the State, and on new material from other collectors as acknowledged in the text. Most of this study material is now deposited in the Invertebrate Collection of the Department of Biological Sciences at the University of Arizona, Tucson; some of it is also at the Museum of Comparative Zoology of Harvard University, Cambridge, Massachusetts.

Our firsthand information was completed with published records, from sources given in the Bibliography. They were accepted as based on Recent (not fossil) specimens only after critical scrutiny, which explains why some of them are ignored. Nevertheless, a few dubious records have been admitted at face value, our Check List accepting (with a query) the following 11 species or subspecies of which we have not seen thus far newly collected living, local specimens: *Sonorella huachucana elizabethae* (only dead shells found by us at type locality), *S. baboquivariensis berryi* (type locality not traced thus far in the field), *Ashmunella mogollonensis, Succinea luteola, Columella columella alticola, Physa humerosa, Gyraulus circumstriatus, Sphaerium striatinum, Pisidium nitidum, P. punctatum,* and *P. insigne.* Some of these are doubtfully Recent in the State, while others may have become extinct there within the past century. Published records of some of them are vague (viz in *Sphaerium* and *Pisidium*), or may have been based on fossils washed up in riparian drift.

Our Check List includes 173 valid species and 46 recognized subspecies in the Recent mollusk fauna of Arizona, their names being printed in **bold face**. Of the total, 140 species are terrestrial and 33 aquatic; 162 are Gastropoda and 11 Pelecypoda. However, all 173 do not have the same value for zoogeographical purposes, since the *indigenous* (pre-Columbian native) species must be distinguished sharply from the *adventive* (post-Columbian introductions by man). Only the 155 natives (128 terrestrial and 27 aquatic) are consecutively numbered in the Check List; the 18 adventives (12 terrestrial and six aquatic) are starred instead. We omit foreign snails found only accidentally thus far, such as the South American *Pomacea bridgesii* (L. Reeve) and *Marisa cornu-arietis*

(C. von Linné), often kept in aquaria or small ponds and sometimes discarded; the African Giant Snail, *Achatina fulica* T. E. Bowdich, one of which was brought alive from Oahu to Mesa in April 1958, but was destroyed soon after (reported by A. R. Mead, 1959); and *Polygyra texasiana texasiana* (S. Moricand) of Texas, of which a colony lived for a while in an irrigated garden in Tucson early in 1970.[1]

Adventive mollusks are of little or no interest to the zoogeographer. Yet they must be recognized as such, particularly when they are fairly well established or feral, being then more often observed by laymen than native snails, or even mistaken for native species. Of the 18 species listed by us as introduced by man in Arizona, mainly during the 20th Century, 12 came from the Old World, and, having no close relatives now living in America, are easily recognized as foreigners. The remaining six (*Lamellaxis gracilis, Succinea campestris, Vallonia pulchella, Pseudosuccinea columella, Biomphalaria havanensis,* and *Anodonta corpulenta*), however, are native elsewhere in the New World, some of them even in the United States. Their adventive status in Arizona is therefore not self-evident, and even sometimes debatable. It must be based on circumstantial evidence, mainly historical, of the type ably discussed for all introduced North American animals by C. H. Lindroth (1957:135-143). Their consistent absence from all local Pleistocene Sites is especially noteworthy, and is often conclusive evidence in this connection, in spite of being of a negative character.

While occasionally found living and freely reproducing in the open, few adventive mollusks have become truly feral in Arizona, that is, permanently established in perpetuating populations away from human environments. The majority are still strictly synanthropic, depending for survival upon artificial habitats, namely gardens, cultivation, greenhouses, nurseries, irrigation canals, impounded water bodies, etc. The distinction between pre-Columbian indigenous and post-Columbian adventive mollusks is of prime importance not only to the student of present and past distribution; it is even more essential for geologists and archaeologists in comparing Late Cenozoic with Recent faunas in the same general area.

---

[1]The devious way in which mollusks may be carried outside their normal range is sometimes puzzling, as the following case shows. On June 23, 1962, the senior author found at a temporary camp site on Eagle Creek, six miles southwest of Morenci, Greenlee Co., a dead but perfect shell of *Polygyra behrii* (W. M. Gabb) that had dropped the night before from the sleeping bag of a Mexican visitor; presumably it had been picked up inadvertently with his camping gear somewhere in Chihuahua or Sonora, where the species is widespread.

Following the concise system of the Check Lists in *The Verte-brates of Arizona* (C. H. Lowe, ed., 1964), suprageneric classification is limited mainly to families. The major headings "Pulmonata Geophila" and "Pulmonata Hydrophila" have no taxonomic signif-icance, being merely a convenient way of dividing the terrestrial from the aquatic pulmonates.

For easy reference, we follow as a rule the sequence, limits, and nomenclature of families, genera, subgenera, species, and sub-species of H. A. Pilsbry's *Land Mollusca of North America* (1939-1948). The reader should consult this indispensable compendium of the malacologist for references to original descriptions and additional information on bibliography, identification, and distribution. How-ever, we have felt free to depart from it on occasion in matters concerning the local fauna. In particular, the treatment of *Sonorella* is a revision of the genus, based on the junior author's original, mostly unpublished investigations of the past six years (W. B. Mil-ler, 1968a). The senior author is responsible for other taxonomic innovations, especially in *Ashmunella, Holospira, Gastrocopta, Chaenaxis,* and *Vertigo.* The accounts of the Arizona *Eremarionta, Sonorella, Oreohelix, Ashmunella,* and *Holospira* include con-densed listings of the species known from other sections of the Southwestern Molluscan Province.

Unfortunately no comprehensive guide to the freshwater mol-lusks of North America is available at present. Merely for conve-nience, we have followed the sequence of aquatic families and genera of J. Thiele's *Handbuch* (1931-1935); but in this case we have taken more liberties with the names of genera and sub-genera. We have also given more complete bibliographic information for the species, since this is now scattered in many publications, some of them not readily accessible. The taxonomy of freshwater mollusks is at present in near-hopeless confusion. Since no agree-ment on the limits or even the names of some genera and species is in sight, we had to rely on our own judgment and ingenuity; what-ever course we decided to follow is bound to be open to criticism.

Our treatment of species and subspecies gives information on nomenclature, type localities (abbreviated T.L.), distribution in Arizona, occurrence elsewhere in the Southwestern Molluscan Province (with emphasis on Sonora), general Recent distribution, presence or absence in Late Cenozoic deposits (especially in the Southwest), and synonymy. Synonyms are usually restricted to names cited in print for Recent or fossil Arizona records, the type localities then being given only for names originally based on Arizona material, or when they clarify present use of the valid name. Purely taxonomic matter is discussed only if essential for an understanding of zoogeography.

Unless they seemed useful in discussing other matters, we have omitted the present location and collection numbers of types or other specimens—information of little if any biological value and, moreover, easily obtainable from other sources, mostly listed in our Bibliography. On the other hand, we attach considerable importance to type localities (T.L.) and have endeavored to determine their precise present-day location and elevations (sometimes exaggerated in print), so that they may be found on modern maps as well as in the field. Due to lack of trustworthy maps in the early days of mollusk collecting, and to local custom of moving roads, abandoning towns, mines, and ranches, or changing their location or names, some published collecting Stations are now vague or misleading. Elevation often was omitted. Since evolution deals with populations, not with single individuals, modern taxonomy should do the same. Therefore, the study of a population of a species is more important than that of its holotype. A rational start of a population study would seem to be at or near the type locality of the species or subspecies; this method was used by the junior author for his revision of *Sonorella* and proved most fruitful.

Perhaps the most useful feature of the Check List will be the accounts of the horizontal and vertical ranges of the species and subspecies within the State. In order to keep these as concise as possible and convenient for future reference, they are arranged on the basis of the 14 counties, in spite of the drawback that these are of very unequal extent, are often too large for the purpose, and rarely correspond to natural biotic areas. The Check List discloses that the known distribution of many species is limited, sometimes restricted to one or a few counties, or to a few localities. In some cases this may only reflect insufficient collecting, the southern tier of counties obviously being better explored than the others, a defect that may be remedied by future fieldwork. Frequently, however, it is a normal, natural situation; for instance, in *Sonorella, Ashmunella,* and *Chaenaxis,* where the species are often strictly isolated within a narrow set of local ecological biotopes. Because of the previous scarcity of published records of Arizona aquatic snails and clams, their distribution has been treated at greater length than that of most terrestrial mollusks.

For most ubiquitous or widespread species we list only the counties where they are known, sometimes adding a few precise localities of special interest, such as mountain ranges and larger cities (Phoenix, Tucson, etc); more details are given for the others.

As noted before, the distributions of the species are based strictly on Recent (presumably living) specimens, either from new firsthand information, or from fully reliable published records. In order to make this clear, the Recent localities are often followed by the usually more numerous records of dead specimens from riparian **drift debris (alluvia, flood debris, or rejectamenta)** thrown by floods or blown by wind on the banks of permanent or temporary rivers, creeks, washes, lakes, or ponds.

In Arizona, shells from drift are often a mixture, difficult to sort, of Recent specimens and Late Cenozoic fossils. Although useful for other purposes, such as the study of variation, they are wholly unreliable for determining Recent distributions. This is well shown by the eventual fate of some species and subspecies originally described from Arizona drift shells. *Sonorella arizonensis,* based on a drift shell from the Santa Cruz River near Tucson, is strictly speaking unrecognizable, since in *Sonorella* the anatomy is essential for precise specific recognition; we list it as a probable synonym of *S. magdalenensis. Holospira ferrissi sanctaecrucis* and *H. ferrissi fluctivaga* do not seem separable from *H. ferrissi ferrissi: sanctaecrucis,* described on a shell from drift of the Santa Cruz near Tucson, was probably washed down from the Canelo Hills or from the Sonora drainage of the river; *fluctivaga,* based on a drift shell from the San Pedro near Mammoth, might have floated down from the Whetstone Mts, the nearest known locality of *Holospira. Chaenaxis tuba,* described from drift of the San Pedro near Benson, is not known to live in that vicinity; its nearest Recent Stations in the San Pedro drainage may have been in the Mustang Mts or Dragoon Mts.

In several genera the discussion of the Recent species is followed by a listing of species known thus far in Arizona from fossils only. Future collecting may possibly add some of these to the Recent list, but most of them became extinct before or within historic (post-Columbian) times.

The Bibliography includes all publications, known to us, dealing with the Recent and Late Cenozoic (Pliocene and Pleistocene) mollusks in the area covered by our Southwestern Molluscan Province, within the boundaries shown in the section on Zoogeography and the accompanying map (Fig. 1). It lists also several works essential for an understanding of the topography, climate, and vegetation of the area, hence essential also for future studies of the ecology of the local mollusks.

Acknowledgments

The present work owes its inception and progress to our close association of several years with the Department of Biological Sciences of the University of Arizona. We are most grateful to the authorities of the University and to the staff of the Department for the many facilities provided, and for their continued interest in our undertaking. To Professor Charles H. Lowe is due the initial incentive to publish our observations, as well as encouragement and constant aid during the preparation of the manuscript. We are also under particular obligation to Professors Albert R. Mead, Joseph T. Bagnara, and Newell A. Younggren for essential assistance. Final completion of the work was assured by financial support from the Roy Chapman Andrews Research Fund of the Arizona-Sonora Desert Museum, sponsored by the Director, Mr. William H. Woodin. The maps were drawn for publication by Dr. James L. Patton, University of California, Berkeley, from original sketches by the senior author.

We were fortunate to obtain help from many friends and acquaintances here listed in alphabetical order, who either joined us in field work, often providing needed transportation, or contributed specimens from a variety of sources often not otherwise accessible. Some of our most valuable information, including the discovery of previously unknown species, resulted from their cooperation. C. F. Adams, J. Ashwanden, K. K. Asplund, J. Ayres, J. T. Bagnara, W. H. Balgemann, G. C. C. Bateman, Mr. and Mrs. Beal, J. E. Beard, J. Beatty, Frank Bequaert, R. L. Bezy, A. R. Brady, P. J. Brady, B. A. Branson, H. E. Broadbooks, J. F. Burger, G. D. Butler, Dorothea Caskey, R. Clark, H. T. Coss, R. Countryman, R. D. Cross, T. L. Cullison, R. J. Drake, J. Edwards, E. H. Erickson, R. C. Ewell, R. S. Felger, Dorothea Franzen, H. C. Fritts, R. L. Gardner, Mr. and Mrs. Gervish, L. H. Gilbertson, S. R. Goldberg, D. Greenfeld, Mr. Greenleaf, W. O. Gregg, R. E. Ground, A. Gubanich, N. G. Guse, E. W. Haury, W. Heath, J. R. Hensley, J. R. Hershey, Mary R. Hestand, B. T. Hinton, Edith Hipple, N. Hodgkin, R. R. Humphrey, R. Hungerford, C. G. Jackson, Mrs. Ardith B. Johnsen, P. H. Johnson, M. Kartchner, B. C. Kell, R. Krizman, J. Landye, M. L. Lindsey, Mrs. Charles (Judith) Lockwood, Barbara Lund, D. Marchant, C. J. May, J. E. May, W. J. McConnell, A. R. Mead, A. L. Metcalf, C. E. Mickel, Nixon Miller, P. F. Min, G. O. Mitchell, L. Moore, N. J. Nerney, Martha Noller, W. L. Nutting, Dr. and Mrs. R. H. Painter, H. L. Parent, J. L. Patton,

A. Phillips, P. E. Pickens, R. Polhemus, R. L. Reeder, J. Reedy, R. C. A. Rice, D. Richman, J. Riddick, M. D. Robinson, K. Roever, J. N. Roney, A. Ross, V. D. Roth, R. H. Russell, P. T. Santana, D. B. Sayner, Mary Schlentz (Mrs. J. C. Truett), W. C. Sherbrooke, N. M. Simmons, A. G. Smith, O. Soule, L. K. Sowls, O. Spitzer, C. B. Stambough, Jr., J. M. Stephenson, D. W. Taylor, W. Thornton, Mr. Treman, D. M. Tuttle, T. R. Van Devender, P. E. Violette, M. L. Walton, M. Wargo, F. G. Werner (and family), C. A. Westerfelt, Jr., and B. Wright.

We are under special obligation to the Department of Geology and Geochronology of the University of Arizona, and in particular to Dr. C. V. Haynes, Jr., Dr. E. H. Lindsay, Dr. P. S. Martin, Dr. P. J. Mehringer, Jr., and Dr. T. L. Smiley, for assistance and continued interest in our project. Their effective cooperation has enabled the senior author to obtain firsthand correctly dated study material of the rich Late Pleistocene mollusk fauna of the upper San Pedro Valley in Cochise County. These fossils, repeatedly mentioned in our Check List, are from a succession of deposits at Lehner Mammoth Site (1.5 miles S of Hereford) and Murray Springs Site (2 miles W of Lewis Springs). The stratigraphy and dating at these Sites have been described by C. V. Haynes, Jr. (1968; 1969:711; 1970:80-82, fig. 3), by C. V. Haynes, Jr. and E. T. Hemmings (1968), and by E. T. Hemmings (1970). Our most valuable material is from the Murray Springs Site in Haynes' Units F_1 and F_2 of the Lehner Formation (Valderon Substage of Wisconsinan), where 40 species of mollusks have been recognized, associated with remains of mammoth and other large mammals now extinct in Arizona, and with Clovis-type artifacts of Early Man. The mammal bones of these Units are about 10,000 to 11,000 years old (before present, B. P.) by radiocarbon dating, thus giving positive proof that species found fossil in situ and also living in Arizona today are true pre-Columbian natives in the State.

PART I

ZOOGEOGRAPHY OF SOUTHWESTERN NEARCTIC MOLLUSKS

I. Zoogeography of Southwestern Nearctic Mollusks

Arizona is predominantly an arid land, in spite of a variety of natural habitats ranging from desert or near-desert (with cacti, agaves, ocotillo, mesquite, yuccas, creosote bush, various thorny shrubs, etc) at low elevations, dry grassland (with or without trees) at moderate elevations, and mesophytic woods (often with lush undergrowth) in the lower mountain canyons, to dense, tall coniferous forest at high elevations, and exceptionally to alpine tundra above timberline. So much of the terrain is level, rocky, sandy, or hard loam, with scant vegetation, that the landscape appears unpromising to the malacologist. Indeed, suitable molluscan Stations are few and far between, often located where least expected.

Gathering material adequate for a fair understanding of the fauna is tedious, time-consuming, and often disappointing. The total of 155 Recent (living) species which we recognize as indigenous in Arizona is, therefore, unexpectedly high. Future collecting will add little to it, except perhaps for some undiscovered *Sonorella*. Further fieldwork will be needed to complete the Recent distribution of certain species in the State, in order to integrate better the Recent fauna with that of the local Late Cenozoic. The information now available seems adequate, however, to attempt an evaluation of the distinctive features of the molluscan fauna and their evolutionary significance.

Our discussions in this section will be restricted to the indigenous (truly native) species and consider only the land snails and slugs, which are the majority (128 land against 27 freshwater species in Arizona). They have been better investigated in the arid Southwest than the aquatic mollusks, most of which were known

thus far from few localities. Moreover, the current confused taxonomy of the aquatic species, due in part to their great intraspecific
variation, makes a worthwhile study of their Recent and past distribution hazardous, if not impossible. It may be noted, nevertheless, that most local aquatic snails and clams are more widespread
than are the majority of land snails, very few of them being restricted to the arid Southwest.

The distribution of mollusks, like that of other animals and of
plants, may be studied either from a purely geographic or from an
ecological viewpoint. Both approaches complement each other and,
in fact, are often difficult to separate in discussions. They are
equally important for a full understanding of the natural history
of mollusks.

In the present work, the Zoogeography section is essentially
spatial; that is, it emphasizes the present-day (Recent) and past (fossil) horizontal and vertical ranges of the native genera and species
now living in the arid Southwest. Ecological considerations will be
incidental, when needed to clarify certain aspects of spatial
distribution, and will be restricted mainly to mentioning Life-
Zones. Perhaps it might be more logical or to the point to call the
spatial approach *historical zoogeography*. Present-day distributions
are rooted in the past, being the final result of the earlier location,
radiating evolution, adaptations, dispersal, and migrations of the
ancestors of the Recent genera and species.

Although disparaged by some latter-day ecologists, C. H.
Merriam's (1890) Life-Zones are still today the most serviceable
tool for the study of ecological problems in arid western North
America. In his masterly detailed analysis of Arizona biota, C. H.
Lowe (1964:15-83) pointed out that the Merriam system, in spite
of its shortcomings, is actually based on the observable ecologic
distribution of animals and plants, and primarily on that of the
dominant perennial vegetation (trees, shrubs, or succulents, etc).
For local naturalists it has the added advantage that it was proposed originally for a most interesting and easily accessible section
of the State (San Francisco Mtn). We have found it particularly
appropriate in dealing with the land snails of Arizona, a State with
peculiar topography of abruptly rising, large and small mountain
Ranges, separated by extensive near-level valleys.

Ecology being currently *the* popular theme, it may be a sur-
prise that we do not intend to discuss it formally for the mollusks
of the arid Southwest. It will be appropriate to explain our views
in the matter, and the reasons for our apparent neglect. Our dis-
quisitions will necessarily focus on the strictly native (not intro-

duced) land and freshwater mollusks, although their ecology is basically similar to that of other animals and of plants. Stripped of conventional technical terminology, ecology is an inquiry into the environment as it affects the basic needs of food, shelter, and reproduction for the survival of the individual and the species (population). To be significant, as well as relevant to evolution, geographical distribution, and paleoecology, it should cover all pertinent environmental factors. It should also be on a specific basis, since even closely related species (in one genus) may have diverse survival requirements. So that a thorough taxonomic knowledge of the local mollusk fauna is an essential prerequisite for ecological studies. Moreover, the inquiry must be based on observations in the field, at the natural habitats (biotopes) of the mollusks. Ecology is primarily the domain of the field naturalist, not of the laboratory biologist; experiments may sometimes illuminate observations in the field, but caution must be used in extending their conclusions to natural conditions.

A satisfactory coverage of ecology requires precise, detailed observations of a wide range of external factors. It must consider the physical and mineral nature of the terrain (topography; exposure; elevation; edaphic or soil features; pH measure of acidity or alkalinity, particularly important for freshwater mollusks; etc), the vegetation (dense, open or scant; dominant species and growth forms; mosses, fungi, and lichens; etc), the soil cover (herbs, grasses, litter, mulch, etc), the climate (general, and microclimate of biotope; diurnal and seasonal variations and extremes of moisture and temperature; etc) and relations with other organisms (associations, consociations, and competition for food and shelter with other mollusks and invertebrates; predators, parasites, and diseases, the potent factors in preventing overpopulation). In addition, for aquatic forms, the type of water (permanent or temporary; stagnant or flowing; pure or polluted; open or with vegetation; etc) is important. It may be noted incidentally that the arid Southwest lacks the predacious snails known elsewhere in North America: the Oleacinidae (*Euglandina*) of Texas and the southeastern States, and the Haplotrematidae (*Haplotrema*) of the Pacific Northwest. All observations should be noted and recorded in sufficient detail, so that they may be repeated and eventual changes at the same locale be assessed correctly. Field observations that cannot be repeated or verified are of little value, which, moreover, is true also of experiments.

An exhaustive study of all factors that relate to survival in natural habitats may reveal how mollusks fit in the general biome of the arid Southwest, of which they are an integral part. It is also essential for a correct understanding of continuing evolution, since new forms arising by genetic change can survive only when ecologically compatible. A clear insight into the relative importance for natural selection and survival of the structural, physiological, and behavioral adaptations of mollusks is therefore vital. This opens up another vast field of research, which it would be premature and pointless to discuss until an adequate body of observed facts is available.

The foregoing outline of what is actually needed for a correct evaluation of molluscan ecology shows that it will be a project of some magnitude, unfortunately devoid of the glamor of some other pursuits in biology. It will be a task for a well-informed and patient investigator, willing to devote several years to it. Like other ecological studies, it is, however, of pressing urgency, before exploitation of natural resources by the steadily growing human population destroys more of the original environment and biota, replacing them with the artifacts of civilization. We regret that we had neither the time nor the opportunity to contribute more than incidental notes on the salient features of the biotopes of some species. The lack of personal observations cannot be remedied from outside sources, since there is at present almost no published ecological information on the local mollusks, beyond general statements not substantiated by detailed, specific field observations.

The only exception is a brief paper by L. H. Gilbertson (1969; excerpts from his unpublished dissertation of 1965), relating his observations at the natural habitat of *Sonorella odorata* in the Santa Catalina Mts. It deals with the natural diet, mating, oviposition, hibernation, estivation, and association with lower fungi of the snail. It should be noted that the biotope of this species is unusual in the genus, as it lives in the Canadian Life-Zone, in a mesic, densely wooded, temperate ambience at high elevation (8,000 feet), with mild temperature and adequate moisture. Most other *Sonorella* prefer the drier and hotter Upper or Lower Sonoran Life-Zones, sometimes in extreme arid biotopes, so that their ecology differs radically from that of *odorata*.

As usual for political boundaries, those of Arizona are artificial and should be disregarded by the naturalist. They are particularly meaningless in studying the processes that underlie the patterns of Recent and past distributions of organisms. Although our Check List is restricted to Arizona, a comprehensive and meaningful analysis of the mollusks of the State must cover a larger area of the Nearctic Region, namely the arid and semiarid parts of North

America here combined into a **Southwestern Molluscan Province,** occasionally referred to for short as the "arid Southwest."

W. G. Binney (1885:18-25) first recognized clearly that, on the basis of distinctive molluscan faunas, the Nearctic Region may be divided into an *Eastern* and a *Western Division,* each sub-divided into discrete faunistic *Provinces.* This system was refined by H. A. Pilsbry (1948:XL-XLVII, map fig. 1) and others, but especially for the Western Division by J. Henderson (1928, 1931). The Divisions and Provinces are seldom encompassed by prominent physical barriers, such as mountain Ranges or rivers, so that their boundaries can only be indefinite, broad transitional belts. The line between the two primary Divisions runs roughly north to south from the Arctic Ocean to the Gulf of Mexico, near the 104th Meridian, the Eastern Division being more extensive than the Western. J. Henderson recognized four Molluscan Provinces in the Western Division, and these are here adopted, though with revised limits.

The *Oregon-Washington Province,* which does not border on our territory, is merely mentioned. Our *California Province,* more restricted than that of Henderson (1928, 1931) and Pilsbry (1948: XLI), covers only the Pacific Coast area, varying in width from some 250 miles in the north (where it reaches Nevada) to about 80 miles in the south (west of the Sierra Nevada, where it borders on the Southwestern Province). It is characterized by several mostly precinctive genera of snails (*Monadenia, Helminthoglypta, Micrarionta,* sensu stricto, *Polygyroidea, Ammonitella, Glyptostoma, Pristiloma,* and *Trilobopsis*) and slugs (*Ariolimax, Hesperarion,* and *Anadenulus*), as well as by the absence of several peculiar Southwestern genera to be discussed later. The *Rocky Mountain Province,* north of Arizona, includes Montana, Idaho, Wyoming, most of Nevada, Utah, Colorado, and northeast New Mexico in the United States, and enters Alberta and British Columbia in Canada. Its distinguishing features are the predominance of *Oreohelix* and the lack of some of the precinctive mollusks of the Southwestern Province, such as *Sonorella, Ashmunella, Holospira,* and *Chaenaxis.*

As shown on our map (Fig. 1), the *Southwestern Molluscan Province* is here enlarged from Henderson's original extent to cover southeast California (Mohave and Colorado Deserts, south of 37° N, west of 114° 30′ W), a small south corner of Nevada (south of 37° N), all of Arizona (south of 37° N), most of New Mexico (south of 37° N, west of ca 104° W), western trans-Pecos Texas (west of 103° W), a small northeast corner and coastal strip of north Baja California (north of 28° N and east of 118° W), north

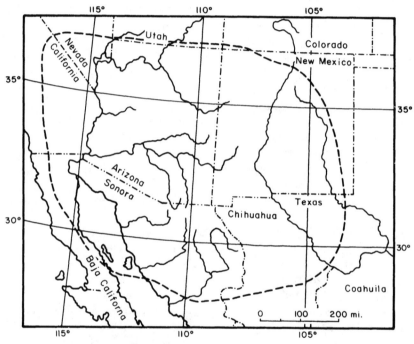

Fig. 1.   Boundaries of the Nearctic Southwestern Molluscan
Province used in this Paper.

and central Sonora (south to 28° N, ca four and a half degrees north of Tropic of Cancer), and north Chihuahua (south to 28° N); a total area of some 500,000 square miles.[2] Binney's (1885) earlier "Central Province" covered most of Henderson's Rocky Mountain and Southwestern Provinces. In 1885 the peculiar Southwestern mollusks were poorly known and, in particular, the distinctive *Eremarionta, Sonorella, Oreohelix,* and *Ashmunella* had not yet been recognized as discrete genera.

The western and northern limits of the Province need no further comment, but the eastern and southern boundaries are more critical and complex. They are to some extent arbitrary, as they are not clearly defined either by physical barriers or by obvious ecological features. They are essentially zoogeographic, that is, based primarily on the presence of certain dominant genera and the absence of others, which produce marked and often rapid changes in the overall composition of the Recent molluscan faunas.

In the eastern section (New Mexico and Texas), between 30° N and 36° N, the Southwestern Province merges fairly gradually with the Eastern Molluscan Division over a transitional strip of generally level, moderately high country with a depauperate molluscan fauna. This limiting strip lacks the distinctive Southwestern genera (*Sonorella, Ashmunella, Oreohelix, Chaenaxis*), which are eventually replaced by typical southeastern elements (*Bulimulus, Polygyra, Helicina,* etc). The fauna is particularly poor in the northern half of the strip, the Llano Estacado of eastern New Mexico and the Texas Panhandle, a southern extension of Henderson's "High Plains Molluscan Province," of Nebraska, eastern Colorado, Kansas, and Oklahoma (covering about the area of L. R. Dice's [1943:26] "Kansan Biotic Province").[3] South of the Llano Estacado and east of the Pecos River (at the west end of the Comanchian Biotic Province of Dice [1943:28]; the Balconian Biotic Province of W. F. Blair [1950:112]), the transitional strip not only lacks the distinctive southwestern genera men-

---

[2]Trans-Pecos Texas includes here El Paso, Hudspeth, Culberson, Jeff Davis, Presidio, and Reeves counties.

[3]The Llano Estacado of Texas and New Mexico lacks most of the southeastern genera and species of central Texas. Except for *Succinea avara, Hawaiia minuscula, Gastrocopta pellucida,* and *Pupoides albilabris,* the 15 species of land snails reported from Palo Duro Canyon, Randall Co., Texas, by Henderson (1908, *Nautilus* 22:9) and W. T. Clarke (1938, *Nautilus* 52:14-15), were dead shells from drift, possibly washed-up fossils, the area having many Pleistocene Sites rich in mollusks. According to Dr. A. L. Metcalf (in litt., 1970), north of the Canadian River, the Panhandle, both in Texas and Oklahoma, has a richer Recent molluscan fauna than its southern section.

tioned above, but in addition harbors many of the widespread mollusks of central Texas (*Polygyra, Microceramus, Euglandina, Helicina, Retinella roemeri, Bulimulus alternatus,* subspecies of *B. dealbatus,* etc). Of these only the two *Bulimulus* and two species of *Polygyra* occur slightly west of the Pecos River, but do not extend west of New Mexico. Our eastern boundary of the Southwestern Molluscan Province thus agrees on the whole with the eastern limits of Dice's (1943) Chihuahuan Biotic Province.

In continental Mexico, the southeasternmost extension of the Southwestern Province merges gradually with the Mexican Plateau Province, the latter mostly an arid, temperate highland ("Tierra Templada") with little relief at fairly high elevations, between the two Cordilleras (Sierras Madre del Norte and del Sur). The transition from the Nearctic to the Neotropical fauna is so gradual and, moreover, it differs so much for every main group of animals that zoogeographers have been at a loss to draw even an approximate, generally acceptable dividing line. As a possible solution to the problem for vertebrates, P. J. Darlington (1957, *Zoogeography:* 456-462, with map, fig. 54) suggested a "Central American-Mexican Transition Area," with a complex, broadly overlapping zone in which the dominance of Nearctic or Neotropical elements varies for the several Classes. The heavy dotted line on his map, marking the divide he suggested for vertebrates between the two Regions, runs near the south end of the Mexican Plateau, west of the Isthmus of Tehuantepec, and about seven degrees south of the Tropic of Cancer. Such an extreme southward extension of the Nearctic Region is hardly tenable for land mollusks. The molluscan fauna of most of Mexico, south of Sonora and Chihuahua, is essentially that of Central America (south to Costa Rica), and predominantly Neotropical, not Nearctic. This is well shown by the mollusks of the transition zone at the southern boundary of the Southwestern Province, a boundary which we have found both logical and convenient to draw at about the 28th Parallel, just south of the range of *Sonorella.* Most of the other dominant or peculiarly Southwestern genera also are now unknown south of this line, where, on the other hand, several typical Mexican and Central American, mostly Neotropical genera appear, such as *Humboldtiana, Polygyra, Euglandina, Drymaeus, Orthalicus,* and *Helicina.* Except for *Humboldtiana* and *Polygyra,* which enter the mountains of Chihuahua and trans-Pecos Texas, these genera are unknown in the Southwestern Province.

On the Lower California peninsula the Southwestern Molluscan Province occupies a small northeastern area of Baja California, north of 28° N, thus including the Vizcaino Desert. The rich molluscan fauna of Baja California Territorio Sur (Arid Tropical Life-Zone of E. W. Nelson, 1922:121, map, Pl. 12; San Lucan Biotic Province of Dice, 1943:39) contrasts sharply with that of the peninsula north of the 28th Parallel. It has many peculiar species of mostly precinctive genera, such as *Berendtia, Pupilla* subgenus *Striopupilla, Bulimulus* subgenus *Leptobyrsus, B.* subgenus *Plicocolumna, B.* subgenus *Puritanina,* and *Coelocentrum* subgenus *Spartocentrum.* Evidently it has been a distinct, active, presumably isolated center of evolution for a long time past. It also harbors several lesser land snails (*Thysanophora hornii, Retinella indentata, Hawaiia minuscula, Gastrocopta pellucida, G. dalliana, G. procera, Pupoides albilabris*) not known at present from arid northeastern Baja California, although some of these may be modern accidental introductions by man.

The State boundaries of Arizona being arbitrarily drawn straight lines on the north, east, and south, it may be expected that the typical Southwestern genera mentioned before are not restricted to the State, although *Sonorella* and *Chaenaxis* may have their centers of density and presumably of evolution there. As Pilsbry (1939:227) recognized, the Colorado River, the only natural obstacle near the western boundary of the Province, has not been a fully impassable barrier for land snails; and this is true also of the Rio Grande at the easternmost extension of the area. The deserts of southeast California, Yuma County, and northwest Sonora, as well as the arid trans-Pecos corner of Texas, have been far more effective in this respect. Nevertheless, the Colorado River may have helped check the eastward advance of the distinctive California genera *Helminthoglypta* and *Sonorelix,* while preventing the extension farther west of certain genera now widespread in Arizona and Sonora (*Sonorella, Chaenaxis, Helicodiscus, Thysanophora, Retinella, Vallonia,* etc), even though adverse climatic factors may have been more powerful deterrents.

The present account of zoogeography is a detailed, critical survey of what is known of the Recent distribution and Late Cenozoic history of the strictly native (pre-Columbian) land mollusks of the Southwestern Province. It is so elaborate that an introductory summary of the main features of the molluscan fauna of the area may be helpful. The most striking features of the Southwestern fauna are its complexity, in spite of a limited number of genera and species, and the unusually high proportion of precinctive, strictly peculiar taxa.[4] The mollusks of the arid Southwest are a heterogeneous assemblage of disparate genera and species, with a variety of patterns of distribution and diversified origins. While such a combination is by no means exceptional in zoogeography, it is nevertheless surprising to find it for a rather small area under ecological conditions basically adverse to molluscan life.

For the orderly and rational discussion of generic and specific distributions we have adopted the following grouping of the native local land mollusks based on their special dominant features.

1. **Precinctive** genera and species, not living now beyond the limits defined before of the Southwestern Province and shown on the map (Fig. 1). They are the overwhelming majority, since 192 (82 pct) of the 234 native species are restricted to the area, where most of them are also autochthonous. In addition, of the 32 genera here recognized for the Province, six (with 130 species) are also precinctive.

2. Essentially **Neotropical** (mainly Mexican) genera and species are the next important element. Of the total 234 species, 39 (16.6 pct) in six genera, have their center of Recent distribution and presumably of past evolution south of the Tropic of Cancer, and later moved northward by natural means to the Southwestern Province in Sonora and south Arizona.

3. A small intrusive **Rocky Mountain** element comprises ten species (4.2 pct of the total 234) in six genera. They appear to be relicts of Late Cenozoic faunas that entered the Province when forced to migrate south by the advancing northern glaciers of successive Ice Ages.

4. A **Circumpolar** element includes eight species (3.4 pct of the total 234) in eight genera, that migrated from northeast Eurasia during the Late Cenozoic, via temporary land connections across the present Bering Strait. Probably few of the Eurasian immigrants that originally reached the present Southwest survive there today. They either retained all their Old World characters or developed only slight differences.

5. A **Pan-American** element is barely represented by five species (2.1 pct of the total 234) that are Recent now over much of the New World, both north and south of the Tropic of Cancer.

---

[4]We prefer the term "precinctive" to "endemic" for taxa restricted to a territory with definite boundaries (a precinct). The term "endemic" is ambiguous, being used also for taxa that are native in an area but not restricted to it. We use the term "autochthonous" for taxa that originated or evolved within the present boundaries of the Southwestern Province.

It will be difficult to unravel their evolutionary history—where they originated and how they spread to their present extensive territories. Three of the five Southwestern species of this group are known as Late Pleistocene fossils, so that these at any rate are ancient natives in the area.

6. A **Californian** element of two species (0.9 pct of the total 234) in one genus seemingly came to the Southwestern Province from the California Molluscan Province of the Pacific Coast, but it is almost negligible. Evidently the deserts of southern California are at present a nearly impassable barrier for mollusks, and probably have been so at times during much of the Late Cenozoic, if not earlier.

7. A **Nearctic** group of 12 widespread North American species (5.1 pct of the total 234) in seven genera probably originated in various areas of the continent, though mostly east of the 100th Meridian. They may have invaded the Southwest independently and at different times during the Cenozoic. The fossil record shows that eight of them were native in Arizona by the Late Pleistocene; the remaining four may yet be found fossil there also.

The reader should be reminded that our discussions are based on present knowledge of Recent and past distributions, from published data and personal observations. This information appears to be sufficiently representative for the Recent fauna of Arizona and other Nearctic areas mentioned by us. The same cannot be said, however, of the Late Cenozoic mollusks of the State, where few fossil Sites have been fully studied by competent malacologists. We are even more poorly informed about the Recent land mollusks of Mexico, particularly in the vast central and northwestern areas which are of paramount importance for a correct evaluation of the fauna of Sonora and Arizona. Little is known there of the lesser species, often difficult to find, and, moreover, unattractive as "collectors' items"; field naturalists being apt to neglect them for the showier and financially more rewarding larger snails. This is evident from the two basic works on Mexican mollusks by P. Fischer and H. Crosse (1870-1894) and by E. von Martens (1890-1901). While our account of the distribution in Mexico of the larger snails (*Eremarionta, Sonorella, Oreohelix, Humboldtiana, Holospira*) may be fairly acceptable, it will no doubt have to be corrected for the lesser species. In particular, some species now regarded as precinctive in the Southwestern Province may yet be found living in Mexico south of Sonora and may have to be transferred to the Neotropical group.

Due to its privileged location near the center of the Southwestern Province, at the crossroads of migrations from several directions, and to its past and present diversified climate and topography, which favored speciation by isolation, Arizona has a richer land mollusk fauna than its neighbors. Mainly due to the many precinctive species of *Sonorella,* the total of 128 Recent native terrestrial species for the State (with 113,956 square miles) tops that of the southeast deserts of California (ca 30 species), New Mexico (86 species, for 121,666 square miles), trans-Pecos Texas (52 species), north Chihuahua (43 species), north and central Sonora (37 species), and northeast Baja California (6 to 10 species).[5] Of the 128 native Arizona species, 88 (68.2 pct) are precinctive in the Southwestern Province. The 40 (31.8 pct) non-precinctives, minus four Succineidae (2.8 pct), may be divided as follows according to their presumed origin: three (2.3 pct) Neotropical, nine (7.2 pct) Rocky Mountain; eight (6.2 pct) Circumpolar, five (4.0 pct) Pan-American, two (1.5 pct) Californian, and 10 (7.8 pct) Nearctic. Seven genera known to be native elsewhere in the Province have not been found in Arizona: *Sonorelix, Mohavelix,* and *Helminthoglypta* of the California deserts; *Humboldtiana* and *Pseudosubulina* of trans-Pecos Texas and Chihuahua; and *Pupisoma* and *Cecilioides* of Sonora.

Distribution of *Eremarionta* and Related Genera

Sonorella, Eremarionta (here raised to generic rank, as explained in the Check List), *Sonorelix, Mohavelix,* and *Helminthoglypta,* five closely related Helminthoglyptidae of the arid Southwest, share many peculiarities in structure. There can be little doubt that they are offshoots of the same ancestral stock, a helminthoglyptid with fully developed accessory reproductive structures. Generic differentiation has resulted from varying degrees of simplification of these structures. In all five genera, the shells are similar, depressed-helicoid in shape, thin in texture, with a thin and slightly reflexed peristome, normally without apertural teeth, superficially sculptured, and usually pale-colored with a single dark peripheral band. They differ essentially in anatomical details of genitalia, and it is often impossible to place the species in the correct genus without dissection. Except for the monotypic *Mohavelix,* they are prolific in species and in often ill-defined sub-

---

[5]The number of species here given for the several political divisions of the Southwestern Province can only be approximate and will no doubt be changed by future investigations. The totals for Chihuahua and Sonora will certainly be increased, especially for the lesser species, of which only 22 are known at present from north Chihuahua and 21 from north and central Sonora.

species, and appear still to be speciating actively at present. As a rule, the species also cannot be recognized positively by shells alone, dissection of genitalia being needed for precise identification. Deficiency in specific and generic shell characters, combined with significant anatomical differences, as observed in these southwestern snails, is exceptional for land mollusks, which can usually be referred to genera and species on shell characters alone (S. S. Berry, 1947b and 1953).

Eremarionta presumably had its center of dispersal in southeast California (map, Fig. 2) where it is now represented by nine described species, after removing some included by Pilsbry in 1939, but segregated in *Sonorelix* by S. S. Berry (1943:8). It is evidently now in active evolution, to judge from the several described subspecies or incipient species listed in the Check List. *Eremarionta rowelli*, the only species known east of the Colorado River, has several poorly defined subspecies, six or seven of them in the Arizona-Sonora sector of the Province. In north Mohave County it extends southeast of the Colorado River to ca 36° N, 114° W, the northeasternmost Station of the genus, but in south Mohave County and north Yuma County it extends only some 10 miles east of the river to 114° 30′ W. It occurs much farther east near and south of the Mexican border: in south Yuma County to the Sierra Pinta (113° 50′ W, ca 80 miles from the river); in Sonora to near Puerto Libertad (112° 30′ W) and on Isla de San Esteban (112° 35′ W), respectively 200 and 270 miles southeast of the Colorado delta. The ranges of *Eremarionta* and *Sonorella* (map, Fig. 2) are not known at present to overlap, although they are sometimes very close. In Mohave County, recently discovered *Eremarionta* and *Sonorella* live in the Rampart Cave area in separate rock slides less than one mile apart; farther south, the Goldroad Station of *Sonorella coloradoensis mohaveana* is only about 25 miles northeast of the Topock Station of *E. rowelli acus*. The Agua Dulce Mts Station of *Sonorella meadi* is about 25 miles southeast of the Sierra Pinta Station of *E. r. rowelli*. In Sonora the Magdalena Stations of *Sonorella magdalenensis* are some 100 miles east of a line drawn through the Sierra San Francisco and Puerto Libertad Stations of *E. rowelli mexicana*.

Sonorelix probably evolved from an *Eremarionta*-like ancestor by secondary simplification of the accessory organs of its reproductive system. It has lost the dart sac and mucus glands, but retains the spermathecal diverticulum and the epiphallic caecum (flagellum). Its distribution largely overlaps that of *Eremarionta*. *S. angelus* W. O. Gregg is a westernmost relict in the extreme southwest

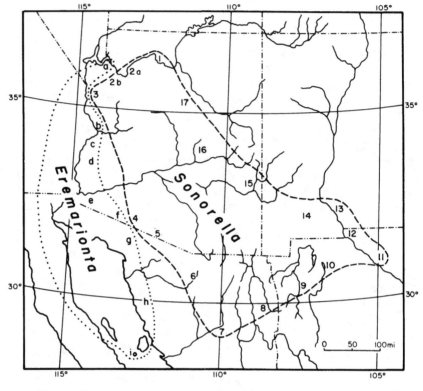

Fig. 2.   Distribution of *Eremarionta* and *Sonorella*. *a* to *i*, eastern
fringe Stations of *Eremarionta rowelli* in Arizona and Sonora: *a*, ca.
1 mi NE of Rampart Cave, Mohave Co., 36° 2′ N, 113° 55′ W; *b*, The
Needles, 4 mi S of Topock, Mohave Co., 34° 40′ N, 114° 40′ W; *c*, 12
mi S of Parker, Yuma Co., 34° 10′ N, 114° 17′ W; *d*, Dome Rock Mts,
Yuma Co., 33° 35′ N, 114° 30′ W; *e*, Tinajas Altas, Yuma Co., 32° 20′ N,
114° 10′ W; *f*, Heart Tank, Sierra Pinta, Yuma Co., 32° 16′ N,
113° 33′ W; *g*, Sierra de San Francisco, Sonora, 31° 40′ N, 113° 15′ W;
h, Puerto Libertad, Sonora, 30° N, 112° 40′ W; *i*, Isla de San Esteban,
Sonora, 28° 40′ N, 112° 35′ W. — 1 to 17, fringe Stations of *Sonorella*:
1 to 5 in Arizona, 1, *S. coloradoensis,* Grand Canyon of Colorado Riv,
Coconino Co., 36° L, 112° W; 2a, *S. coloradoensis,* NW of Chino near
Seligman, Coconino Co., 35° 30′ N, 113° W; 2b, *S. boreoccidentis* [MS,
W.B. Miller, 1971], at Rampart Cave, Mohave Co., 36° N, 113° 55′ W;
3, *S. coloradoensis mohaveana,* Goldroad near Oatman, Mohave Co.,
35° N, 114° 20′ W; 4, *S. meadi,* Agua Dulce Mts, Pima Co.,
32° N, 113° 20′ W; 5, *S. ambigua,* Santa Rosa spur of Nariz Mts, Pima
Co., 31° 49′ N, 112° 38′ W. 6 to 8 in Sonora, 6, *S. magdalenensis,* near
Magdalena, 30° 35′ N, 110° W; 7, *S. magdalenensis* and *S. sitiens
montezuma,* Sierra Pajaritos, E of Urés, 29° 25′ N, 110° 10′ W,
southernmost Station of genus; 8, *S. mormonum huasabasensis,* Huasabas,

corner of the Mohave Desert at Solemint near Newhall, Los Angeles County (34° 15′ N, 118° 30′ W). *S. peninsularis* (H. A. Pilsbry) is known from Baja California at Bahia de los Angeles (ca 29° N, 113° 30′ W; W. O. Gregg, in litt.) and at Rancho Mezquital (28° 20′ N, 114° W; collected by R. L. Bezy and W. C. Sherbrooke, July 28, 1965; dissected by the junior author); it is probably widespread in the central Baja California peninsula. W. O. Gregg (in litt.) reports two specimens of a population of *Sonorelix* from the southern tip of the peninsula which do not appear to be significantly different from *S. peninsularis*. The majority of known species of *Sonorelix,* however, are found in the Mohave Desert and the Colorado Desert of California.

Mohavelix micrometalleus (S. S. Berry, 1930) has undergone a more extensive simplification of its reproductive accessory organs. The dart sac, mucus glands, and spermathecal diverticulum have been lost, and the epiphallic caecum is minute. Because a similar simplification has also occurred in *Sonorella, Mohavelix* was originally described by S. S. Berry as a subgenus of *Sonorella.* Evidence available to date indicates that *Mohavelix,* known only from the El Paso Mts, Kern County, California, a low Range in the northern Mohave Desert, did not evolve from the same ancestor as *Sonorella.* Accordingly, *Mohavelix* was raised to generic rank, in order to avoid an implied sharing of a common ancestral generic founder with *Sonorella* (W. B. Miller, 1968b:51).

The recently discovered *Helminthoglypta micrometalleoides* W. B. Miller (1970) of the El Paso Mts exhibits numerous characters that strongly suggest a relatively recent common ancestry with *Mohavelix.* Its shell is nearly indistinguishable from that of *Mohavelix,* having similar dimensions, flattened shape, unusually wide umbilicus, thin peristome, and a pale, usually bandless color. Furthermore, it lives on the same mountain, only about three miles from known *Mohavelix* populations. It is attractive, there-

---

29° 56′ N, 109° 15′ W. 9 and 10 in Chihuahua, 9, *S. mormonum* and *S. pennelli,* near Colonia Juárez, 30° 15′ N, 108° W; 10, *S. nelsoni,* near Laguna de Santa María, 31° 10′ N, 107° 15′ W. 11 and 12 in Texas, 11, *S. orientis,* Sierra Blanca, Hudspeth Co., 30° 15′ N, 105° 20′ W, easternmost Station of genus; 12, *S. orientis,* Franklin Mts, El Paso Co., 31° 50′ N, 106° 30′ W. 13 and 14 in New Mexico, 13, *S. orientis,* San Andres Mts, Doña Ana Co., 32° 40′ N, 106° 40′ W; 14, *S. hachitana flora,* Florida Mts, Luna Co., 32° 5′ N, 107° 40′ W. 15 to 17 in Arizona, 15, *S. franciscana,* San Francisco Riv, Greenlee Co., 33° 10′ N, 109° 5′ W; 16, *S. ashmuni* and *S. anchana,* Sierra Ancha, Gila Co., 33° 40′ N, 110° W; 17, *S. coltoniana,* Walnut Creek, 8 mi. E of Flagstaff, Coconino Co., 35° 5′ N, 111° 30′ W.

fore, to hypothesize that *Mohavelix* evolved from an ancestral *Helminthoglypta* population which originally ranged widely over the El Paso Mts during pluvial periods, but is now restricted to occasional isolated rock slides by the prevailing drought. The isolation process, coupled with reduced population size and concomitant extensive inbreeding, could have been conducive to saltational speciation of the type discussed by Harlan Lewis (1966, *Science,* 52 [3719]:167-172). Further research should determine the extent of relationship between the El Paso Mts populations, such as comparisons of proteins, chromosomes, etc. For the present, however, it is more plausible to consider all populations of the two genera in the El Paso Mts as derived from a common ancestral *Helminthoglypta* population, rather than to postulate a *Sonorella* invader from Arizona which left no trace of intervening colonization for over 200 miles.

Helminthoglypta is almost restricted to California, with only two species in southwest Oregon and two species and several subspecies in northwest Baja California. By contrast, there are 47 recognized, described species, with numerous subspecies, in California. As a genus it has adapted well to the more humid regions, such as the northern coast, the Monterey Peninsula, the Sierra Nevada, and the coast ranges between the Mohave and Colorado Deserts and the coast. Relatively few desert species have been described, and one of them, *H. caruthersi,* has not been found again in recent years, in spite of several attempts by W. O. Gregg and the junior author. They share certain morphological characteristics, such as a non-malleate shell and a peculiarly swollen penis. The desert *Helminthoglyptae* are restricted to the edges of the Mohave Desert, the northeasternmost being *H. fisheri* in the Panamint Mts, Inyo County, and the southwesternmost *H. fontiphila* in Soledad Canyon, Los Angeles County; the greatest concentration of species is found in the mountains around Victorville, San Bernardino County. It is tempting to speculate that the ancestral *Helminthoglypta* evolved in the desert mountains from an isolated *Eremarionta*-like population and subsequently underwent rapid adaptive radiation in the more humid regions of California as the climate became more favorable. It appears more likely, however, that *Helminthoglypta* is relatively unchanged from the original North American invader, being very similar in anatomy to many Old World Bradybaenidae.

Distribution of *Sonorella*

To the malacologist, Arizona is the land of *Sonorella*. As now delimited by the junior author (1968a) and explained in the Check List, the genus occurs over most of Arizona (except a strip north of the Grand Canyon, an extensive northeast corner, and the small southwest *Eremarionta* area), the southwest corner of New Mexico, trans-Pecos Texas (El Paso and Hudspeth counties), northeast Sonora (south to the Sierra Pajaritos, east of Ures, 29° 25′ N), and the northwest corner of Chihuahua (southeast to Colonia Juárez, 30° 10′ No, 108° W); (map, Fig. 2). It does not occur in California or Baja California.

The boundaries of *Sonorella* shown on the map are probably nearly definitive for Arizona. Present knowledge of the mollusks of north Mexico is, however, unsatisfactory, so that new localities and even new species of *Sonorella* will no doubt be discovered there; but it is unlikely that the genus occurs much beyond the south boundary on our map. Farther south and southeast the molluscan fauna is strikingly different, with *Polygyra, Euglandina, Drymaeus, Helicina,* and other genera, all unknown from the Southwestern Molluscan Province, but characteristic of the Mexican Plateau.

Sonorella is precinctive and autochthonous in the Southwestern Province and mainly a snail of Arizona, where the genus speciated and radiated most actively. It now occurs over nearly two-thirds of the State (about 80,000 of a rough total 114,000 square miles). The junior author's unpublished revision (W. B. Miller, 1968a) recognizes 68 valid species (with 19 subspecies), 57 of them in Arizona (three in common with Sonora), three in New Mexico, one in trans-Pecos Texas (in common with New Mexico), eight in Sonora (three in common with Arizona), and three in Chihuahua. Most of the species (47) live in the south third of Arizona (the Basin-and-Range Section) south of the Gila River, and particularly in the southern mountains from the Tucson area to New Mexico. By contrast, only 11 species are known north of the Gila River and some of these also occur farther south. The present center of density of the genus is in the Santa Rita Mts (with seven species), Huachuca Mts (with six species), and Chiricahua Mts (with six species and one more in the adjoining Dos Cabezas Mts).

It is generally thought that *Sonorella* is a simplified helminthoglyptid. If, as some believe, its ancestor was an *Eremarionta,* the simplification involved loss of the dart apparatus, mucus glands, spermathecal diverticulum, and most of the epiphallic caecum.

On the other hand, if the ancestor was a *Sonorelix,* the simplification involved only the loss of the spermathecal diverticulum and the shortening of the epiphallic caecum.

It should be pointed out, however, that total loss of the dart apparatus and accessory structures does not appear to be too difficult a genetic feat. There is some evidence that it occurred in *Eremarionta* populations at least once (W. O. Gregg, in litt., 1969, for *E. argus*), and possibly twice, namely in producing also *Mohavelix* (W. B. Miller, as mentioned before).

In accordance with established procedure in zoogeography, it is generally considered that the center of radiating dispersal of a genus is usually indicated by its center of density, that is, the territory with the most species at present. For *Sonorella* this would be the area of the Huachuca Mts, Santa Rita Mts, and Chiricahua Mts. Not only do those Ranges have more distinct species, but they are also situated rather close to the geographic center of distribution of the genus.

Nevertheless, even though the Santa Rita, Huachuca, or Chiricahua Mts may have been once the center of radiation, it is difficult to visualize them as the site where the original helminthoglyptid ancestor evolved into a primitive *Sonorella.* No species now living in these Ranges even remotely resembles *Eremarionta* or *Sonorelix.* Furthermore, if an ancestral *Eremarionta* did exist there once, most probably some *Eremarionta* populations might have survived in the territory between these Ranges and the lower Colorado River; none are to be found there now east of the Ajo Mts.

A more likely hypothesis is that *Eremarionta* populations radiated at one time eastward from their present limits in the vicinity of Parker, Quartzsite, and the Sierra Pinta, in Yuma County, and extended perhaps as far east as the Phoenix Mts, to form "marginal populations." The ancestral *Sonorella* then originated from one of these marginal populations, possibly by saltational chromosomal reorganization (Harlan Lewis, 1966). From this founder, during periods favorable for areal expansion, *Sonorella* populations spread as far as the Santa Rita, Huachuca, and Chiricahua Mts, where they found ideal conditions for further rapid radiating speciation.

The area between Phoenix and the known *Eremarionta* localities has not yet been explored for populations of either *Sonorella* or *Eremarionta,* so nothing can be said about their distribution there. However, two species of true *Sonorella* have been found recently in the Phoenix area: *Sonorella allynsmithi* W. O. Gregg and W. B. Miller (1969), closely resembling *Eremarionta* in shell and genitalia and with similar ecological niches (extremely hot and arid rock piles), in the northern Phoenix Mts; and *S. superstitionis taylori* W. B. Miller (1969) in the southern Salt River Mts, not closely related to *allynsmithi,* though with the same xerophilous ecology. *S. r. rooseveltiana* and *S. rooseveltiana fragilis,* both close relatives of *allynsmithi* and of the same supraspecific complex, occur farther east, along the Salt River. Between the Salt River Mts and the Santa Rita Mts, no *Sonorella* of this complex has been found thus far.

Concerning the mechanism of speciation in *Sonorella,* the selective advantages of the main morphological differences between species are not obvious, the most striking differences being in shell size and reproductive structures. An outstanding example is the case of three species of the *granulatissima* Complex found in proximity in the Huachuca Mts. *S. granulatissima, S. parva,* and *S. danielsi* all live in essentially identical habitats and have probably been exposed to much the same selection pressures. In *S. parva* the verge is considerably larger than in *granulatissima,* but any selective advantage of the one over the other is not apparent, since both species appear to survive equally well. In such cases, the most probable mechanism of speciation is "genetic drift," which operates as follows. An original, polymorphic, large, ancestral population becomes divided into small, isolated subunits, possibly following a climatic change and radiating dispersal. If the subunit populations become small enough, it will be increasingly probable that some of the alleles present in the original population will be lost. Accordingly one of the isolated groups may have only combinations of alleles causing large verges, while another group may have a combination for small verges. As long as the new combinations do not prove lethal, both isolated populations could survive to become founders of new species.

The subspecies now recognizable in several species, though sometimes poorly defined, suggest that active evolution with invasion of new territory may be going on even today. The present boundaries of the genus may be dictated primarily by the rate and speed of past speciation and the relatively short intervening periods of pluvial conditions favorable for radiating since the genus' inception. For example, its present absence north of the Mogollon Rim in New Mexico and in northeast Arizona may be due to the fact that the genus simply has had no time to migrate there, rather than to the present-day adverse ambience. Time lag in dispersal may conceivably account also for other irregularities in specific distributions. Thus, only one species (*S. coloradoensis*) has managed to migrate north beyond the western outliers of the Mogollon Rim. It now lives, perhaps as an isolated relict, on both walls, a few hundred feet below the top, in the Grand Canyon of the Colorado River, at present the northern boundary of the genus. Moreover, it is surprising that no *Sonorella* was found thus far in the massive and very high San Francisco Mtn, although two species are known from Oak Creek Canyon and one from Walnut Creek Canyon, both only 25 miles farther south, at 5,000 to 6,500 feet.

Since all *Sonorella* are large (adults 12 to 30 mm in diameter), fairly heavy, and well protected in their natural haunts, live snails could hardly be transported to new Stations by wind or rain, or by adhering to moving objects—modes of transport often used by small land mollusks. Dispersal over the large area now occupied by the genus must therefore have been mainly automotive (H. B. Baker, 1958), that is, by snails crawling away from an original herd. Being soft-bodied animals, protected against heat and drought only by a superficial film of slimy secretion, they can travel only at night, with suitable damp weather. Moreover, progress must have been slow for crawling snails, bound to the ground when in motion.

In land snails, as in other organisms, successful colonization of new areas with self-perpetuating populations is the keystone of dispersal. It is not the mere appearance or existence of individuals, but the steady survival of a population under a particular type of environment that is essential for evolution. Under present-day climatic and topographic conditions of the arid Southwest, it would seem that dispersal to new territory would be difficult for *Sonorella*. It could only be a random, hit-and-miss undertaking; a straying individual could hardly ever reach a suitable new Station, and the probability of two snails from one colony reaching the same Station would be very low; at best a self-fertile or gravid adult snail might conceivably start a colony at a new locale.

It seems therefore more in accordance with the natural conditions of the area to assume that *Sonorella* reached its present-day wide distribution by a series of substantial migrations of several individuals during the successive cool and moist Pluvial Intervals of the Pleistocene Era, when the topography of Arizona may also have been at times more suitable for colonization by large snails. Even so, it must have been a protracted venture, that lasted throughout the Pleistocene and may have started earlier, no doubt with occasional regresses as well as advances. Meanwhile, it provided the time needed for profuse speciation by the combined action of mutation, isolation, and genetic drift.

The hazards and vicissitudes of evolution and dispersal in *Sonorella* account for the motley pattern of present-day specific distributions. Pilsbry's early attempts to deal with this problem predicated that it was primarily a matter of topographical or physical isolation. He thought that presumably each of the many mountain Ranges of the area, now separated by broad, arid valleys and bajadas, might have acquired in situ its own one or more precinctive species. This extreme isolationist approach even led him to describe putative subspecies for individual canyons of the more extensive Ranges. However, his attitude gradually softened as his exploration of the mollusks of the Southwest progressed and the anatomy of the snails became better known. He recognized eventually that some of his named, supposedly distinct, forms were only spatially isolated populations without consistent distinctive features of either shell or anatomy. The weeding process he initiated was further extended by the junior author in his revision of the genus (W. B. Miller, 1968a), on which our listing of the species and subspecies in the Check List is based.

Species strictly segregated in relatively small areas are actually the exception rather than the rule in *Sonorella*. Some mountains have only one species restricted to a single Range: *S. papagorum,* on Black Mtn near San Xavier Mission; *S. eremita,* only in one rock slide at the northwest end of San Xavier Hill of the Mineral Hill Group; *S. meadi,* in the Agua Dulce Mts; *S. tortillita,* in the Tortolita Mts; *S. micromphala,* on the west slope of the Milk Ranch Point extension of the Mogollon Rim, above Pine; *S. bartschi,* in the Mule Mts; and *S. allynsmithi,* in the Phoenix Mts. Excepting *S. micromphala,* these isolated precinctives are decidedly xerophilous, living in the Lower Sonoran Life-Zone under conditions of extreme aridity at low elevations (1,500 to 3,900 feet).

A few mountains harbor two or three species, each usually in a separate part of the Range: *S. imperatrix* in the north and *S. imperialis* in the south section of the Empire Mts; *S. delicata* and *S. caerulifluminis* in the north and *S. waltoni* in the south part of the Arizonan extension of the Peloncillo Mts; and *S. grahamensis* at lower and *S. imitator* at higher elevations in the Pinaleno Mts. None of the three precinctives of the Dragoon Mts, *S. dragoonensis, S. apache,* and *S. ferrissi,* live in close proximity.

Some of the smaller mountain Ranges have two species, often one of them precinctive and the other widespread: *S. superstitionis* and *S. galiurensis,* in the Superstition Mts; *S. xanthenes* (precinctive) and *S. ambigua* (widespread), on Kitt Peak of the Quinlan

Mts; *S. insignis* (precinctive) and *S. walkeri* (widespread) as subspecies *cotis*, in the Whetstone Mts; *S. anchana* (precinctive) and
S. ashmuni (widespread), in the Sierra Ancha; and *S. franciscana*
(precinctive) and *S. caerulifluminis* (widespread), in the Blue Mts.
The Mustang Mts have at present one living precinctive, *S. mustang;* but on the north slope of East Dome, at ca 5,200 feet, it is
associated with many fossil *S. huachucana,* a species still living in
the Huachuca Mts, farther south, and in the Patagonia and Santa
Rita Mts, farther west. Possibly *S. huachucana* became extinct
in the Mustang Mts recently, perhaps even during the past century,
due to human activity (mainly deforestation followed by rapid
erosion, which removed the needed shelter).

Sonorella is at its best in the extensive, massive, and lofty
mountains of southeast Arizona, where a vertical succession of Life-
Zones and many deep canyons offer a variety of suitable exposures
and habitats, the number of species in each Range being fairly
proportional to its height and extent. The Chiricahua Mts are
peculiar in having only precinctives, six in the main Range (*S.
optata, S. binneyi, S. neglecta, S. virilis, S. bowiensis,* and *S.
micra*) and one more in the closely connected Dos Cabezas Mts (*S.
bicipitis*). In the Huachuca Mts four of the six species are precinctive
(*S. granulatissima, S. danielsi, S. parva,* and *S. dalli*) and two (*S.
huachucana* and *S. sitiens*) have spread farther west, but not east.
The Santa Rita Mts, including the northern foothills and the Patagonia Mts, boast seven species, four precinctive (*S. tryoniana, S.
santaritana, S. clappi,* and *S. rosemontensis*) and three more widespread (*S. huachucana, S. walkeri,* and *S. magdalenensis*). The
Baboquivari Mts proper (excluding the Coyote and Quinlan Mts)
have three species, one precinctive (*S. vespertina*), the others widespread (*S. baboquivariensis* and *S. sitiens*). Of the two species of
the Santa Catalina Mts, one (*S. odorata*) is probably precinctive,
the other (*S. sabinoensis*) has strayed to the nearby Tucson Mts.
The Rincon Mts have three precinctives (*S. rinconensis, S. bagnarai,* and *S. bequaerti*).[6] Even the arid Tucson Mts and their

---

[6]Pilsbry and Ferriss (1919a:287 and 288) recorded and described briefly as
S. odorata shells from the Rincon Mts, collected presumably by J. H. Ferriss at
his Stations 20 (north slope of the Rincon Mts) and 22 (Spud Rock Ranger Station,
"found deep in a rock slide in a quaking asp thicket"). In 1923:87, Pl. 6, fig. 8,
they published a brief account of the genitalia of a snail from Station 22. In 1939:369
Pilsbry proposed the name *S. odorata* form *populna* for the snails of Station 22.
Attempts to find *Sonorella* in the Spud Rock area by the senior author (1963)
and by the junior author and J. Bagnara (1968) have failed. The actual occurrence of
S. odorata in the Rincon Mts is questionable.

loosely connected nearby hills have three species, all widespread (*S. magdalenensis, S. sabinoensis,* and *S. baboquivariensis*).

Six species, *S. sabinoensis, S. ambigua, S. magdalenensis, S. walkeri, S. sitiens,* and *S. huachucana,* are exceptional in that they cover large areas, in some cases at several widely separated localities. *S. magdalenensis* is the most interesting, being known at present from an alignment of moderately high mountains over a strip of territory, ca 200 miles long from north to south, but only 30 miles wide from east to west, mostly along the edges of the Santa Cruz Valley. Originally described from near Magdalena, Sonora, the junior author's fieldwork and anatomical studies (W. B. Miller, 1965:51; 1967c) revealed that it extends south in Sonora to the Sierra Pajaritos, near Ures (29° 25' N, 110° 10' W), and north in Arizona to Tumamoc Hill at the western city limits of Tucson (32° 14' N). In the intervening area it was given some six varietal, subspecific, or specific names, now synonymized in the Check List with *magdalenensis. S. sabinoensis,* restricted to the Santa Catalina Mts, Tanque Verde Hills, and Tucson Mts, usually in the Upper Sonoran Life-Zone, has an interesting clinal distribution, which, however, is not clearly correlated with an environmental gradient: nominal *S. s. sabinoensis* lives in the central canyons of these Ranges; eastward it gradually changes to *S. s. buehmanensis,* most distinctive at its T.L. in Buehman Canyon; westward it grades into *S. s. dispar* originally described from Pima Canyon, and extends to the northern Tucson Mts as *S. s. tucsonica.*

Pilsbry and Ferriss (1915b:395) and Pilsbry (1939:288) noted that, in Walnut Branch of Agua Caliente Canyon, Santa Rita Mts, three species, *S. walkeri, S. santaritana,* and *S. clappi,* lived "sometimes all under the same rock, sometimes in separate rock piles," at ca 6,000 feet; in upper Madera Canyon of the same Range we found *S. walkeri* and *S. clappi* living together in a small rocky area near the bottom of the creek at 5,700 feet. On the east slope of Marble Peak, Santa Catalina Mts, *S. sabinoensis* is the normal species of the Upper Sonoran Life-Zone, and *S. odorata* that of the Canadian Life-Zone; but in the intervening Transition Life-Zone we found both in the same rock pile at 6,800 to 7,000 feet. On Kitt Peak we observed *S. xanthenes* and *S. ambigua* sealed side by side to the same boulder at 6,200 to 6,400 feet. In the Sierra Ancha we dug *S. anchana* and *S. ashmuni* on two occasions from the same small rocky bank in a deep moist canyon at 6,500 to 7,200 feet. On Sonoita Creek, near Sanford Butte, at the south end of the Santa Rita Mts, we saw *S. tryoniana* and *S. huachucana* living in the same rocky bank at 3,800 to 3,850 feet. In Sonora the junior author found *S. nixoni* and *S. walkeri* in the same rock slide in Los Pilares de

Nacozari at ca 4,800 feet. Such close association of two or three related species at the same niche or biotope should, it seems to us, be recognized as truly sympatric. Presumably in *Sonorella* the structural specific differences in the genitalia effectively prevent interbreeding. Moreover, it must be assumed that each of the associated species has slightly different ecological requisites (for food, moisture, temperature, reproductive behavior, etc) which preclude de facto competition.

Speculation on the origin and evolution of *Sonorella* is hampered by the scarcity of fossil evidence. Thus far there is only one definite record of a fossil species with a reliable Cenozoic date. M. F. Skinner (1942:152) reported fossil *S. huachucana* (identified by H. A. Pilsbry), a species now living in the Huachuca, Patagonia, and Santa Rita mountains, from the matrix of Late Pleistocene vertebrates in Papago Spring Cave of western Canelo Hills, 5.5 miles southeast of Sonoita, Santa Cruz County. It was associated there with fossil *Holospira ferrissi, Helicodiscus singleyanus, Thysanophora hornii,* and *Retinella indentata paucilirata,* all species still living in the vicinity. *S. huachucana* is now extinct in the western Canelo Hills, which no longer seem to offer habitats suitable to *Sonorella.* Its nearest present (Recent) two Stations are Saddle Mtn, eight miles south of Papago Spring Cave, and Lookout Knob, some 10 miles to the southeast. As mentioned before, only fossil *S. huachucana* are now found in the Mustang Mts, at a Station of living *S. mustang* separated from the Huachuca Mts by the five- to six-miles wide Babocomari Valley. The fossil occurrences in the western Canelo Hills and in the Mustang Mts show that at one time (perhaps Middle Pleistocene) *S. huachucana* had a wider and perhaps more continuous distribution than at present, the Recent Stations in the Huachuca Mts, Saddle Mtn, eastern Canelo Hills (Lookout Knob), southern Patagonia Mts (Mt Washington area), and northwest Santa Rita Mts (Agua Caliente Canyon) being relict populations.[7]

---

[7]In 1919 J. H. Ferriss collected in the eastern Canelo Hills (originally cited without more precise Station) dead and live snails described as *Sonorella elizabethae* H. A. Pilsbry and J. H. Ferriss (1919b:20). The T.L. was better defined later by the authors (1923:65; with dissection by Pilsbry) as "Station 276, a large peak east of Huachuca-Duquesne road." A careful study of J. H. Ferriss' itinerary, using a modern topographic map, shows that his Station 276 was Lookout Knob, in the eastern Canelo Hills, ca 5 miles south of Canelo P.O. (not Mt Hughes, given as T.L. by Pilsbry, 1939:283). However, repeated search on Lookout Knob by several collectors in 1965 and 1967 has failed to turn up live *Sonorella,* although dead *S. huachucana* were common at 5,500 to 5,700 feet. Possibly the species became extinct there during the past 50 years, as it is now in the western Canelo Hills (near Papago Spring Cave) and in the Mustang Mts.

The absence of *Sonorella* at other fossil Sites, such as those in the San Pedro Valley where lesser snails are abundant, may be due primarily to their thin, fragile, weakly calcified shells decaying rapidly by weathering and breaking easily when carried by floods. In addition, the habitats of *Sonorella* are well protected and usually far from potential fossil Sites. Most species favor fairly steep slides of large, bare rocks, where they can crawl through the interstices to the proper depth for protection and then seal to a rock when inactive; the slides are often with scant or no plant cover. A few species prefer better covered sloping talus of loose material, in which they may lie loose with the aperture closed with a temporary epiphragm and are sometimes associated with *Ashmunella* or *Oreohelix*.

Distribution of *Humboldtiana*

Although *Humboldtiana,* a genus of Helminthoglyptidae, is not known at present from New Mexico, Arizona, Sonora, and farther west, it is a most revealing biotic element in the eastern trans-Pecos and Chihuahua sections of the Southwestern Province. It contains the largest land snails of the area. The adult shell, 19 to 32 mm high and 21 to 34 mm in diameter, is robustly helicoid, globose, with blunt spire, of moderately thick texture, with a broad, rounded-oval, toothless aperture, a simple, not reflected outer peristome, and a closed or narrowly rimate umbilicus. It is normally marked with one to three, equal or unequal, irregular, dull brown to black spiral bands (exceptionally lacking), on a much lighter, pale yellowish background. The number, usually three, and width of the bands vary within the same population and lack specific or subspecific significance. Although most species have some reliable specific shell characters, such as shape or sculpture, dissection of genitalia is needed for more precise identification, as shown by Pilsbry (1927b:166-186, Pls. 11-14; 1939:395-410) and by J. B. Burch and F. G. Thompson (1957).[8]

The genus is essentially Mexican, with centers of density and almost certainly of origin and evolution in the two Cordilleras and the intermontane highlands of the Mexican Plateau (map, Fig. 3). Here it is widespread at 5,500 to 12,000 feet elevation, but mostly above 7,000 feet, preferring the often cooler and generally drier "Tierra Templada" and avoiding the hot and humid tropical low-

---

[8]The anatomy of only five of the eight species here recognized in the Southwestern Province has been described thus far (*texana, chisosensis, ultima, högeana,* and *torrei*); it is unknown for the three species of the Davis Mts.

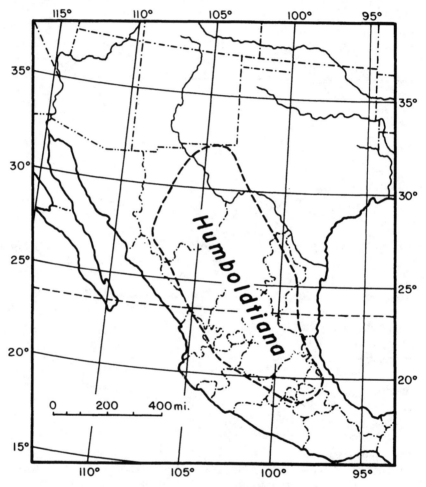

Fig. 3.   Distribution of *Humboldtiana*.

lands near the Coast. The westernmost Station, nearest the Pacific, in Jalisco, is ca 100 miles from the ocean; while in the east that of *H. pilsbryi* A. Solem (1954), northwest of Ciudad Victoria, Tamaulipas, is ca 125 miles from the Atlantic. Most of the 23 species now recognized (with a few subspecies) are from the Mexican States of Chihuahua, Coahuila, Hidalgo, Durango, Nuevo Leon, San Luis Potosi, Tamaulipas, Vera Cruz, Zacatecas, Queretaro, Guanajuato, and Mexico (D.F.); we have also seen immature shells of a species, perhaps undescribed, from Jalisco (north of Ajijic near Lago de Chapala, 5,300 feet, collected by W. B. Nutting, June 1964). Possibly the genus occurs also elsewhere in Mexico, but probably not in Sonora and Baja California, which are too arid. The southernmost record is from Mexico City (19° 26′ N, 99° 7′ W), and the northwesternmost from the south end of the Guadalupe Mts, Culberson County, in trans-Pecos Texas (ca 31° 40′ N, 140° 40′ W).

The following two species occur in Chihuahua, within the Southwestern Province, one of the two reaching also trans-Pecos Texas, presumably as a distinct subspecies: *Humboldtiana högeana* (E. von Martens, 1892:148, Pl. 7, figs. 20-21, as *Helix humboldtiana* var. *högeana*); T.L.: near the city of Chihuahua. *H. torrei* H. A. Pilsbry, 1935:1, Pl. 1, fig. 10; T.L.: San Antonio and Santa Rosalia (now Ciudad Camargo, 27° 40′ N, 105° 10′ W).

In the United States *Humboldtiana* is definitely known only from trans-Pecos Texas, with the following seven species (one a subspecies of a Chihuahua species):*H. texana* H. A. Pilsbry, 1927b: 179, Pl. 11, fig. 1, Pl. 12, figs. 9-11, and Pl. 14, fig. 5; T.L.: Housetop Mtn, ca 5,400 feet, 15 to 20 miles east of Marathon, ca 30° 9′ N, 102° 57′ W; and nearby, e. g. at Altuda, ca 5,000 feet, 13 miles east of Alpine, ca 30° 20′ N, 103° 25′ W (earliest United States record of the genus, R. E. C. Stearns, 1891:96, as *Helix humboldtiana*); all known Stations in Brewster County, at 3,900 to 5,500 feet. *H. chisosensis* H. A. Pilsbry, 1927b:182, Pl. 12, figs. 7-7a, and Pl. 14, fig. 8; T.L.: Oak Creek Canyon, cliff of Pulliam Bluff, left of Naill's Ranch, 4,500 feet (shown by H. A. Pilsbry, 1927b:184, fig. 11), in the Chisos Mts of the Big Bend National Park, Brewster County; widespread in the Chisos Mts at 4,500 to 7,550 feet; synonym: *Humboldtiana edithae* J. J. Parodiz, 1954:107, Pl. 9, 3 figs. at left; T. L.: Emory Mtn, Chisos Mts, 7,000 feet (based on a two-banded dead shell; a better preserved shell from Emory Mtn, seen by the senior author, has the normal three complete dark bands and is not separable from paratypes of *H. chisosensis*). *H. ultima* H. A. Pilsbry, 1927b:184, Pl. 12, figs. 12-14, and Pl. 13, fig. 4; T.L.: originally given as "canyon at high elevation southeast of Orange, New

Mexico, along the Texas-New Mexico border, at Station 241, east of the crest of the Range"; recorded also in 1927 from "Station 240, west of the crest of the Range"; as shown below, both Stations were actually in Texas, not in New Mexico; later records are definitely also from Culberson County, Texas: P. B. King (1948:145) reports finding *H. ultima* "living in sheltered places along cliffs high up on south wall of Pine Springs Canyon"; and A. R. Mead et al collected it in April 1969 at 5,300 feet in South McKittrick Canyon on the southeast slope of the Guadalupe Mts, one mile south of the New Mexico State line. *H. ferrissiana* H. A. Pilsbry, 1923b:32; holotype figured by H. A. Pilsbry, 1935c:2, Pl. 1, fig. 5, and 1939:400, 2 figs. 269c; T.L.: Mitre [or Miter] Peak, Davis Mts, ca 7,500 feet, Jeff Davis County. *H. palmeri* W. J. Clench and H. A. Rehder, 1930:12, Pl. 2, figs. 1-4; T.L.: head of Madera Canyon, north side of Mt Livermore, Davis Mts, 7,350 feet, Jeff Davis County. *H. cheatumi* H. A. Pilsbry, 1935:93; holotype figured by H. A. Pilsbry, 1939:402, 2 figs. 269a; T.L.: small canyon tributary to Limpia Canyon, Davis Mts, elevation not given, Jeff Davis County; known also from the north side of Blue Mtn, Davis Mts, 7,300 feet. *H. högeana praesidii* H. A. Pilsbry, 1939:402, 3 figs. 269f; T.L.: San Carlos Mine, Sierra Vieja, Presidio County, elevation not given; the Site of the former mine is three miles west of Vieja Peak, at ca 4,000 feet.

Pilsbry believed that the type locality and his other original Station of *Humboldtiana ultima* were in New Mexico, but a careful study of his itinerary shows that both Stations were in Texas, well south of the New Mexico-Texas line. In November 1922 Pilsbry and Ferriss started from Orange in Otero County, New Mexico, to explore the higher peaks of the southern section of the Guadalupe Mts. The now deserted settlement of Orange, no longer on most modern maps, is shown on P. B. King's (1948:6, fig. 2) geologic map about three miles north of the New Mexico-Texas line and 12 miles northwest of the west corner of Culberson County, Texas. Traveling southeasterly, they reached the PX Trail, shown on the Guadalupe Peak Quadrangle of the Geological Survey maps as entering Culberson County, Texas, about two miles south of the New Mexico State boundary, so that the remainder of their journey, including Stations 240 and 241 of *H. ultima,* being southward, was entirely in Texas. Eventually they reached the head of Pine Spring Canyon (mentioned by Pilsbry, 1940:997, as T.L. of *Ashmunella kochii amblya,* "above Walter Glover's house") and the south flank of Guadalupe Peak ("Signal Peak," cited as a Station of *Bulimulus pasonis* by Pilsbry, 1946:19, where he places

it in Texas). Moreover, Pilsbry (1946:124-125) gives the T.L. of
Holospira montivaga as "Guadalupe Mts east of Orange, from a ter-
raced butte in a deep dry canyon (our Station 240), about two
miles south of the PX Trail over the mountains"; (Station 240
is one of the two original localities of *H. ultima.*) He also gives
the T.L. of *Holospira montivaga* form *breviaria* as "eastern slope
of Guadalupe Mts near south end and above Walter Grover's ranch
house in Pine Creek Canyon" (Pilsbry here again places both localities in New Mexico, although they are clearly in Texas).[9] The
two later collecting Stations of *H. ultima* cited above are also
unquestionably in Texas. There is at present no real evidence that
Humboldtiana occurs in New Mexico. It is, of course, not impossible that *H. ultima* may yet be found north of the New Mexico-
Texas line, but this seems improbable since the highest and most
massive section of the Guadalupe Mts, with suitable habitats for
Humboldtiana, is restricted to Texas.

Humboldtiana is of minor importance to us, since it is restricted to the southeast corner of the Southwestern Molluscan
Province. Its main interest is as an outstanding Mexican immigrant.
All eight species of the genus in the Province are precinctive, so
that the genus must have reached the area early during the Late
Cenozoic, after which it must have speciated actively within a
limited territory. Most Texas species are now isolated in rather
small, disjunct mountain Ranges; only the Davis Mts share three
presumably distinct species, which, however, do not seem to be
truly sympatric, since no two are thus far known living in the
same niche (as is known for some *Sonorella*). It is puzzling that the
genus did not spread farther north or west. Perhaps it was stopped
by the longer and more severe winters farther north, or it may have
had formerly a more extensive range.

As is often the case for land mollusks, further speculation on
the evolutionary history of *Humboldtiana* is precluded by the
lack of fossil evidence. P. B. King (1948:145) listed *H. ultima*
among nine species of snails dug from a flat-lying reddish clay at
the base of a deposit washed down from the Guadalupe Mts, ca a
mile west of Pine Spring, on the north side of Pine Spring Canyon,
Culberson County; however, these *Humboldtiana* were perhaps not
true Pleistocene fossils, as the author thought, but rather Recent

---

[9]C. C. Hoff (1961) pointed out previously that Pilsbry's Stations 236 to 241
of 1922 were in Texas, not in New Mexico. This is further substantiated by A. L.
Metcalf's (1972) critical analysis of Pilsbry's manuscript Field Journal of his New
Mexico-Texas expedition of October 30 to November 13, 1922, now at the
Academy of Natural Sciences of Philadelphia.

or sub-Recent shells, since snails of this genus often burrow in the soil, where they may die in situ.[10]

Distribution of *Oreohelix*

The distribution of *Oreohelix,* in the family Oreohelicidae (map, Fig. 4), one more of the few Southwestern genera of large snails (adult shell 9 to 30 mm in diameter), contrasts sharply with that of *Sonorella.* The genus is primarily native in the Rocky Mountain and Great Basin States, and is the dominant element of the Rocky Mountain Molluscan Province. It occurs, with a wealth of species, subspecies, and minor variants, through east Washington (northwest to mouth of Entiat River, Chelan County, 47° 35′ N, 120° 10′ W), Idaho, Montana, Wyoming, Utah, and Colorado. It extends north into southeast British Columbia (north to Donald, ca 51° 30′ N, 117° W) and south Alberta (northeast to Medicine Hat, 50° 2′ N, 110° 40′ W; not in Saskatchewan, as given by Pilsbry); west into northeast Oregon (westernmost Station, Rufus, Sherman County, 45° 40′ N, 120° 50′ W), south Nevada (west to 30 miles north of Las Vegas, 36° 35′ No, 115° 10′ W), and to one Station in southeast California (*O. californica* S. S. Berry, 1931a and 1931b, at west side of Clark Mtn, San Bernardino County, ca 35° 30′ N, 115° 43′ W, at 7,000 feet). Northeastward it enters western South Dakota (Black Hills, Lawrence County, 44° 10′ N, 103° 50′ W).

In the Southwestern Province it is discontinuous in the northeast half (west to Mingus Mtn, 7,000 feet, 34° 42′ N, 112° 6′ W) and southeast corner (west to Huachuca Mts, ca 31° 25′ N, 110° 22′ W) of Arizona, and in the western half of New Mexico (Recent east to Sierra Blanca, Lincoln County, 33° 23′ N, 105° 52′ W). It barely enters northeast Sonora (in San José Mts, 31° 15′ N, 110° W) and northwest Chihuahua (Colonia Juárez area, ca 30° 15′ N, 108° W; Sierra de Almoloya, five miles northwest of Cueva Diablo, Distrito Jimenez, ca 27° N, 105° 21′ W).[11] Most forms known from the Chiricahua Mts in Arizona, the Big Hatchet and Mogollon mountains of New Mexico, and the mountains of Sonora and Chihuahua, belong in the subgenus *Radiocentrum.*

---

[10]We agree with J. B. Burch and F. G. Thompson (1957:2) that *Humboldtiana tuckerae* W. C. Mansfield (1937, *Dept. Conservation Florida, Geolog. Bull.* 15:66, Pl. 1, figs. 9 and 12), described from the Lower Miocene of Florida, is not a member of the genus.

[11]*Oreohelix avalonensis* "H. Hemphill" H. A. Pilsbry (1905:284, pl. 11, figs. 4-7), described from Santa Catalina Island, off the coast of South California, is a problematic species, disregarded for the present discussion. It is not impossible, however, that it may be a disjunct, perhaps now extinct, relict from an earlier more extensive range of the genus.

In his latest revision of the genus, Pilsbry (1939:412-553) recognizes 35 valid Recent species, the majority from the Rocky Mountains and the Great Basin States. Only 21 of these are in the Southwestern Province: one in California, two in Nevada, nine in Arizona (two of them also in the Rocky Mts, three also in New Mexico), nine in New Mexico (three of them also in Arizona), one in Sonora (also in Arizona), and three in Chihuahua. Most of these species are precinctive in the Southwestern Province.

In October 1971 an undescribed species of the subgenus *Radiocentrum* was found by the junior author living in Baja California Territorio Sur. It extends the Recent range of *Oreohelix* some 400 miles to the southwest of the previously known limits, shown on our map (Fig. 4), beyond the boundaries of the Southwestern Province.

The shell of most *Oreohelix* varies much in size (for same number of whorls), shape, sculpture, width of umbilicus, and color pattern, often within the same species. Hence an oversupply of subspecies, varieties, and minor named forms whose true affinities usually can be recognized only by dissecting the genitalia.

The genus appears to be much older than *Sonorella* and *Ashmunella*, with which it now occurs sometimes at the same Stations in east Arizona and west New Mexico. That it was more widely distributed during the Cenozoic than at present is shown by several fossil species of the Upper Cretaceous, Paleocene, Eocene, and Miocene of Alberta, Wyoming, and New Mexico, partially listed by J. Henderson (1935:138-140). These fossils agree in generic shell characters with Recent *Oreohelix* and some even show the peculiar embryonic sculpture of the subgenus *Radiocentrum*. T. D. A. Cockerell (1914:103) suggested that *Radiocentrum* might be the ancestral form of the genus, although it now includes only a few of the Recent species. As Pilsbry noted (1939:417), "it is extraordinary that so ancient a genus is now in a stage of prolific speciation." He then (1939:445) referred the *Oreohelix* from the Pleistocene loess in eastern Iowa (ca 41° 40′ N, 91° 30′ W) to the Recent *O. strigosa cooperi* (W. G. Binney, 1858), whose nearest known living Stations are in South Dakota, some 600 miles farther west.[12]

---

[12]The Pleistocene *Oreohelix* of Iowa was at first reported as *Patula strigosa,* "the variety known as *P. cooperi,*" by B.? Shimek (March 1888, *American Geologist* 1 [3]:150) from Iowa City, Johnson County, and later (November 1888), *Bull. Lab. Nat. Hist. Univ. Iowa* 1:203) also from Polk County. Pilsbry (1898a:141) cited it, without description, as *Pyramidula strigosa iowensis,* a *nomen nudum* used by B. Shimek in 1901 (*American Geologist* 28 [6]:345). Pilsbry (1916e:357) adopted this name for his *Oreohelix cooperi* form *iowensis,* but he later (1939:443 and 445, fig. 294 of holotype) made *iowensis* a synonym of *O. strigosa cooperi,* thus reverting to B. Shimek's original identification of 1888. The Recent *O. strigosa depressa* (T. D. A. Cockerell, 1890) was reported from the Pleistocene of Utah by E. J. Roscoe (1951, *Proc. Utah Acad. Sci. Arts Lett.* 25 for 1948-1949:135-136).

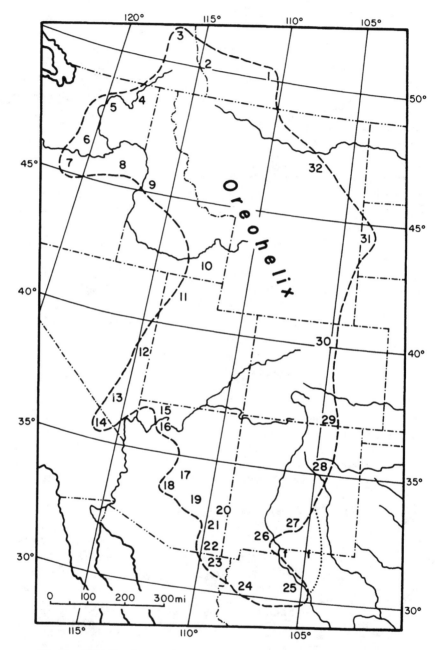

Fig. 4.   Distribution of *Oreohelix*. 1 to 32, Recent fringe Stations:
1 and 2 in Alberta, 1, 15 mi SE of Medicine Hat, 50° N, 110° 40′ W;
2, Waterton Lake, 49° 8′ N, 114° W. 3, in British Columbia, Donald (RR

J. P. E. Morrison (1943:104) reported *O. s. cooperi* also from the Pleistocene of Illinois at Copperas Creek, Rock County (ca 41° 20′ N, 91° W), just east of the Mississippi, now the easternmost known Station of the genus. In Arizona Pilsbry and Ferriss (1923:26) found only dead *O. concentrata* form *huachucana* at the two known Stations of the genus in the Mustang Mts, where we collected it also fossil at 5,200 feet, in a gravelly deposit on the north slope of East Dome, together with the fossil *Sonorella huachucana* mentioned before. We have seen fossil *O. c.* form *huachucana* from the Little Rincon Mts, 25 miles northwest of Benson, Cochise County, and it is known fossil also from the Dragoon Mts (recorded by Pilsbry and Ferriss, 1915b:372, as a variety of *O. strigosa*). *O. yavapai fortis* T. D. A. Cockerell (1927:101) and its synonym *O. yavapai vauxae* W. B. Marshall (1929:1), of the Grand Canyon of the Colorado River, are known only from presumably Pleistocene or sub-Recent fossils.

---

station), 51° 30′ N, 117° W, northernmost Station of genus. 4 to 6 in Washington, 4, Colville, Stevens Co., 48° 30′ N, 118° W; 5, Entiat Riv at mouth in Columbia Riv, Chelan Co., 47° 35′ N, 120° 10′ W; 6, 15 mi N of Ellensburg, Kittitas Co., 47° 10′ N, 120° 30′ W. 7 and 8 in Oregon, 7, Rufus, Sherman Co., 45° 40′ N, 120° 50′ W; 8, Milton, Umatilla Co., 45° 56′ N, 118° 25′ W. 9 and 10 in Idaho, 9, Race Creek near Lucille, Idaho Co., 45° 30′ N, 116° 20′ W; 10, Malad City, Oneida Co., 42° 10′ N, 112° 15′ W. 11 in Utah, Tooele, Tooele Co., 40° 30′ N, 112° 20′ W. 12 and 13 in Nevada, 12, White Pine Mts, White Pine Co., 38° 55′ N, 115° 30′ W; 13, 30 mi N of Las Vegas, Clark Co., 36° 35′ N, 115° 10′ W. 14 in California, Clark Mtn, San Bernardino Co., 35° 30′ N, 115° 43′ W. 15 to 22 in Arizona, 15, Snake Gulch, Mohave Co., 36° 30′ N, 112° 40′ W; 16, Supai, Coconino Co., 36° 5′ N, 112° 40′ W; 17, Oak Creek Canyon, Coconino Co., 35° N, 111° 40′ W; 18, Mt Mingus, Yavapai Co., 34° 42′ N, 112° 6′ W; 19, Sierra Ancha, Gila Co., 33° 40′ N, 111° W; 20, Y Salt House Branch of Eagle Creek, Greenlee Co., 33° 20′ N, 109° 30′ W; 21, Mt Graham, Pinaleño Mts, Graham Co., 32° 40′ N, 109° 52′ W; 22, Huachuca Mts, Cochise Co., ca 31° 30′ N, 110° 25′ W. 23 in Sonora, San José Mts, 31° 15′ N, 110° W. 24 and 25 in Chihuahua, 24, Rio Piedras Verdes near Colonia Juárez, 30° 15′ N, 108° W; 25, Sierra de Almoloya, 27° N, 105° 21′ W. 26 to 28 in New Mexico, 26, Sierra Co., 33° N, 107° 40′ W; 27, Sierra Blanca, Lincoln Co., 33° 23′ N, 105° 52′ W; 28, San Miguel Co., 35° 50′ N, 105° 30′ W. 29 and 30 in Colorado, 29, Las Animas Co., 37° 20′ N, 104° 30′ W; 30, Longmont, Boulder Co., 40° 15′ N, 105° W. 31 in South Dakota, Black Hills, Lawrence Co., 44° 10′ N, 103° 50′ W. 32 in Montana, Musselshell Co., 46° 30′ N, 108° 30′ W. *ff*, Fossil Stations in trans-Pecos Texas. Omitted from map: Saskatchewan (no precise Station known); Santa Catalina I, off south California (the questionable *O. avalonensis*); Baja California Sur (undescribed species discovered in 1971).

Some of the *Oreohelix* of New Mexico appear to be extinct also: *O. socorroensis socorroensis* H. A. Pilsbry (1905:279) of the Negra Mts, Socorro County (Black Mts? or Oscura Mts?); *O. socorroensis magdalenae* H. A. Pilsbry (1939:515) of Magdalena, Socorro County; *O. metcalfei florida* H. A. Pilsbry (1939:513) of the Florida Mts, Luna County; and *O. hachetana cadaver* H. A. Pilsbry (1915a:332) and *O. ferrissi morticina* H. A. Pilsbry (in Pilsbry and Ferriss, 1915a:334), both of the Big Hatchet Mts, Hidalgo County. *O. yavapai compactula* T. D. A. Cockerell (Aug. 1905:46), based on Pleistocene fossils of Pecos Canyon, a few miles above Valle (or Valley) Ranch, San Miguel County, appears to be identical with the Recent *O. yavapai neomexicana* H. A. Pilsbry (March 1905:282).

In trans-Pecos Texas *Oreohelix* is known at present from fossils only. In El Paso County two distinct species occur in Late Pleistocene (Illinoian or Early Wisconsinan) canyon sediments at ca 4,900 feet, five miles east of Vinton, near the northwest end of the Franklin Mts. Sometimes they are associated with living *Ashmunella pasonis*. They were first collected by W. O. Gregg, later also by M. L. Walton, A. L. Metcalf, ourselves, and others. We accept the conclusions reached in a recent study of these fossils by A. L. Metcalf and W. E. Johnson (1971:100-103, figs. 2a-b, of *O. s. socorroensis*, and 2c-d, of *O. f. ferrissi*). In the fossils which they regard as identical with Recent nominate *O. ferrissi ferrissi* H. A. Pilsbry (1915), of the Big Hatchet Mts, New Mexico, the shell is depressed, with several low spiral ridges, a strong peripheral carina, and the embryonic radial riblets of the subgenus *Radiocentrum*. The other species, more convex, with more rounded whorls, without noticeable spiral sculpture, with a weak peripheral carina, and without the embryonic *Radiocentrum* riblets, they refer to nominate *O. socorroensis socorroensis* H. A. Pilsbry (1905), a fossil species of the indefinite New Mexico "Negra Mts." They also mention a fossil shell, similar to *O. s. socorroensis*, found by H. A. Harris in cave sediments of the Hueco Mts, seven miles north of the Hueco Tanks, El Paso County. Fossil *Oreohelix* of Bell Canyon, Culberson County, at the south end of the Guadalupe Mts, called *O. yavapai compactula* T. D. A. Cockerell by P. B. King (1948:145), were more probably *O. s. socorroensis*. They extend the former Pleistocene range of the genus southeastward to near 105° W.

Our analysis of the Recent distribution and known past history of *Oreohelix* suggests that the genus may have originated in or near the Rocky Mountain and Great Basin area, probably toward the close of the Mesozoic Era or in the Early Tertiary. It then speciated profusely and extended greatly beyond its present Recent range. Presumably throughout the Cenozoic Era, it migrated from time to time southward into the area of the present-day Southwestern Molluscan Province, as well as eastward at least as far as Iowa and Illinois. Early southwestern immigrants may have been the ancestors of the local Recent precinctive species, while some later invaders became established without evolving into distinctive new species. Moreover, the relatively few Recent species of the area, their restricted ranges and often disjunct distributions, as well as the several known extinct colonies, suggest that *Oreohelix* is at present somewhat marginal in the Southwestern Province, or perhaps locally even on the verge of extinction.

Distribution of *Ashmunella*

Ashmunella, in the family Polygyridae, another distinctive Southwestern genus of large snails (9 to 24 mm in diameter), is clearly derived from a *Polygyra*-like ancestor. It has a convex or flattened, helicoid, unbanded shell, with a rounded to carinate periphery, a wide umbilicus, and with or without apertural teeth. Precinctive in the eastern section of the Southwestern Province, it is restricted to a relatively small area, with a present center of density in the southern two-thirds of New Mexico, extending westward to a small southeast corner of Arizona, eastward over much of trans-Pecos Texas, and southward in Mexico over the northern half of Chihuahua (map, Fig. 5). In the west it reaches near 110° 30′ W in the Blue, Chiricahua, and Huachuca Mts of Arizona, and in the east to ca 103° 30′ W in the Davis Mts of Texas. The northernmost Stations are in San Miguel County, New Mexico (near 35° 40′ N), for *A. thomsoniana* (C. F. Ancey, 1887), and the southernmost in Chihuahua, 16 miles southwest of Buenaventura (ca 29° 40′ N, 107° 30′ W), 7,500 feet, for a species related to *A. intricata* H. A. Pilsbry (1948), collected by R. H. Russell, Aug. 1971.

Of the 32 species here recognized as valid, 19 occur in New Mexico (one also in Arizona), six in Arizona (one also in New Mexico), five in Texas (two also in New Mexico), and five in Chihuahua. Most species live at moderate elevations (4,000 to 6,000 feet), but a few reach 9,000 to 12,000 feet (e.g. in New Mexico). They prefer well-sheltered and shaded biotopes in sloping talus of loose soil and rock debris.

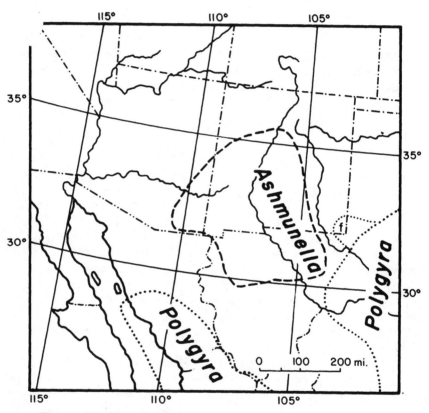

Fig. 5. Distribution of *Ashmunella* and southwestern extension of
Polygyra.

The Texas species, mostly described since Pilsbry's last revision of the genus (1940), are important in defining the eastern limits of the genus. As explained before in the discussion of *Humboldtiana,* Pine Spring Canyon, above Walter Glover's house, in the Guadalupe Mts, the T.L. of *A. kochii amblya* H. A. Pilsbry (1940), was not in New Mexico, as the author thought, but near the south end of the Guadalupe Mts, Culberson County, Texas. *A. edithae* H. A. Pilsbry and E. P. Cheatum (1951) is also from the Guadalupe Mts, Culberson County, and *A. pasonis* (R. J. Drake, 1951, as *Polygyra*) is from the Franklin Mts, El Paso County. *A. bequaerti* W. J. Clench and W. B. Miller (1966) and *A. mudgei* E. P. Cheatum (1971:107, figs. 1-4; here made a subspecies of *A. bequaerti*) are both from the Davis Mts, Jeff Davis County. *A. carlsbadensis* H. A. Pilsbry (1932), described from Dark Canyon, Guadalupe Mts, Eddy County, New Mexico, was also found in 1969 by A. R. Mead et al living at 5,300 to 5,400 feet in South McKittrick Canyon of the Guadalupe Mts, Culberson County, ca one mile south of the New Mexico-Texas line.

The map (Fig. 5) shows that the known ranges of *Polygyra* and *Ashmunella* nowhere actually overlap, although they come close in the Pecos River and Chisos Mts area of trans-Pecos Texas. Here the Davis Mts localities of *A. bequaerti* and *A. mudgei* are only some 75 miles from the nearest Station of *Polygyra texasiana tamaulipasensis* (I. Lea, 1858), at Sanderson, Terrell County (ca 30° 10′ N, 102° 25′ W), and some 100 miles from that of *Polygyra chisosensis* H. A. Pilsbry (1936) in the Chisos Mts, Brewster County. In Chihuahua, the southernmost *Ashmunella* is some 200 miles northeast of the nearest *Polygyra* (*P. behrii* W. M. Gabb, 1865) in Sonora and Chihuahua.

Ashmunella agrees with other southwestern large snails in speciating actively even at present, as shown by the many described subspecies. Pilsbry (1940:912-978) recognized 25 subspecies for his 27 species: in New Mexico, three for *A. rhyssa,* two for *A. thomsoniana,* one for *A. ashmuni,* one for *A. pseudodonta,* one for *A. townsendi,* one for *A. danielsi,* four for *A. tetrodon,* two for *A. cockerelli,* and one for *A. kochii;* and in Arizona, five for *A. proxima,* five for *A. levettei,* and one for *A. chiricahuana.* Some of the named Arizona variants were so poorly defined that they are now relegated to the synonymy. Pilsbry (1940) started the process by eliminating seven of his earlier subspecific or varietal names. He also recognized (1940:917) that: "In most of the larger mountain Ranges the *Ashmunellae* are perplexing on account of the multiplicity of local races and forms of which the taxonomic status remains uncertain. . . . A further source of difficulty has been the remarkable parallel development of similar toothless species in different Ranges, such as *A. chiricahuana, A. varicifera, A. mogollonensis, A. ashmuni robusta*. By their anatomical structure these species prove to be most nearly related, not to one another but to dentate species of their respective Ranges. The inference that existing toothless species are secondarily so, and derived from dentate ancestors, seems to be justified." This appears to be a case of subspecific evolution in which the same or similar genes remain phenotypically active in spatially and presumably reproductively isolated gene pools (species).

In spite of active evolution, *Ashmunella* seems to have been unable to colonize much territory in the past and may even be losing ground today. Fossil evidence for or against this hypothesis is very scant and, moreover, uncertain, since *Polygyra* and *Ashmunella* cannot be distinguished by shells alone. It is possible that some American fossils now referred to *Polygyra* may have been *Ashmunella*. However, it is fairly certain that the following fossils of New Mexico, within the Recent range of the genus, are correctly placed in *Ashmunella*. T. D. A. Cockerell (1901, *Science*, N.S., 14:1009) described briefly *Ashmunella antiqua* from the Pleistocene at Las Vegas, San Miguel County; this was synonymized with Recent *A. thomsoniana* by Pilsbry (1940:919). *Ashmunella thomsoniana pecosensis* T. D. A. Cockerell (1903a:105), from a "deposit of uncertain age," at Valley Ranch near Pecos, San Miguel County, was retained as a valid extinct subspecies by Pilsbry (1940:922). A. L. Metcalf (1967:49, figs. 3, 1-2) described and figured an unidentified *Ashmunella* found in some numbers in Dona Ana County at three Late Pleistocene Sites in "Tortugas alluvium"; he also mentions (1967:50) finding an *Ashmunella* in Pleistocene alluvium of the upper Gila River north of Duncan, Greenlee County, Arizona.

Distribution of *Holospira*

Of the genera discussed thus far, *Eremarionta* and *Sonorella* in the Helminthoglyptidae, *Oreohelix* in the Oreohelicidae, and *Ashmunella* in the Polygyridae are autochthonous denizens of the Western Molluscan Division of the Nearctic Region, where they now dominate the fauna of cool to warm temperate areas and where they must have arisen from Late Mesozoic (possibly Cretaceous) or Early Cenozoic ancestors. With *Holospira,* in the

Urocoptidae, we enter a different molluscan world. The Urocoptidae now prevail in the fauna of southern North America and the Antilles, with many species and several distinctive genera. They possibly originated in the Neotropical Region, where they certainly underwent a stupendous evolution, at least since the Late Mesozoic. In the Nearctic Region they appear to be later, perhaps Early Cenozoic invaders, being now represented there by only three genera in the southern United States. Two genera of Urocoptinae, *Cochlodinella* and *Microceramus,* reach south Florida, the latter also south Texas. The Holospirinae were more successful, having spread as far north as the 43rd Parallel, presumably from the Eocene onward. At present *Holospira* is fairly well represented from southeast Arizona to central Texas.

The shells of *Holospira* are large enough to be noticed by casual collectors, particularly since they often proliferate and congregate in populous colonies. Hence the plethora of about 120 described species, many poorly known, but providing nevertheless adequate data for the general distribution of the genus (map, Fig. 6). The bulk of them (ca 100) occupy four main areas in Mexico: the northern States (Sonora, Chihuahua, Coahuila), with 22 species; the western States (Sinaloa, Durango, Zacatecas), with four; the eastern States (Nuevo Leon, Tamaulipas, San Luis Potosi, Veracruz), with 17; and the southern States (Hidalgo, Mexico, D.F., Morelos, Queretaro, Puebla, Guerrero, Oaxaca), the present center of density of the genus, with 44. The lack of records from Nayarit, Jalisco, Colima, and Michoacan could be due to scarcity of suitable limestone outcrops, but more likely to insufficient collecting. It is fairly certain, however, that the genus does not occur in the Baja California peninsula, Chiapas, Tabasco, Campeche, Yucatan, and farther south, in spite of published records to the contrary.[13] The southeasternmost known occurrence is at present in Oaxaca (ca 29° N, 111° W). The headquarters of the genus are south of the Tropic of Cancer, where it occurs on the slopes of the Sierras Madre Occidental and Oriental and is more generally distributed over the Cordillera del Sur. It selects limestone terrain with sunny, dry exposure, usually at moderate elevations (4,000 to 7,000 feet; exceptionally it reaches 9,500 feet). North of the Tropic of Cancer, in a more arid environment, it may live at lower elevations, as, for in-

---

[13]The type locality "Yucatan" originally given for *H. yucatanensis* was erroneous, as it lives in the Big Bend area on both sides of the Rio Grande in Texas and Coahuila. The senior author's listing of *H. berendti* from Chiapas (1957, *Bull. Mus. Comp. Zool.* 116 [4]:227) was an oversight, duplicating the record of *Epirobia berendti* from that State. Other erroneous records were due to misuse of the generic name.

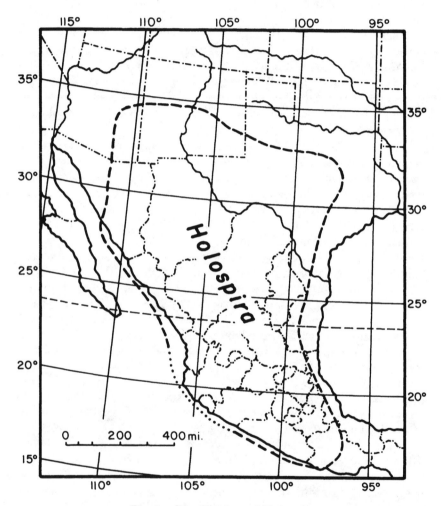

Fig. 6.   Distribution of *Holospira*.

stance, at 900 to 1,400 feet near Hermosillo, Sonora, near the northwest limit of the genus (ca 29° N, 111° W).

North of the Mexican border, *Holospira* is poorly represented by some 19 valid species, several subspecies, and minor variants, sporadic over a narrow north-south, but wide west-east, strip of Arizona, New Mexico, and Texas. The six species we recognize in Arizona are restricted to a small southeast corner of the State, extending from the Mexican border northward to Dos Cabezas (32° 11′ N, 109° 39′ W) and westward from New Mexico to the Empire Mts (northwest to ca 31° 55′ N, 110° 36′ W) and to the western Canelo Hills (southwest to ca 31° 32′ N, 110° 35′ W). Five species are listed by Pilsbry (1946:123-135) from adjoining New Mexico (northeast to Sierra County, ca 33° N, 107° 40′ W, the northernmost Recent Station of the genus). The Arizona and New Mexico species are closely related and placed by us in subgenus *Eudistemma* W. H. Dall (1895).[14] In this northwest part of the range, the genus occurs only in the Upper Sonoran Life-Zone at moderate elevations (4,800 to 6,500 feet).

Farther east, in Texas, the senior author recognizes eight valid species, some of them more widespread than usual in the genus, which extends there from the upper Rio Grande in the trans-Pecos area, east to Travis County and Hays County, with the northeast limit near San Marcos, Hays County (ca 30° N, 98° W, farther east than in south Mexico). Mostly from shell characters, but with some support from meager anatomical data, the Texas species appear to belong to four distinct lineages, here given subgeneric status. *H. montivaga* H. A. Pilsbry (1946) and *H. oritis* H. A. Pilsbry and E. P. Cheatum (1951), of the Guadalupe Mts, trans-Pecos Texas, resemble the forms of Arizona and southwest New Mexico and may be placed provisionally with these in subgenus *Eudistemma*.[15] *H. goldfussi* (K. T. Menke, 1847), placed by Pilsbry in the subgenus *Holospira* (sensu stricto), is isolated in the "Hill Country" of central Texas. *H. roemeri* (L. Pfeiffer, 1848), the only member of the subgenus *Metastoma* H. Strebel (1880), is more widespread than usual

---

[14]The taxonomy and nomenclature of *Holospira* are discussed in the Check List. *Haplostemma* W. H. Dall (1895), *Distomospira* W. H. Dall (1895), *Tristemma* P. Bartsch (1906; not of J. F. Brandt, 1835), and *Malinchea* P. Bartsch (1945) are here considered synonyms of *Eudistemma*. All five names were originally based on species from Arizona or southwest New Mexico.

[15]*H. montivaga* and its form *breviaria* H. A. Pilsbry (1946) were originally described from type localities in Culberson County, Texas; not in New Mexico, as thought by the author, as explained before for *Humboldtiana ultima*.

in *Holospira,* occurring from the mountains of the upper Rio Grande (in Texas and adjoining New Mexico counties) to New Braunfels, Comal County, and from the Mexican border to 32° 20′ N. The other four, *H. hamiltoni* W. H. Dall (1897), *H. mesolia* H. A. Pilsbry (1912), *H. yucatanensis* P. Bartsch (1906), and *H. pasonis* W. H. Dall (1895) of trans-Pecos Texas (two of them also in Chihuahua and one in Coahuila), represent the mainly Mexican subgenus *Haplocion* H. A. Pilsbry (1902) in the United States. In trans-Pecos Texas *Holospira* occurs at moderate elevations (2,800 to 5,000 feet), but farther east it lives as low as 700 feet (*H. goldfussi* and *H. roemeri* near San Marcos, Hays County).

The presence of several lineages (subgenera) in the Recent fauna of Arizona, New Mexico, and Texas suggests that *Holospira* had a wider northward range during the Early Cenozoic. This is supported to some extent by the fossil record. *Pupa* ? *leidyi* F. B. Meek (1873, in F. V. Hayden, *6th Ann. Rept. U.S. Geol. Surv. Terr. for 1872*:517), based on two Early Tertiary (probably Eocene) fossils from 12 miles south of Fort Bridger, Uinta County (41° 10′ N, 110° W), Wyoming, was said to look like *Holospira,* "so that the name may have to be changed to *Holospira leidyi*"; also that "it rather resembles *H. remondi,* an existing species found at Sonora, Mexico." Size (length, 0.55 inches = ca 14 mm) and shape fit *Holospira;* but the aperture was missing in both shells, and none of the truly diagnostic features of the genus were mentioned. It was not figured and the types are lost, according to C. A. White (1883, *U.S. Geol. Surv. 3rd Ann. Rept. for 1881-1882*:456). Pilsbry (1902, in 1902-1903:68, where the Wyoming and New Mexico fossil *Holospira* are cited as cospecific; in 1946:113, the two fossils are cited as distinct species) and J. Henderson (1935:148) listed *leidyi* as a true *Holospira;* while this appears to be correct, it should be confirmed from newly collected fossils from the original Site. D. W. Taylor (1962, in M. C. McKenna, P. Robinson, and D. W. Taylor, *Amer. Mus. Novit.* 2102: 16, fig. 5) described and figured a *"Holospira* ? sp." from the Eocene Bridger Formation at Tabernacle Butte, 25 miles north of Farson, in southeast Sublette County (ca 42° 30′ N, 109° 2C′ W), Wyoming; although the fossil lacks the diagnostic features cf the last whorls and aperture, the size, shape of spire, type of sculpture, and hollow columella leave little doubt of its being a true *Holospira;* it may even be cospecific with the enigmatic *H. leidyi* of F. B. Meek, in our opinion. Perhaps this is also the case for the undescribed and unfigured Early Eocene fossils from Pass Peak Formation near Bondurant, in northwest Sublette County (ca 43° 20′ N, 110° 20′ W),

Wyoming, placed tentatively in *Holospira* by J. A. Dorr, Jr (1969, *Contrib. Mus. Paleont. Univ. Michigan* 22 [16]:216, 217).

The generic identity of supposedly fossil *Holospira* reported from British Columbia and Utah is uncertain. *Holospira adventicia* L. S. Russell (1955, *Bull. 136 Ann. Rept. Nation. Mus. Canada for 1953-1954*:109, fig. 1 and Pl. 1, 2 figs. 7), based on poorly preserved shells from the Late Eocene or Early Oligocene Kishenehn Formation near Couldrey Creek, 2 miles north of the Montana border, Fernie District (ca 49° 5′ N, 114° 35′ W), southeast British Columbia, was placed tentatively in the genus; the photographs of the type show no positive *Holospira* features, the adnate, heavy, interrupted peristome and the strong parietal lamella suggesting rather a pupillid. Equally poor Paleocene or Early Eocene fossils from three Sites in the Flagstaff Formation of Sanpete County and Juab County (39° 20′ to 39° 40′ N, 111° 30′ to 111° 55′ W), Utah, called *"Holospira"* or *"Holospira* cf. *leidyi"* by A. La Rocque (1956, *7th Confer. Internat. Assoc. Petrol. Geol.*:143, Pl., fig. 24; 1960, *Mem. Geol. Soc. America* 78:49, Pl. 4, figs. 15-16) can hardly be placed definitely in either family or genus.

As matters now stand, of the fossils mentioned thus far, those from the Wyoming Sites may be accepted as valid evidence that *Holospira* lived during Early Eocene times as far north as the 42nd to 43rd Parallels, some 700 miles north of its Recent limits in New Mexico. However, the evidence is far from conclusive for a former extension of the range to the 49th Parallel in British Columbia.

Fossil true *Holospira* were first reported in New Mexico from the Early Eocene or Paleocene Puerco Formation of the Nacimiento Group, near Nacimiento (present-day Cuba, ca 36° N, 107°W), Sandoval County, by C. A. White (1886:27, Pl. 5, figs. 8-10; figs. 8-9 and fig. 10 seem to be of two species). He referred them to F. B. Meek's *"Pupa ? leidyi,"* a name used also for these fossils by C. Schuchert (1905, *U.S. Nat. Mus. Bull.* 53 [1]:561), J. H. Gardner (1910, *Journ. Geol.* 18:731), and J. B. Reeside, Jr. (1924, *U.S. Geol. Surv. Profess. Paper* 134:42). T. D. A. Cockerell at first (1906:459, footnote) called White's fig. 10 of 1886 *Holospira leidyi.* He changed it later to *Holospira grangeri* T. D. A. Cockerell, when he described that fossil species (1914:102, Pl. 8, fig. 5) from the Eocene Torrejon Formation of the Nacimiento Group at the East Fork of the Torrejon Arroyo (Encino Wash of present maps), near Torreon, some 24 miles southwest of Cuba (ca 35° 45′ N, 107° 15′ W), Sandoval County. He also stated that he was at a loss to separate the fossil *H. grangeri* from the Recent *H. ferrissi* of Arizona. The two New

Mexico fossil *Holospira* Sites are some 200 miles north of the Recent limits of the genus.[16]

In Arizona, M. F. Skinner (1942:152) found *H. ferrissi* in the matrix of Late Pleistocene vertebrate fossils of the Papago Spring Cave, in western Canelo Hills, 5.5 miles southeast of Sonoita, Santa Cruz County, at 5,200 feet. This snail does not now live at the cave, but is Recent only five to eight miles to the southeast in the Canelo Hills. *Holospira* has not been found thus far at any Pleistocene Site in the San Pedro Valley, where other land snails are common.

True fossil *Holospira* are known also at two Mexican Cenozoic Sites. The only figured type of *Holospira eva* Julia Gardner (1945, *Geol. Soc. America Mem.* 11:266, Pl. 10, fig. 21) is from the Upper Oligocene, ca 9 km due south of Mendez, Tamaulipas (U.S. Geol. Surv. Sta. 13583). Another species from the Mid-Oligocene, mentioned by the author but not described nor figured, is said to be abundant ca 2 km south of Rancho Miralejas near Carlos Cantú, China, Nuevo Leon. Pilsbry (1953:139) studied one of the fossils from the Nuevo Leon lot (not the type lot of *H. eva* from Tamaulipas, as he thought) and found that it had internal lamellae typical of the subgenus *Holospira* (sensu stricto). Since Recent species of the genus are known from nearby localities in Tamaulipas and Nuevo Leon, the Mexican fossils do not add significantly to the known range of *Holospira*.

Distribution of Lesser Land Mollusks

The genera of large, conspicuous snails discussed thus far are the bulk of the native land mollusk fauna of the Southwest (79 species, about two-thirds of the total 128, in Arizona), and illustrate the composite nature of the local fauna. The 64 precinctive and autochthonous *Eremarionta, Sonorella,* and *Ashmunella* are the dominant element in the State; the remaining 15 large snails, only in part precinctive, evolved from Rocky Mountain (nine *Oreohelix*) or Mexican Plateau (six *Holospira*) ancestors. The heterogeneity of the fauna is further enhanced by 49 lesser native species to be considered presently. Except for one *Bulimulus* and a few slugs, their shells are small to minute, at most 10 mm in greatest dimension when full grown.

Due to small size and hidden habitats, these lesser species are apt to be overlooked or even ignored by zoogeographers and ecol-

---

[16]We are greatly indebted to Dr. Artie L. Metcalf, University of Texas at El Paso, for his efforts in tracing the precise location and present names of the New Mexico fossil Sites of *Holospira*. We also wish to express to him our appreciation of much pertinent criticism and of permission to use unpublished information credited to him (in litt.) in our text.

ogists, although their study is as rewarding as that of the showier kinds. Not only do they bolster the precinctive character of the fauna, but they add components from other sources which throw further light on present distribution and past history. It may be noted also that, like other small invertebrates, they may occur in large populations and are then ecologically important consumers of vegetation, decomposers of decaying organic matter, and food sources for vertebrates and other animals. To geologists and archeologists they may be invaluable as frequent and often abundant fossils associated with Late Cenozoic vertebrate remains, or with artifacts of Early Man, sometimes furnishing useful clues to past chronology and environments.

Whatever their origin or past history might have been, in the arid Southwest most lesser land mollusks are more widespread than the larger snails, particularly in areas with scant plant cover. Many of them also range at present beyond the accepted limits of the Southwestern Molluscan Province. Small size and light weight allow for wide passive, mechanical dispersal, mainly by wind (eolian transport), and to a lesser degree by rain or floods. Mechanical transport is most common for immature and adult *Helicodiscus, Retinella, Vallonia,* and pupillids, which often close the shell temporarily with a mucous secretion that glues the edges of the mouth to leaves and debris in the loose, poorly sheltered litter, their favorite biotope in wooded grassland, on rocky ledges, or on canyon slopes. In the arid Southwest powerful atmospheric disturbances, such as high winds, twisters, and dust storms, are frequent on the plateaus and in the intermontane valleys, while cloud bursts associated with violent winds are seasonal features of the canyons. The term *"anemochores,"* used for plants normally dispersed by wind, seems suitable also for snails and other small invertebrates commonly spread by air currents, especially in arid country. The term "aerial or eolian transport" seems unsuitable or ambiguous in this connection, being often applied to passive, or perhaps sometimes deliberate, phoresis by birds or flying insects of live terrestrial or aquatic mollusks (*Succinea, Physa,* lymneids, ancylids, small clams, etc), which possibly also is effective in the Southwest; such mollusks might be called *"zoochores,"* a term now applied to plants that are normally dispersed by animals.[17]

---

[17]D. T. Jones (1940, *Bull. Univ. Utah* 31 [4]:22) noted that Valloniidae and Pupillidae are especially prone to be scattered by wind storms, as they may seal the shell to litter. Transport by wind of small snails has also been observed in Europe (E. Frömming, 1954:35). The effectiveness of air currents for the dispersal of small animals, minimized by some zoogeographers, is ably stressed by P. J. Darlington, Jr. (1938, *Quart. Rev. Biol.* 13:274-300; 1957, *Zoogeography*:20, where mollusks are mentioned).

Frequent transport by wind might also explain, at least in part, why the lesser snails of the Southwest have failed to produce by isolation the wealth of specific and other variants so striking a feature of the larger genera of the area. When live small snails with new mutations or unusual gene pools are blown by wind to a new station or biotope, later gusts of air easily carry there other live individuals of the same species with a different or more normal gene pool, so that both types of stragglers or their offspring may inter-breed before the unusual strain becomes reproductively isolated. Repeated haphazard dispersal by wind thus tends to prevent specia-tion by isolation, by enhancing the stabilizing effect of gene flow. Studies of lesser snails show that, even in species with notable intra-specific variation, such as pupillids with variable apertural teeth, several variants often occur within the same population. Sometimes one particular, more or less fixed, variant may be dominant over a wide area, and be recognizable as an incipient subspecies. More often, however, two or more fairly stable variants live together or in prox-imity. In either case, gusts of wind may readily shuffle the variants. In our discussion of *Holospira* in the Check List, we suggest that the apparent lack of clear-cut specific and subspecific differentiation may be due, even in these larger snails, to their preferred habitat in scantily covered limestone areas, where their immatures especially are exposed to passive dispersal by wind.[18]

Earlier in this section on Zoogeography, we proposed seven groups of species for the land mollusks of the arid Southwest, based on prominent peculiarities of their Recent and past distribution. The species which we include in these several groups will now be discussed in detail in the same order.

1. A high proportion of strictly **precinctive** land mollusks (190 species, 86 of them in Arizona) is the most prominent feature of the fauna of the Southwestern Province. Of the 32 genera, in 12 families, now recognized as native there, six, in three families (with 130 species, 66 of them in Arizona), are precinctive. Five of these are large snails discussed before: *Eremarionta* (nine species, one in Arizona), *Mohavelix* (one, not in Arizona), *Sonorelix* (ten, none in Arizona), and *Sonorella* (68, 57 in Arizona), in Helminthoglyptidae; and *Ashmunella* (31, six in Arizona) in Polygyridae. The sixth, *Chaenaxis* (monotypic), restricted to southern Arizona and Sonora, so far as known, is one of the lesser snails. Being one of the most

---

[18]We omit from the treatment of the lesser snails the native Succineidae (three *Succinea* and one *Catinella*). No worthwhile discussion of their distribution could be attempted with their present confused taxonomy.

distinctive elements of the Southwestern fauna, its taxonomy and distribution are given in detail in the Check List. It ranges from Yuma County (113° to 114° W) in the west to near New Mexico (ca 109° 30' W) in the east, and from Yavapai County (ca 34° N) in the north to southern Sonora (ca 28° N) in the south (map, Fig. 7). Although not definitely known thus far as fossils, all six genera are presumably autochthonous, having most probably originated and certainly evolved further in the area they now occupy.

An additional 70 precinctive Southwestern land snails (20 of them in Arizona) belong to nine other, more widespread genera in six families: *Helminthoglypta* (nine species, none in Arizona) and *Humboldtiana* (eight, none in Arizona) in Helminthoglyptidae; *Oreohelix* (14, five in Arizona) in Oreohelicidae; *Holospira* (25, six in Arizona) in Urocoptidae; *Bulimulus* (subgenus *Rabdotus:* five, one in Arizona) in Bulimulidae; *Pallifera* (one in Arizona and Sonora) in Philomycidae; and *Gastrocopta* (five in Arizona and Sonora), *Pupilla* (one in New Mexico and trans-Pecos Texas), and *Vertigo* (one in Arizona and Chihuahua) in Pupillidae. The majority are large snails discussed before (*Helminthoglypta, Humboldtiana, Oreohelix,* and *Holospira*). These four genera appear to have originated outside the present-day Southwestern Province, but most of their local species (except two non-precinctive species of *Oreohelix*) are probably nevertheless autochthonous, having evolved within their Recent ranges from ancient, presumably Cenozoic immigrant ancestors. This is reasonably certain for the precinctive species of *Oreohelix* and *Holospira,* the fossil record showing that both genera lived in the Southwest at least since the mid-Tertiary. It is also acceptable for the few precinctive *Humboldtiana* of trans-Pecos Texas and Chihuahua, which are well differentiated from their Mexican congeners farther south.

Due to lack of trustworthy fossil evidence and incomplete knowledge of Recent distribution, the autochthony of most precinctive Pupillidae remains doubtful. *Gastrocopta pilsbryana, G. quadridens,* and *Vertigo hinkleyi* have not been found fossil and have no known Recent or extinct close relatives. Their very restricted distribution suggests, however, that they are autochthonous in the Southwest, even though some of their present-day Stations may be relictual. *Pupilla sonorana* also has some earmarks of autochthony. It is known Recent from New Mexico (Otero County, 6,600 to 8,500 feet, Pilsbry, 1948:909, and A. L. Metcalf, 1967:43) and from trans-Pecos Texas (South McKittrick Canyon, Guadalupe Mts, Culberson County, A. R. Mead et al, Apr. 1969), but not from Arizona or Sonora. It is not definitely known fossil, but H. G.

Richards (1936:371) reported it as a sub-Recent fossil at the Linde-meier Site of Early Man, north of Fort Collins, Larimer County, Colorado (possibly a misidentification of *Pupilla blandii*). A. L. Metcalf (1967:43) found it in Recent alluvium of the Rio Grande floodplain near the Robledo Mts, Dona Ana County, New Mexico. A. L. Metcalf and W. E. Johnson (1971:99) report it from Late Pleistocene canyon colluvium of the Franklin Mts in Vinton and McKelligon canyons, El Paso County, Texas.[19]

Three of the five precinctive *Gastrocopta* (*G. ashmuni*, *G. cochisensis*, *G. dalliana*) belong in the subgenus *Immersidens*, mainly a Neotropical group, south to Brazil and Argentina, which enters the Nearctic Region only in the Southwestern Province. Possibly some of them may yet be found living in Mexico or Central America south of their present known range, in which case they should be regarded as part of the Neotropical element of the local fauna.

Gastrocopta ashmuni and its sinistral form *perversa* are wide-spread and characteristic for Arizona, New Mexico, trans-Pecos Texas, Chihuahua, and Sonora, from 105° W to 113° 30′ W and from 28° N to 36° 30′ N (detailed distribution in the Check List). It is surprising, therefore, that thus far they have not been found in the local Pleistocene. *G. ashmuni* was reported by H. G. Richards (1936:371) as a sub-Recent fossil from northern Colorado, some 300 miles north of the nearest known Recent Stations in Arizona and New Mexico; but this record, based on a single shell, will need scrutiny before being accepted as evidence for a former more northern extension of the species. The known distribution of *Gastrocopta cochisensis* is too sporadic and fragmentary for a useful discussion.

The distribution of *Gastrocopta dalliana* and its subspecies *bilamellata* has some unusual aspects. Together they extend over much of Arizona, Sonora, northern Chihuahua, and Baja California Territorio Sur (details for both in the Check List), from 109° W to 114° W and from 26° N to 35° 30′ N. Neither occurs in New Mexico or California. We suspect, however, that part of this wide range may be due to accidental post-Columbian transport by man. The most reliable difference between the two subspecies is in the parietal edge of the peristome (lip), in comparing shells from eastern and western populations, as in Pilsbry's figs. 488 and 489 (1948:902).

---

[19]It is as yet uncertain that *Pupilla sonorana* and *P. blandii* are specifically distinct, although A. L. Metcalf and W. E. Johnson treat them as such, having found them as sympatric fossils in the Franklin Mts. The trivial name *"sonorana"* is a misleading misnomer, since the species is not known from Sonora or from the Lower or Upper Sonoran Life-Zones in New Mexico or Texas.

In southeastern Arizona (Cochise County), which is relatively damp (often wet in summer) and cool (often cold in winter), all populations are nominate *G. d. dalliana,* with the upper ends of the peristome widely separated on the body-whorl and a very thin, weak connecting parietal callus. In southwest Arizona (Yuma County and adjacent areas west of 112° W), the most arid section of the State with prolonged droughts (more moist in winter, usually very dry in summer) and high temperatures (warm in winter), all populations are *G. d. bilamellata.* In this the peristome of full-grown, adult shells covers the body-whorl with a strong, adnate, slightly raised parietal callus, so that the mouth of the shell is encircled by a continuous, somewhat expanded edge. Obviously this is an adaptation to extreme aridity, enabling the snail to close the mouth of the shell tight against water loss, when it is glued by mucus to litter or stone for hibernating or estivating diapauses. Such an external, obvious structural adaptation to an arid environment is exceptional in land snails, which as a rule adjust to prolonged drought and excessive heat by physiological or behavioral arrangements (reduced metabolic activity, moving to appropriate shelter, etc).

Both subspecies of *dalliana* have spread beyond their preferred or typical ranges (mainly by the action of wind, as explained before), so that they now meet or sometimes overlap in south-central Arizona (between 110° W and 112° W). The junction of the two is particularly evident in the canyons and foothills that border the Santa Cruz Valley in Pima and Maricopa counties: some populations there are pure *d. dalliana* (in more humid, sheltered biotopes); others are pure *d. bilamellata* (at drier, more exposed Stations); more often they are a mixture in varying proportions of both subspecies and of transitional shells, so that definite identification of certain populations may be arbitrary. The transitional specimens were described by H. A. Pilsbry (1948:902) as a distinct subspecies, *G. dalliana media,* with the following statement: "The localities of *media* are west of those known for *G. dalliana,* but its range seems to be overlapped by the eastern extension of *bilamellata.* Its structural characters are between the two."

The peculiar distributional relations of *d. dalliana* and *d. bilamellata* could be interpreted as involving two distinct, reproductively isolated sibling species, the transitional specimens being then merely individual variants of both where they occur together. This fails to explain, however, why a similar profuse individual variability is not found in the pure populations of *dalliana* or of *bilamellata* at the eastern and western ends of their combined range. An alternative explanation would be to regard the entire combined range as a distributional cline, in which the two types merge gradually along an environmental west-east gradient from extreme to moderate aridity; but this does not seem to account for the simultaneous occurrence of pure and mixed populations of both types, as well as of transitional snails, in the central area of the range. A third possible solution of the problem, tentatively accepted by the senior author, recognizes *dalliana* and *bilamellata* as two subspecies derived from a common ancestral stock, that is, two geographic groups of populations with recognizable phenotypic differences, adapted to different degrees of aridity, but not reproductively isolated; they may have been at one time spatially isolated, but at present, due to passive dispersal by wind (as explained before), they are often sympatric and then interbreed freely, producing the mixed populations of both types and transitional hybrids. This interpretation not only agrees better with the facts, but has the added advantage that it may be tested experimentally.

The Bulimulidae are the most prevalent and most diversified family of Recent Neotropical land snails. It is surprising that they have not been more successful in invading and settling the Nearctic Region, where they are now restricted to a few species of *Bulimulus* (in the Southwest and the warm temperate Southeast) and *Drymaeus* (in Florida). The prolific genus *Bulimulus* produced five precinctive Southwestern species of the subgenus *Rabdotus* (sometimes treated as a genus): *B. nigromontanus* is fairly widespread in northern Sonora and enters Arizona barely in Santa Cruz County; *B. sonorensis* (doubtfully distinct from *B. nigromontanus*) is restricted to Sonora, and *B. baileyi* to Sonora, Chihuahua, and Sinaloa; *B. pasonis* and *B. pilsbryi* are associated in trans-Pecos Texas with peculiar subspecies of *B. dealbatus* (*B. d. neomexicanus*) and *B. alternatus* (*B. a. schiedeanus*), two species that extend eastward beyond the Southwestern Province. *B. dealbatus,* still in active evolution at present, is a dominant faunal element in much of the southeastern United States, east to 85° 30′ W (Cleburne County, Alabama), and north to ca 40° N (Marion County, Missouri). *B. alternatus* occurs in south Texas (north to San Antonio, 29° 25′ N, 98° 30′ W) and the Mexican States of Tamaulipas, Nuevo Leon, San Luis Potosi, Jalisco, Puebla, Durango, Coahuila, and Chihuahua, but not in Sonora in the Southwest or eastward in Louisiana.

Perhaps it might be more logical and agree better with the facts to regard *Vallonia perspectiva,* in Valloniidae, as an aberrant precinctive and not as a widespread Nearctic element, as we shall discuss it later. It appears to be a unique case of a snail definitely autochthonous in the Southwestern Province that eventually migrated eastward, where it is now established, perhaps temporarily, at several Stations, or possibly survives there only as a local relict of a wider, more continuous former range.

Although *Gastrocopta cristata* is not strictly precinctive in the Southwest, since it lives also outside the Province as far east as central Texas, it is best discussed here as an autochthonous element that probably originated in the area where it now retains its center of distribution and from where it spread farther east. In Arizona it is widespread and one of the most common Recent snails at moderate elevations (Lower Sonoran Life-Zone, usually at 2,500 to 4,500 feet), from Maricopa County (northwesternmost Station on Hassayampa River, ca 33° 50′ N, 112° 40′ W) and Yavapai County, northeast to Apache County (50 miles north of St Johns, ca 35° N, 109° 20′ W), and southeast to Cochise County (alive on the floodplain of the San Pedro River at Hereford, ca 31° 20′ N, 110° 5′ W) (details in the Check List). It also lives in the valleys of New Mexico (especially near Las Vegas), in trans-Pecos Texas (El Paso, A. L. Metcalf, 1967:41; alive at Marathon, Brewster County, L. Beatty, 1962; Davis Mts, Jeff Davis County, senior author), and in Sonora (drift of Rio Yaqui, near Ciudad Obregón, ca 28° N, 108° 50′ W, southernmost Station of the species; drift of Rio Bavispe, 12 miles south of Agua Prieta; both B. A. Branson et al, 1964:104). The general distribution elsewhere is imperfectly known, as the species is often confused with the closely related *G. procera* (discussed later), and as some published records are of drift shells, possibly washed-up fossils. It is definitely Recent in Kansas (east to Douglas County, ca 39° N, 95° 20′ W), in Oklahoma (east to Le Flore County, ca 35° N, 94° 30′ W), and in west and central Texas (east to Robertson, Brazos, and Austin counties, ca 29° N to 30° N, 97° W); farther south and southeast it is known only from dead shells in riparian drift. D. W. Taylor's (1960:69) record from north Nebraska (Brown County, ca 42° 30′ N, 100° W), far beyond the normal range, is puzzling and needs further investigation. Occasional sporadic occurrences in the eastern United States, such as the three Stations in Maryland (Talbot and Cambridge counties, W. Grimm, 1959, *Nautilus* 73:22), if based on correct identifications, must be due to transport by man, as the collector suggested. South of the Mexican border it is known only in Sonora. Like its sibling species *G. procera,* it is unusually prone to be spread by man, adapting readily to artificial environments in otherwise unsuitable territory. Its occurrence in the Pleistocene of Arizona, New Mexico, Oklahoma, Kansas, and Texas proves, however, that it is truly native in most of its Recent range.

It is unknown, either Recent or fossil, in California, the Rocky Mountain States, and east of Texas.

Pallifera arizonensis is, so far as known at present, restricted to Arizona and Sonora. Its true relationships and its status within the genus, as well as its actual distribution, are too uncertain for a fruitful discussion.

2. The most conspicuous **Neotropical** elements of the Southwestern mollusk fauna are the several large precinctive snails discussed before in *Humboldtiana* (seven species), *Holospira* (25 species), and *Bulimulus* (four species). These genera originated and speciated profusely south of the Tropic of Cancer, particularly in Mexico and Central America, where they still retain their centers of density and of speciation. Eventually they invaded the Southwestern Province, probably from the Late Mesozoic or Early Tertiary onward, which provided them with the lapse of time required for the evolution of precinctive species now fully adapted to subtropical and more arid environments. In the Nearctic Region they are at present restricted to relatively small southern areas; but fossil evidence shows that *Holospira* and *Bulimulus,* at any rate, were more widespread formerly, when suitable topography and climate allowed them to settle farther north.

Some species of lesser snails in *Gastrocopta* and *Vertigo,* possibly of Neotropical ancestry, but so far as known precinctive now in the Southwestern Province, have been discussed before. However, three other more widespread lesser snails appear to be also essentially Neotropical, with main ranges south of the Tropic of Cancer; they have invaded the southern Nearctic Region with little or no noticeable change of specific characters: *Gastrocopta prototypus* in Pupillidae, *Thysanophora hornii* in Thysanophoridae, and *Radiodiscus millecostatus* in Endodontidae.

Gastrocopta prototypus, including here *G. oligobasodon* and *G. prototypus basidentata* (synonymy explained by the senior author in the Check List), belongs in the subgenus *Immersidens,* of which three species were discussed before as local precinctives. Its general Recent distribution is fragmentary, with few published records (all known Stations given in the Check List). It is at present disjunctive, consisting of three widely separated areas in southeast Arizona (alive and in drift), south New Mexico (in drift only), south Mexico (presumably alive), and Guatemala (presumably alive). In Arizona it is known alive at four Stations (Patagonia Mts, Santa Cruz County; Huachuca Mts and Chiricahua Mts, Cochise County), and from riparian drift at ten (in Gila, Pima, Santa Cruz, and Cochise counties). Dead shells were found in drift in New Mexico at two

Stations (in Luna and Doña Ana counties). Two Mexican Stations are in Jalisco and Michoacan. There is one Station in Guatemala. Most fringe records are of riparian drift shells: the northwesternmost in Gila County, Arizona (ca 34° N, 110° 30′ W), the northeasternmost in Doña Ana County, New Mexico (ca 32° N, 107° W), and the southeasternmost in Guatemala (ca 14° 30′ N, 90° 30′ W). Since drift shells may be washed-up fossils, they are unreliable for defining Recent distributions, particularly for *G. prototypus,* which is known from the Late Pleistocene of southeast Arizona. Whether its Recent distribution is actually discontinuous, as appears from the records known at present, remains uncertain until the lesser snails of north Mexico are better known.

Thysanophora hornii is an unusual Neotropical invader which deserves a detailed study. *Thysanophora,* usually placed in the Sagdidae, but recently in its own family Thysanophoridae, is a prolific, strictly American genus mainly of Mexico, Central America, Colombia, Venezuela, and the Greater Antilles. One species, *T. hornii,* reaches the southern United States in Texas, New Mexico, and Arizona (all records in the Check List). In Arizona, it extends north to Mt Logan (36° 20′ N, 113° 10′ W, northernmost Station of species) and Shinumo Canyon (36° 15′ N, 112° 20′ W), and west to Palm Canyon, Kofa Mts (33° 20′ N, 114° 5′ W, westernmost Station of the species), and Arch Canyon, Ajo Mts (32° 2′ N, 112° 42′ W). In New Mexico, it is known from four southern counties (Hidalgo, Luna, Catron, and Sierra; north to Cuchillo Mts, 33° 30′ N); a dead shell was found in a cave in Dark Canyon, Eddy County, by W. H. Balgemann. In Texas, it is restricted to the southern counties, viz El Paso, Culberson (north to South McKittrick Canyon, Guadalupe Mts, 5,300 ft, ca 31° 50′ N, 104° 46′ W, senior author, 1969), Jeff Davis, Brewster, Presidio, Val Verde, Edwards, Uvalde, Live Oak, San Patricio, Hidalgo, and Cameron. In Mexico, it is known from Sonora (at 3,550 to 5,400 feet; northwest to Magdalena, ca 30° 35′ N, 111° W), Baja California Territorio Sur (San José de Comondú, 1,500 feet, 26° 5′ N, 111° 50′ W, westernmost Station of the species, junior author, Dec. 1970), northwest Chihuahua (Sierra de la Breña, 7,000 feet, 30° 10′ N, 108° 10′ W, H. A. Pilsbry, 1940:987, and junior author, 1966), Sinaloa (Rio Fuerte near San Blas, 26° 5′ N, 108° 46′ W), Jalisco, Nuevo Leon, Tamaulipas, and San Luis Potosi (south to ca 20° N).

T. hornii adapts to a wide variety of environments, from very dry (arid) to moderately humid (mesic). In the Southwestern Province, it is usually a xerophile, indifferent to nature of terrain, soils, or type of vegetation, and often found where other snails are unable

to survive. It is most prevalent in bajadas and canyons of the
Lower Sonoran Life-Zone at 2,800 to 6,000 feet, but it may live
also at higher elevations in the Upper Sonoran and Transition
Life-Zones (e. g. in upper Pinery Canyon, Chiricahua Mts, at 6,500
to 7,000 feet; on the south slope of Marble Peak, Santa Catalina
Mts, at 7,000 feet). The senior author found it in Arizona living
at the Saguaro National Monument, east of Tucson, under decaying
flat joints of prickly pear, fully exposed to heat and sun, at ca
3,500 feet. On the other hand, in southeast Texas (Hidalgo
County) it was thriving at near sea level, in well-shaded leaf litter
of damp, wooded bottomland, on the lower Rio Grande in the
Santa Ana National Wildlife Refuge. The short hairs of the shell
of young snails usually disappear later in dry habitats, but are often
retained by adults of moist biotopes. Perhaps *T. hornii* is a thermophile rather than a xerophile, partial more to high temperature than
to low precipitation, its present northern limits being set by frequent
or normal long and severe winters.

It is puzzling that *T. hornii* is seldom found fossil. The only
reliable fossil record for Arizona is from the matrix of Late Pleisto-
cene vertebrates in Papago Spring Cave of the western Canelo Hills,
Santa Cruz County (M. F. Skinner, 1942:152), where it now lives
nearby. It has not been found thus far in our large collections of
Late Pleistocene mollusks of the San Pedro Valley, Cochise County.
In trans-Pecos Texas, P. B. King (1948:145) reports it, with extinct
Oreohelix, from presumably Pleistocene deposits of Bell Canyon,
one mile north of the Hegler Ranch, Guadalupe Mts, Culberson
County, where it now lives. A. L. Metcalf and W. E. Johnson (1971:
100) found it in cemented canyon sediments, possibly of Illinoian
or Early Wisconsinan age, in the Franklin Mts, El Paso County.

According to Pilsbry (1948:665), *Radiodiscus* is "probably an
austral genus spreading north from South America," where several
species are known as far south as Tierra del Fuego. *R. millecostatus,*
the only Nearctic species, is clearly an invader from Mexico, known
there at present from a few northern localities in Tamaulipas,
Michoacan, and Chihuahua (Sierra de la Breña, in litter at 7,000
feet, 30° 10′ N, 108° 10′ W, collected by the junior author), but
not thus far in Sonora. In south Mexico (Puebla; Mexico, D.F.) and
Costa Rica it is represented by *R. m. costaricanus* H. A. Pilsbry
(1926c:132). As it easily escapes collectors due to its minute size,
its Neotropical range is no doubt imperfectly known. In the United
States it has been taken mostly at high elevations at only a few
Stations in Arizona (all records in the Check List; north to Reservation Creek, Apache County, 33° 42′ N, 109° 29′ W) and one in New

Mexico (Mogollon Mts, ca 33° 20′ N, 108° 40′ W). The discontinuous distribution suggests that it is a relict in the Southwestern Province from a former more extensive range during warmer interglacial Pleistocene periods. It is not known as yet as a fossil.

3. Intrusive elements from the **Rocky Mountain** Molluscan Province. The most interesting snails in this group are the two *Oreohelix* (*O. strigosa* and *O. subrudis*) which Arizona and New Mexico now share with the Rocky Mountain States. As shown before in this section, the present centers of density and presumably also of dispersal of *Oreohelix* are in the Rocky Mountains of Wyoming, Montana, Utah, and Colorado. Since it is known that the genus was at one time much more widely distributed, extending east at least to Iowa and Illinois, it can hardly be surmised when and where it originated. It may have reached the Southwest as early as the Paleocene or even the Late Cretaceous, thus accounting for the several precinctive forms of the area; but the two species which Arizona and New Mexico now share with the Rockies could well have been later, possibly Pliocene or Pleistocene immigrants.

The following lesser land snails appear to be also relatively late natural invaders from the Rocky Mountains: *Microphysula ingersollii* in Thysanophoridae; *Discus shimekii* in Endodontidae; *Pupoides hordaceus, Pupilla blandii, P. hebes,* and *P. syngenes* in Pupillidae; and *Vallonia cyclophorella* and *V. gracilicosta* in Valloniidae.

Microphysula ingersollii, probably autochthonous in the Rocky Mountains, is known Recent in British Columbia (north to Field, 51° 25′ N, 116° 30′ W), Washington, Oregon, Idaho, Montana, Wyoming, Utah, Colorado, New Mexico (east to Beulah, San Miguel County, ca 35° 50′ N, 105° 30′ W), and Arizona (known Stations in the Check List, at 5,000 to over 9,000 feet; south to Miller Peak, Huachuca Mts, Cochise County, 31° 23′ N, 110° 17′ W). It is not known fossil anywhere, possibly due to the fragile shell, or to its being restricted to moist biotopes at high elevations.

Discus shimekii, originally described from Pleistocene fossils of Iowa, is known fossil also from Illinois, Missouri, Nebraska, and perhaps Alberta and Ontario. It is Recent in Yukon Territory (north to Lake Lindeman, at the head of the Yukon River, ca 64° 30′ N, 140° 50′ W), Alberta, Oregon, Montana, Wyoming, northeast California, Utah, Colorado, South Dakota (east to Lawrence County, ca 44° 20′ N, 103° 50′ W), north New Mexico (east to Taos County, ca 36° 40′ N, 105° 30′ W), and east Arizona (at three Stations; southernmost for the species in Carr Canyon, Huachuca Mts, ca 31° 20′ N, 110° 15′ W); usually now only at high elevations, 7,000 to 12,000 feet. Arizona specimens are often referred to a distinct subspecies, *D. shimekii cockerelli* (H. A. Pilsbry, 1898:85, as *Pyramidula cockerelli,* from Labelle, Taos County, New Mexico, and from Custer County and Saguache County, Colorado; Saguache County selected as T.L. by Pilsbry, 1948:619). Commenting on specimens from Alberta, S. S. Berry (1922, *Victoria Mem. Mus. Ottawa Bull.* 36:10; Pl. 1, 3 figs. 8) quotes Pilsbry (in litt.): "The *D. shimekii* from Canada are about intermediate between typical *shimekii* and *cockerelli.* The last should have a noticeably wider umbilicus. In your shells the umbilicus is equal to that of the most openly umbilicate of the original *shimekii.* The distinction of the race *cockerelli* is probably of doubtful expediency, though usually it is more depressed and more open beneath." It thus appears that the subspecies *cockerelli* is scarcely recognizable, as B. Shimek concluded long ago (1901, *Bull. Lab. Nat. Hist. State Univ. Iowa* 5:139).

The subgenus *Ischnopupoides* of *Pupoides* is strictly American, with two closely related Nearctic species, *P. hordaceus* (to be discussed presently) and *P. inornatus* (not in Arizona and not known alive in the Southwestern Province). A few other species live in Mexico and west South America (Ecuador, Peru, Bolivia, Argentina, Chile). Thus far *P. hordaceus* was somewhat of a problem, as it was known only from dead or fossil shells, casting some doubt upon its status as a Recent snail. The problem has recently (1969) been solved with the discovery of a living colony in north Arizona (see the Check List). The junior author's dissection of some of these snails shows that the anatomy, particularly of the genitalia, differs only in minor details from that of the type species of *Pupoides,* *P. albilabris.*

In the Recent fauna, *P. hordaceus* is primarily a Rocky Mountain snail, ranging slightly farther east and south, particularly in the Southwest. Its greatly discontinuous range includes southeast Wyoming (a fresh, dead shell from a rocky crevice in arid country at Guernsey Reservoir, 4,400 feet, Platte County, 42° 15′ N, 104° 45′ W, northernmost Station of the species, D. E. Beetle, 1960, *Nautilus* 73:155), southwest Utah (Cannonville and Henry Mts, Garfield County, ca 37° 32′ N, 112° 8′ W, for the synonym *Pupoides eupleura* R. V. Chamberlin and E. Berry, 1931, *Bull. Biol. Soc. Washington, D.C.* 44:7, fig. 1), southwest Colorado (Gypsum Creek, San Miguel County, ca 38° N, 108° W), central Kansas (Hutchinson, Reno County, 38° N, 98° W, easternmost Station; the Cheyenne County record by D. S. Franzen, 1947, *Trans. Kansas Acad. Sci.* 49[4]:416, Pl. 1, fig. 5, is omitted, as the figure lacks the diagnostic riblets of *hordaceus*), New Mexico (widespread; south to Mesilla, in drift of the Rio Grande), and Arizona (all Stations in the Check List; westernmost Station of the species, Mt Trumbull, 36° 20′ N, 113° 5′ W; southernmost, Walnut Gulch near Tombstone, 31° 40′ N, 110° 2′ W). It is not known from Sonora or elsewhere in Mexico.

P. hordaceus occurs in Pleistocene or sub-Recent deposits in Kansas (Upper Pleistocene, Jones Sink, Meade County, D. S. Franzen and A. B. Leonard, 1947, *Univ. Kansas Sci. Bull.* 31, Pt. 2 [15]:372-374, Pl. 21, fig. 2), New Mexico (Las Vegas, Pilsbry, 1948:925; Tortugas and Picacho alluvium in Sierra County and Dona Ana County, A. L. Metcalf, 1967:43, figs. 3, 4), Texas (Pleistocene or later: Silver Falls, Crosby County, 33° 40′ N, 110° 10′ W, J. C. Frye and A. B. Leonard, 1957, *Bur. Econ. Geol. Univ. Texas Rept. Investig.* 32:35; Odessa Meteor Crater, Ector County, F. Wendorf, 1961, *Paleoecology of Llano Estacado* 1:106 and 108), and Arizona.

Of the five North American species of the Holarctic genus *Pupilla,* all Recent in the Southwestern Province, one is circumpolar and a natural, ancient immigrant from Eurasia (*P. muscorum,* to be discussed later); another is perhaps a local, autochthonous precinctive (*P. sonorana,* if specifically distinct from *P. blandii*). The remainder, *P. blandii, P. hebes,* and *P. syngenes,* are essentially Rocky Mountain snails that evolved independently in the Nearctic Region, presumably from very early *muscorum*-like Old World immigrants. All three species are very variable, seemingly still in active evolution, their identification being sometimes difficult, particularly when fossil. Pilsbry (1948:928) noted that they "are scarcely differentiated from Old World forms. *P. blandii* and especially *P. sonorana* are hardly separable from the Alpic *triplicata* (S. Studer, 1820), all three having teeth like the Miocene species. *P. hebes* resembles the European *P. cupa* (J. de Cristofori and G. Jan, 1832) and *alpicola* (J. de Charpentier, 1837)."

P. blandii is known Recent from Alberta (northernmost Station of the species, Red Deer, 52° 17′ N, 113° 46′ W), Idaho, Montana, Wyoming, Nevada (west to Mt Charleston, Lincoln County, ca 38° N, 115° W), Utah, Colorado, west half of New Mexico (east to Rio Arriba County, ca 36° 30′ N, 107° W), and Arizona (west to Mt Logan, Mohave County, ca 36° 20′ N, 113° 10′ W; and south to Mt Graham, Graham County, 32° 40′ N, 109° 52′ W). In Arizona and New Mexico it lives at 5,000 to 9,000 feet; dead shells at lower elevations in riparian drift (in Pima and Cochise counties) are mostly washed-up fossils. East of the Rocky Mountains (e. g. at the T.L., Fort Berthold, North Dakota, ca 47° 30′ N, 102° W), it occurs only as dead shells in drift. It does not seem to have spread west to British Columbia or to any Pacific Coast State. It is a common Pleistocene fossil in Iowa, Missouri, Nebraska, Kansas, Oklahoma, Texas, New Mexico, and Arizona. Its extreme variability accounts for the several named forms, none of which can claim subspecific status: *Pupa sublubrica* C. F. Ancey, 1881; form *obtusa* T. D. A. Cockerell, 1888; variety *edentata* H. Squyer, 1894; subspecies *pithodes* H. A. Pilsbry and J. H. Ferriss, 1917; and subspecies *charlestonensis* H. A. Pilsbry, 1921. Sinistral shells, so common for most species of *Pupilla,* are unknown for *blandii.*

Pupilla hebes, one of the most distinctive snails of the Rocky Mountains, extends farther south (to north Mexico) than *P. blandii.* Such differences in distribution are, however, of little significance and perhaps merely due to scanty collecting. The nominate dextral form is Recent in southeast Washington (Lime Point, Asotin County, ca 46° 15′ N, 117° 15′ W, northwesternmost Station of the species), Idaho (north to the Salmon River, ca 46° N, 116° 30′ W), Montana (Bearmouth, Granite County, ca 4,000 feet, 46° 40′ N, 113° 20′ W, northernmost Station of the species; Birch Creek, 24 miles northwest of Dillon, Beaverhead County, ca 5,000 feet, 45° 15′ N, 112° 35′ W; both collected by R. H. Russell, 1964-1965), Wyoming (Chimney Rock, Albany County, 41° 40′ N, 106° W), Nevada (T.L.: White Pine, White Pine County, 38° 55′ N, 115° 30′ W; Austin, Lander County, 39° 30′ N, 117° W, southwesternmost Station of the species), Utah, Colorado (east to Tolland, Gilpin County, 39° 40′ N, 105° 35′ W, easternmost Station of the species), and Arizona (above 5,000 feet; west to Hualapai Mts, Mohave County, 8,000 feet, ca 35° N, 114° W). In Mexico, dextral shells only are known from north Chihuahua (on Rio Piedras Verdes, 3.25 km below Pacheco, in talus at 5,900 feet; Sierra de la Breña, 17.5 km from Pearson on road to Pacheco, at 7,000 feet, ca 30° 10′ N, 108° 10′ W; both Pilsbry, 1953:164). A sinistral form *nefas* occurs in southeast Arizona (Santa Catalina, Rincon, Santa Rita, Huachuca, and Chiricahua mountains), usually with dextral shells. Pilsbry (1948:939) states that the sinistral form "evidently appeared as a mutation somewhere in southeastern Arizona, and as yet has spread over only a few mountain Ranges of that region. Nothing has been seen of it in the extensive regions north and northwest, inhabited by dextral *hebes.*" Both dextral and sinistral shells are fossil in the Late Pleistocene of the San Pedro Valley, at the Lehner Mammoth and Murray Springs Sites.[20]

---

[20]The *P. hebes* from drift at Tecumseh and Lawrence, Douglas County, Kansas, recorded by G. D. Hanna (1909, *Nautilus* 23:94), may have been *P. muscorum.*

The range of *Pupilla syngenes* is more restricted and more discontinuous than that of most of its congeners. It is Recent in Wyoming (five miles east of Thermopolis, Hot Springs County, 43° 30′ N, 108° 10′ W, at 4,400 feet), east Montana (northeasternmost Station of the species, drift of Beaver Creek at Wibaux, Wibaux County, 47° N, 104° 10′ W), Utah (Zion National Park, Washington County 37° 10′ N, 113° W; Moab, Grand County, 38° 35′ N, 109° 30′ W), Colorado (Tolland, Gilpin County, and Eldora, Boulder County, both ca 39° 40′ N, 105° 35′ W), New Mexico (east to San Rafael, Valencia County, 35° 10′ N, 108° W), and Arizona (all Stations in the Check List; westernmost at Mt Logan, Mohave County, 36° 20′ N, 113° 10′ W; southwest to Brown Canyon, Huachuca Mts, Cochise County, 31° 27′ N, 110° 19′ W, southernmost record of the species, by Pilsbry and Ferriss, 1923:62, omitted by Pilsbry, 1948; northeast to a "branch of Chinle Creek," Apache County, ca 36° 30′ N, 109° 40′ W; southeast to Mt Graham, Graham County, 32° 40′ N, 109° 52′ W). It lives normally at 4,500 to 9,000 feet and is found at lower elevations in riparian drift only. It is also known as a Late Pleistocene fossil in the San Pedro Valley, Cochise County.

Originally based on sinistral shells from Arizona, without precise locality, *P. syngenes* has three named variants, none of them here recognized as subspecies: the dextral *P. s.* form *dextroversa* H. A. Pilsbry and E. G. Vanatta (1906:606); the albino *P. s.* mutation *nivea* H. A. Pilsbry (1921:169); and the aberrant *P. s. avus* H. A. Pilsbry and J. H. Ferriss (1911:196, 5 figs. 9; Pilsbry, 1948:942). *P. s. avus* was described with the following comment: "The special characters of this race, being those of senility, are unequally developed in different individuals. . . . Finding these shells associated with about an equal number of *P. s. dextroversa* of about the same size, we at first were disposed to think them all one race in which the shell was indifferently dextral or sinistral; but on closer study it appears that the dextral forms never have the last whorl and aperture abnormal, nor are the teeth so deeply immersed or the parietal lamella so long, while almost every sinistral shell collected in this colony is markedly distorted. It seems, therefore, that although the two forms are of common origin and live together, the different direction of the coil probably prevents interbreeding, thus segregating the sinistral stock, which in this colony is in a late stage of degeneration." *P. syngenes* may be at present in the throes of explosive evolution, with an accelerated mutation rate producing a multiplicity of aberrations, some of which could eventually evolve into new subspecies or species; a type of speciation perhaps more frequent in the past and at present among land mollusks than suspected. Possibly *P. syngenes* itself may have been the product of a similar evolutionary process, in which selection pressure and adaptation played only a minor role.

The Valloniidae are a strictly Holarctic family of few genera, but with many Recent and extinct species, mostly in *Vallonia*. This genus should be of considerable zoogeographic interest because of its great antiquity, dating as far back as the Paleocene, and because it contains three Recent circumpolar species, *V. pulchella* (O. F. Müller, 1774), *V. excentrica* V. Sterki (1893), and *V. costata* (O. F. Müller, 1774). It is regrettable, therefore, that the taxonomy of the Nearctic species is at present so confused that it seems impossible to give a satisfactory account of their Recent and past distribution. With some misgivings, we have followed in the main Pilsbry's (1948:1019-1039) latest revision, which recognizes eight Recent Nearctic species, and is hardly more than a rewrite of V. Sterki's earlier work (July 1893:247-261; Sept. 1893, *Proc. Acad. Nat. Sci. Philadelphia* 45:238-279).

Fortunately, four species are clearly recognizable in the South-western Province. One of these, *V. pulchella,* is only adventive there and may therefore be dismissed without further comment. Of the three native species, *V. cyclophorella* V. Sterki (1892) and *V. gracilicosta* O. Reinhardt (1883), to be discussed presently, are Late Cenozoic immigrants from the Rocky Mountains. The third, *V. perspectiva* V. Sterki (1893), will be considered later as an autochthonous, but widespread Nearctic element.

Pilsbry (1948:1032) records also *V. albula* V. Sterki (1893) from two New Mexico Stations (Arroyo Pecos near Las Vegas, San Miguel County; Santa Fe, Santa Fe County). *V. albula* ranges from south Canada and northeast United States south to New York and west to Manitoba and South Dakota; records from farther west are either erroneous or dubious. Pilsbry did not attempt to separate *albula* and *gracilicosta* in his key (1948:1023), and stated (1948:1032): "In Canada, New England and New York this species [*albula*] is readily separable from other vallonias of those regions, but in the Mountain States it is often not easy to decide between *albula* and *gracilicosta,* the most obvious distinction being that *albula* has a narrow lip." This difference is by no means obvious in published figures, and the suspicion arises that *albula* and *gracilicosta* are perhaps not specifically distinct, in which case our discussion of the distribution of *gracilicosta* may need emending.

Vallonia cyclophorella is Recent in southwest North Dakota (east to Billings County, ca 46° 50' N, 103° 30' W), northwest

South Dakota (east to Perkins County, ca 45° 30′ N, 102° 30′ W), Wyoming, Montana (north to Helena, Lewis and Clark County, 46° 30′ N, 112° W), Idaho (north to Nez Perce County, ca 46° 20′ N, 117° W), Washington (northwest to Grand Coulee, Grant County, ca 47° 30′ N, 119° 70′ W), Oregon (west to Klamath Lake, Klamath County, ca 42° 20′ N, 122° W), Wyoming, Nevada, California (south to Mill Creek, San Bernardino County, ca 34° 10′ N, 117° 10′ W), Utah (widespread), Colorado (widespread; east to El Paso County, ca 38° 45′ N, 104° 20′ W), New Mexico (widespread; southeast to Cloudcroft, Sacramento Mts, Otero County, ca 32° 50′ N, 105° 40′ W; northeast to Colfax County, ca 36° 30′ N, 105° W), and Arizona (widespread at 7,500 to 10,000 feet; northwest to Hualapai Mts, Mohave County, ca 35° N, 114° W; south to Miller Peak, Huachuca Mts, Cochise County, ca 31° 23′ N, 110° 17′ W). It does not now live in Canada, Alaska, Mexico, or Texas; Pilsbry's (1948:1035) record from a canyon 15 miles southeast of Amarillo, Potter County, Texas, was based on washed-up fossils. The species is a frequent Pleistocene to sub-Recent fossil in Texas, New Mexico, and Arizona.

Vallonia gracilicosta (including *V. costata* var. *montana* V. Sterki, 1893, and *V. sonorana* H. A. Pilsbry, 1915b:345, 4 figs. 5) has much the same general distribution as *V. cyclophorella*. It is reported Recent from North Dakota (T.L.: "Little Missouri"; according to Pilsbry, 1948:1030, from Medora, Billings County, ca 46° 50′ N, 103° 30′ W, where H. B. Baker found it later in drift; east to Minnewaukan, Benson County, 48° N, 99° 20′ W, collected by Mina L. Winslow, July 1919), Montana (widespread; north to Blaine County, ca 48° 30′ N, 109° W), southeast Idaho (north to Bannock County, ca 42° 40′ N, 112° 30′ W), Wyoming (widespread), South Dakota (widespread; east to Deuel County, 44° 50′ N, 96° 30′ W), northwest Nebraska (Agate, Sioux County, ca 42° 30′ N, 103° 50′ W), Colorado (widespread; east to El Paso County, ca 38° 45′ N, 104° 30′ W), Utah (widespread), east-central California (Yosemite National Park, Mariposa County, 37° 30′ N, 119° 30′ W), New Mexico (northeast to Colfax County, 36° 30′ N, 105° W; southeast to Sacramento Mts, Otero County, 32° 50′ N, 105° 40′ W), Oklahoma, and northwestern Texas (living in Gray County, in Panhandle, A. L. Metcalf, in litt., 1970).

V. gracilicosta is not known thus far living in Arizona or Mexico; but it should be looked for in the White Mts and Blue Mts of northeast Arizona. A. La Rocque (1953, *Nat. Mus. Canada Bull.* 129:337) includes Alberta, Manitoba, and Ontario in the range from earlier published records; but these provinces were omitted by Pilsbry (1948:1029-1030), presumably because the records might have been based on misidentified *V. costata,* a circumpolar species found in southern Canada and northern United States, but unknown from the Southwest. In Arizona *V. gracilicosta* is known at present only from Pleistocene fossils, either in situ or washed up in drift. It apparently was more widespread in the Late Cenozoic, being known from the Pleistocene of Ohio, Iowa, Missouri, northern Minnesota (fossils in drift at Thief River Falls, Pennington County, 48° 10′ N, 96° W), Kansas, Oklahoma, Texas, New Mexico, and Arizona. *V. sonorana,* recorded only from New Mexico (T.L.: Big Hatchet Mtn, 8,360 feet, Hidalgo County, ca 31° 30′ N, 108° 20′ W), appears to be a synonym of *V. gracilicosta,* rather than of *V. albula* as Pilsbry (1948:1033) suggested, although the distinction between *albula* and *gracilicosta* may be untenable.

3. **Circumpolar** (Holarctic) species. In his review of Recent and past faunal connections between Eurasia and North America, C.H. Lindroth (1957:114-122) lists 19 species of Recent land Pulmonates which he regards as definitely or probably indigenous (pre-Columbian native) in both the Nearctic and Palearctic Regions. *Columella columella,* which he omits, should in our opinion be added to his list, but we do not regard as truly circumpolar four of the species he included. We accept therefore 16 species as originally common to Eurasia and North America without the intervention of man.[21]

It should be emphasized that the only circumpolar elements discussed here are the land mollusks of the original (pre-Columbian) Holarctic fauna. C. H. Lindroth (1957) mentions 33 additional terrestrial mollusks introduced in North America by man during the last four (post-Columbian) centuries, and his list is by no means complete. These modern human immigrants should be clearly recognized as adventives, even when they are now well established (feral) in America; they are of no further interest to the zoogeographer. The most valid proof that a Recent mollusk is truly indi-

---

[21]In addition to the seven circumpolar species of the Arizona fauna discussed here, C. H. Lindroth lists the following eight that reached North America by natural means, but are not now part of the Southwestern native fauna; *Zonitoides nitidus, Columella edentula, Vertigo alpestris, V. pygmaea, Vallonia pulchella* (only as an adventive in Arizona), *V. excentrica, V. costata,* and *Zoogenetes harpa.* Some others, included doubtfully by C. H. Lindroth, are in our opinion post-Columbian introductions by man either to the Old or New World: *Cepaea hortensis* (pre-Columbian native for Pilsbry, 1939:8; a doubtful American adventive for C. H. Lindroth, 1957:114 and 234-237, with whom we agree), *Zonitoides arboreus* and *Hawaiia minuscula* (both discussed by us later), and *Subulina octona* (native in Florida for C. H. Lindroth, 1957:116; a widespread tropical and subtropical tramp, probably originally from tropical America, almost certainly introduced by man in Florida, in our opinion).

genous in America, and not a modern human introduction, is its occurrence in situ in some North American Cenozoic deposit. Such proof is now available for 13 of the 16 species we here accept as circumpolar. The exceptions are the two *Vertigo* and the *Zoogenetes,* which, moreover, are not known from the Southwestern Province. Although not known fossil in the New World, we regard these as circumpolar because their peculiar ecology and extreme boreal distribution preclude the possibility of transport by man to their present American Stations.

Paleontological evidence, based mostly on mammals, indicates that northern Eurasia was the main Holarctic center of evolution and dispersal for circumpolar land animals (G. G. Simpson, 1947, *Bull. Geol. Soc. America* 58:613-688; *Evolution* 1:218-220. C. H. Lindroth, 1957:303. P. J. Darlington, Jr., 1957, *Zoogeography:*365-367. C. A. Repenning, in D. M. Hopkins, 1967:309). They migrated from northeast Asia eastward to the New World by way of mostly Cenozoic north-Pacific land connections across the present-day Bering Strait, and spread from Alaska to their present ranges farther south. This explains why most circumpolar mollusks are cold-adapted (boreal) types. The land bridges were chiefly one-way corridors, favoring eastward migrations. At any rate, there is no convincing evidence that, at any time during the Cenozoic, land mollusks spread by natural means westward from the New World to Asia. Westward migrations from north Europe to America by way of a transatlantic Iceland-Greenland bridge appear to have been exceptional also, and there is no fully trustworthy evidence for them among land mollusks.

In the north Pacific area now occupied by the Chuchi Sea and the Bering Sea, north and south of the Bering Strait, a combination of fairly permanent physiography (particularly an extensive shallow Continental Shelf, even now at most 300 feet below sea level) and frequent oscillations of the sea level, promoted a succession of regressions and transgressions of the ocean during the Tertiary and Quaternary periods. According to D. M. Hopkins (1967:451-481), the geologic history of the area shows that from the Early Tertiary onward, land connections across the north Pacific were made repeatedly, joining northeast Siberia and Alaska by means of a temporary isthmus ("Beringia"). Temporary land bridges were most frequent during the Late Cenozoic (Pliocene and Pleistocene), being then regulated primarily by the successive glacial and interglacial periods of the northern continents. (See maps of "Beringia" at the height of Illinoian and Wisconsinan Glaciations, in D. M. Hopkins, 1967:462, figs. 2-3). However, in spite of frequent opportunities for eastward dispersal, these Late Cenozoic trans-Pacific corridors resulted in comparatively few successful mollusk migrations, to judge from the small, almost insignificant proportion of circumpolar species in the Recent Nearctic land mollusk fauna as a whole. The overwhelming bulk of this fauna now consists of precinctive, autochthonous New World genera and species. Moreover, most American circumpolar mollusks are restricted even today to northern, boreal areas. Few have ventured farther south, where (as in Arizona) they are usually confined to high elevations with a cooler climate.

With few exceptions, the pre-Columbian immigrants from the Old World seem to have undergone little evolution since they reached America. It is true that American populations of circumpolar species have sometimes been separated by taxonomists as "subspecies," allegedly distinct from their Eurasian ancestral counterparts. But in most cases these were based only on geography, or on such tenuous differences as to be unconvincing without knowing the origin of the specimens.

The disparity between precinctive and autochthonous New World genera and species and pre-Columbian Eurasian immigrants is most striking in the arid Southwest. The Southwestern Molluscan Province has only the following eight native circumpolar land mollusks, confined mostly to high elevations and often only as disjunct relicts: *Euconulus fulvus* in Euconulidae, *Nesovitrea hammonis* in Zonitidae, *Deroceras laeve* in Limacidae, *Pupilla muscorum, Columella columella,* and *Vertigo modesta* in Pupillidae, *Cochlicopa lubrica* in Cochlicopidae, and *Vitrina pellucida* in Vitrinidae. Several of these extend southward into Sonora and Chihuahua, and may have been formerly more widespread on the Mexican Plateau, when extensions of the Pleistocene glaciations induced them to migrate farther south.

Euconulus fulvus is Recent over much of the Holarctic Realm, north to Greenland, Iceland, Scandinavia, Siberia, Alaska, Manitoba (north to Churchill, 58° 45′ N, 94° 5′ W), Labrador, and Newfoundland. Southward, it extends in Eurasia to the mountains of Europe and North Africa in the west and to Kamchatka in the east; in North America to North Carolina in the east and in the west to trans-Pecos Texas, New Mexico, Arizona, and the Pacific States (south to the San Bernardino Mts, California, ca 34° 8′ N, 116° 30′ W). South of the Mexican border it is known from northwest Chihuahua (Sierra de la Breña, 30° 10′ N, 108° 10′ W, by junior author), but thus far not from Sonora. Its present southeast limits are in trans-Pecos Texas (collected alive by the senior author on Casa Grande Peak, Chisos Mts, 6,800 feet, Brewster County, ca 29°

20' N; and in South McKittrick Canyon, Guadalupe Mts, 5,300 feet, Culberson County, ca 31° 50' N), and southeast New Mexico. In the boreal areas it may live at low elevations to near sea level; but in the South only at 5,000 to 10,700 feet and mainly in the Transition and Canadian Life-Zones. Pilsbry (1899:116; 1946:238, figs. 118A-C; fig. C, shell from Chiricahua Mts) refers the records from Kamchatka, Alaska, western Canada, and western United States (east to trans-Pecos Texas) to a distinct subspecies, *E. f. alaskensis,* while retaining in nominate *E. f. fulvus* those from Canada and the United States east of the Rocky Mountains, as well as those from most of Eurasia; but in our opinion it is not possible to separate these two forms consistently on shell characters. *E. fulvus* is known fossil from the Middle Pliocene and the Pleistocene on both sides of the Atlantic; in North America, from Quebec, Ohio, Iowa, Indiana, Illinois, Missouri, Kansas, Oklahoma, Mississippi, Texas, and Arizona (common in Late Pleistocene of San Pedro Valley at Lehner Mammoth and Murray Springs Sites, associated with mammoth remains 10,000 to 11,000 years old).

As revised by H. B. Baker (1930) and Pilsbry (1946:253-304), the genus *Retinella* comprises many, often ill-defined species, most of them Holarctic, although a few occur south of the Tropic of Cancer. In North America, Pilsbry recognizes 22 species (and seven subspecies) in four subgenera (and two sections), but most of these supraspecific taxa are based on differences so tenuous as to verge on the esoteric. Identification of *Retinella,* particularly fossils, is frustrating. Pilsbry claims only that H. B. Baker's key "*may* assist in identification."

So far as known, only two species occur in the Southwest: the widespread, strictly American *R. (Glyphyalinia) indentata* (T. Say, 1823), to be discussed later, and the circumpolar *Nesovitrea hammonis* (K. A. af Stroem, 1765). Typical *N. hammonis* occurs throughout Eurasia, from Iceland, the Faeroes, British Isles, continental Europe (except the South), and north Asia to Kamchatka and Japan. Its Nearctic representative, *N. hammonis electrina* (A. A. Gould, 1841), is Recent from Alaska (Kodiak Island and north to Point Barrow, 71° 23' N, 156° 28' W), British Columbia, Alberta, Manitoba, Ontario, Quebec, Labrador, Newfoundland, and Nova Scotia, south to Virginia and Missouri, and west to Kansas, north Nebraska, Colorado, Wyoming, Utah, New Mexico, and northeast Arizona (see the Check List). We follow C. H. Lindroth (1957: 117) in recognizing *electrina* as subspecifically distinct from *hammonis,* although the distinction is perhaps purely geographic. Even Pilsbry (1946:258), who gives *electrina* specific status, recognizes that "the separation of the American race specifically is more a matter of convenience than of definable constant diversity; the subspecific rank seems more fitting." To retain *electrina* as a distinct species would mask the interesting fact that it is a Late Cenozoic invader from the Old World. That it has not changed much since it reached North America is attested by the Late Pliocene to Pleistocene fossils found in Texas, Arizona, Kansas, Oklahoma, Mississippi, Indiana, Missouri, Illinois, and Quebec, which do not differ appreciably from Recent Holarctic shells.

The slug *Deroceras laeve* has an extraordinary, puzzling distribution, partly indigenous and partly adventive, therefore difficult to evaluate correctly. Pilsbry (1948:522) gave his reasons for regarding it in North America as a truly indigenous, pre-Columbian circumpolar element. This is supported by the senior author's finding Late Pleistocene internal shells, not separable from those of Recent Arizona slugs, in the San Pedro Valley, associated with mammoth remains 10,000 to 11,000 years old. It is the only native circumpolar member of the Limacidae, although there are five other Recent species of *Deroceras* in North America (three strictly American; two introduced from the Old World). *D. laeve* is indigenous over much of temperate Eurasia, from Iceland, Scandinavia, and Siberia (north to 70° N) to Kamchatka; in America from Alaska (west to the Aleutian Islands), Alberta, Manitoba (north to Churchill, 58° 45′ N, 94° 5′ W), Northwest Territories (north to Southampton Island, 64° 8′ N, 83° 10′ W, where it is the only land mollusk), Ontario, South Baffin Island, north Labrador, and Newfoundland, southward to Florida and to the Mexican border in Texas, New Mexico, Arizona, and California. It is not known from Sonora or Chihuahua, but occurs farther south in Mexico (Mexico, D. F.; Puebla; Vera Cruz), Guatemala, and Nicaragua; whether it is always native there is uncertain. It is remarkable in America for its unusually wide horizontal, vertical, and ecological range—from sea level in the north and east, to nearly 12,000 feet in some New Mexico mountains (C. C. Hoff, 1962:55), and from Arctic tundra to southern mesic forests. It was introduced by man to Hawaii, Australia, South Africa, and no doubt elsewhere. In America it is a Pliocene to sub-Recent fossil in Quebec, Ohio, Missouri, Illinois, Mississippi, Louisiana, Kentucky, and Arizona. Like most cosmopolitan synanthropic mollusks it has an excess of synonyms, some 15 listed by Pilsbry (1948:539) for North America alone. *D. aenigma* A. B. Leonard (1950, *Kansas Univ. Paleont. Contrib. Mollusca* 3:38; Pl. 5, fig. E), based on Pliocene internal shells of Kansas and also reported from the Pleistocene of Nebraska, Oklahoma, and Texas, seems doubtfully distinct from *D. laeve.*

Pupilla, a strictly Holarctic genus mainly of boreal or cold temperate areas, has a checkered geologic history. The fossil record shows that it probably originated in north Asia, later migrated to central Europe, perhaps in the Late Oligocene, and eventually to North America in the Pliocene or Pleistocene. Whether Recent or fossil, all species vary much in size, shape, coiling, and apertural armature of shell, making correct specific identification often hazardous, particularly of fossils. Their taxonomy is in urgent need of critical revision on a worldwide basis. We follow Pilsbry (1920-1921: 152-225; 1948:926-942) in recognizing six American species (some with several named "forms" or "subspecies"). The most distinctive, *P. sterkiana* (H. A. Pilsbry, 1889), precinctive in Baja California, is perhaps an offshoot of a very early (Middle Tertiary?) immigrant from Asia. As mentioned before, *P. sonorana* is a Recent precinctive of New Mexico (if specifically distinct from *P. blandii*) and trans-Pecos Texas. *P. blandii, P. hebes,* and *P. syngenes* were discussed before as invaders from the Rocky Mountain Province, where they presumably originated from an ancient, now extinct Eurasian ancestor.

P. muscorum, circumpolar in the Recent fauna, does not seem to have changed much since it reached the New World from Asia. In both North America and Eurasia it occurs in four tooth-mutations, but the toothless form is the more common in America, where some shells resemble *P. hebes* in this respect. In the Old World it lives from Iceland and the British Isles throughout most of Europe, North Africa, west, central, and north Asia, east to Sakhalin Island and south to Iran. In America it is Recent from Alaska (north to Anuk, 62° 19′ N, 163° 51′ W), British Columbia, Alberta (north to Laggan), Manitoba (north to Churchill, 58° 45′ N, 94° 5′ W), Ontario, Quebec, Anticosti Island, and Newfoundland, south to New Jersey (Atlantic City, 39° 21′ N, 74° 27′ W) and Maryland (Carroll County and Frederick County, 39° 30′ N, 77° W, W. Grimm, 1959, *Nautilus* 72 [4]:124) in the east (the southeast Stations probably due to human transport), and west to Oregon (nine miles east of Milton, Umatilla County, 45° 56′ N, 118° 25′ W), Colorado, central New Mexico (southeast to White Oaks, Lincoln County, 33° 42′ N, 105° 30′ W; Duran, Torrance County, 34° 35′ N, 105° 30′ W), Utah, and northern Arizona. Pilsbry (1948:934) noted that it "has a far greater zonal or climatic range, as well as a wider geographic distribution in the Palearctic Region than in America," where "it has not yet become adapted to warm climates such as the circum-Mediterranean zone which it inhabits in the Old World." In this connection, it is relevant that in the northern, cooler section of the Nearctic Region it occurs to near sea level, while in Arizona and New Mexico it now lives only at high elevations (6,700 to 12,000 feet) in the Alpine, Hudsonian, and Canadian Life-Zones. It is not known Recent south of the Mexican border. Evidently it was more widespread, even at low elevations, during the Late Cenozoic Era, since it is a common Pliocene to Pleistocene fossil in Ohio, Missouri, Iowa, Nebraska, Kansas, Oklahoma, New Mexico, Texas, Arizona, and Wyoming. The few specimens taken in river drift in Kansas by G. D. Hanna (1909:94) were no doubt washed-up fossils.

During the Late Cenozoic two species of *Columella* entered North America from Eurasia via the Bering Strait isthmus ("Beringia"). Typical *C. edentula,* widespread in the Palearctic Region from Iceland to Japan, occurs now in America from Alaska, the Magdalen Islands, Labrador, and Newfoundland, south to Oregon, Montana, Colorado, Minnesota, Iowa, Pennsylvania, Virginia, West Virginia, and Alabama. It is not known from Arizona, or elsewhere in the Southwest. Eastern specimens from the United States and Canada, as far west as Minnesota, are sometimes called "form" *simplex* (A. A. Gould, 1840). Snails from Arizona, referred at one time to *C. edentula,* were *C. columella alticola,* as Pilsbry (1948:1003) recognized. In a recent revision of the genus, L. Forcart (1959) divides the circumpolar *C. columella* (J. P. R. Draparnaud, 1805) in three subspecies: *C. c. columella* in boreal Eurasia, from Scandinavia to Siberia; *C. c. inornata* (A. G. L. Michaud, 1831) (synonym: *C. c. gredleri* S. Clessin, 1872) in the Central European Alps; and the North American *C. c. alticola* (E. Ingersoll, 1875), of which he did not study specimens. Since Pleistocene fossils of *C. columella* are known from both the Old and New Worlds, the species should be added to C. H. Lindroth's (1957:119) list of truly pre-Columbian circumpolar animals.

In the Recent American fauna, *C. c. alticola* is strictly western, in Alaska (west to the Aleutian Islands, north to Port Clarence, 65° 15′ N, 166° 30′ W), British Columbia (Field, 51° 25′ N, 116° 30′ W), Alberta (Banff, 51° 15′ N, 115° 29′ W), Manitoba (north to Churchill, 58° 45′ N, 94° 5′ W, W. J. Wayne, 1959, *Nautilus* 72:92), Wyoming, Utah, Colorado, Arizona, and New Mexico (southeasternmost record, Willow Creek, Mogollon Mts, ca 33° 10′ N, 108° 30′ W), usually at high elevations, particularly in southern areas.[22] It is

---

[22] J. Oughton's (1940, *Nautilus* 53 [4]:128) record of *C. c. alticola* from the south end of Baffin Island (60° 50′ N, 66° 20′ W) was no doubt based on misidentified *C. edentula,* as appears from Pilsbry's notes on Oughton's paper.

recorded as a Pleistocene fossil in Kansas, Oklahoma, Iowa, Ohio, Illinois, Missouri, and Mississippi, but not in Arizona. However, the senior author found in the Late Pleistocene of San Pedro Valley, Arizona, at Lehner Mammoth Site, specimens of the extinct *Columella hasta* (G. D. Hanna, Oct. 1911, *Proc. U.S. Nat. Mus.* 41 for 1912 [1865]:372, fig. 1), originally described from the Pleistocene of Kansas; this is perhaps a giant subspecies of *C. columella.*

Vertigo, a prolific genus of Pupillidae with distinctive features of shell, animal, and anatomy, should be particularly interesting to the zoogeographer. It is restricted to most of the southern Holarctic Realm, extending north to a few degrees below the Arctic Circle. It occurs to near sea level in the boreal and cold temperate areas, but reaches high elevations (10,000 feet or more) in the south. It may have had two disconnected active centers of evolution, one in Eurasia, the other in North America. Of a total of some 80 species, 35 are recognized as Nearctic in Pilsbry's latest revision (1948: 943-1000). In America, where it is an old element (two extinct species known from the Eocene of Wyoming), it remains in active evolution, to judge from the many described, but often weak, subspecies (nine in *V. gouldii,* six in *V. modesta,* eight in *V. californica*).

Most Nearctic species are precinctive, and some are restricted to small, disjunct areas (for instance, *V. hinkleyi,* known only from the Huachuca Mts and the Dos Cabezas Mts in Arizona, and from the Sierra de la Breña in Chihuahua). It is surprising that only three of the many species are circumpolar: *V. pygmaea* (J. P. R. Draparnaud, 1801), of Eurasia and northeast North America, sometimes suspected of being only a post-Columbian introduction by man in the New World; *V. alpestris* (J. Alder, 1839), widespread in Eurasia, but known in America from only two sub-Arctic localities (south end of Baffin Island and Fort Severn on Hudson Bay), a puzzling distribution possibly due to insufficient collecting; and *V. modesta* (T. Say, 1824), with its synonyms *V. borealis* A. Morelet, 1858, *V. arctica* von Wallenberg, 1858, and *V. krauseana* O. Reinhardt, 1883. Of the three circumpolar species, *V. modesta* is the only one known fossil thus far in North America (Pleistocene of Illinois and Ohio).

Vertigo modesta ranges in several subspecies in northern Eurasia from Iceland, boreal and montane Europe, and Siberia to the Commander Islands; in America from the Aleutian Islands, continental Alaska, British Columbia, Manitoba, Hudson Bay, Labrador, Greenland, Newfoundland, and northeast United States, south to Connecticut in the east and to south California, Colorado, Arizona, and New Mexico in the west. It is known as a

Late Cenozoic fossil in Eurasia, and in North America from Indiana, Illinois, Iowa, Missouri, Kansas, and Wyoming, but thus far not from Arizona. Two Recent subspecies are recognized in Arizona, both at high elevations (6,000 to 12,000 feet; all known local Stations in the Check List), but, as Pilsbry (1948:989) admits, their distinction is "rather finely drawn." *V. m. corpulenta* is known only in the Huachuca Mts. *V. m. ingersolli* is more widespread in the northern and eastern highlands and in the southeast mountains of the State.

Cochlicopa lubrica (*Cionella lubrica* of most American malacologists) has one of the widest horizontal and vertical distributions among terrestrial snails. It is Recent over most of Eurasia and much of North America, from near sea level in Alaska, the eastern United States, and Canada to over 10,000 feet in the Southwestern Province. In the Palearctic Region, it ranges from near the Arctic Circle in Iceland, the Faeroes, and Scandinavia, through the British Isles, most of continental Europe, and Siberia, east to Kamchatka and Japan, and south to northwest Africa, Transcaucasia, north China, and Korea; also in the Azores and Madeira.

In America it lives from extreme north Alaska (Point Barrow, 71° 23′ N, 156° 28′ W) and British Columbia to Hudson Bay, Labrador, and Newfoundland, southeast to Missouri and north Alabama, and west to Washington, Oregon, Idaho, Nevada, Utah, Colorado, Arizona (all Stations in the Check List), New Mexico (east to the Sacramento Mts, 32° 50′ N, 105° 40′ W), and trans-Pecos Texas (McKittrick Canyon, Guadalupe Mts, Culberson County, ca 31° 50′ N, 104° 40′ W); it is not known from Greenland or California. In Mexico it occurs in north Chihuahua (Sierra de la Breña, ca 30° 10′ N, 108° 10′ W, 7,000 feet) and Nuevo Leon (Sierra Madre near Galeana, ca 30° N, 108° W, 7,800 feet, southernmost Station of the species). It is a Pleistocene fossil in the Old and New Worlds; in America, in Illinois, Missouri, Mississippi, Kansas, Oklahoma, New Mexico, Arizona, and Texas. Variation is about the same in fossil and Recent shells, with slender and obese shells, as well as intermediates, often occurring together in the same population or deposit.[23]

---

[23]The nomenclature of the variants of *Cochlicopa lubrica* was fully discussed by Pilsbry (1907-1909:321). *Achatina lubrica* variety *exiguus* K. T. Menke (1830, without definition) is an invalid nomen nudum. The first definition and valid use of the name *exiguus* in the genus was as *Bulimus (Cochlicopa) subcylindricus* variety *exiguus* by A. Moquin-Tandon (1855) for Recent snails at two localities in France (Metz and Grenoble), for dwarfed, somewhat more slender shells. This European *exiguus* is a synonym of the earlier *Cochlicopa lubrica lubricella* (C. Porro, 1838), also based on Recent European snails, and this name should be used for slender specimens if one were needed.

The shell of *C. lubrica* is quite variable, often within the same Recent or fossil population, in size, shape, proportions, thickness and color, shape of aperture, apertural lip, etc. Nearctic and Palearctic specimens are usually regarded as strictly cospecific, although two names, *Zua buddii* D. Dupuy (1849) and *Bulimus lubricoides* W. Stimpson (1851), were proposed (as nomina nuda) for American specimens merely on geography. Pilsbry (1948:1049) accepts the subspecies *C. lubrica morseana* W. Doherty (1878) for the snails of the eastern United States, from Michigan and New York south to Alabama. It is doubtful that all southeastern lots of shells could be separated consistently from all northeastern and western populations without knowing their origin.

The family Vitrinidae is primarily Palearctic, with many Recent species and several supraspecific groups in Eurasia, where it is also known fossil (Upper Oligocene to sub-Recent). *Vitrina*, sensu lato, of J. Thiele (1931:599-600), has its centers of present density, and presumably of past speciation and dispersal, in the Palearctic Region. Most species occur in continental Eurasia, from the Arctic Ocean to the Mediterranean, Asia Minor, and Kamchatka; a few reach the Atlantic islands (Azores, Madeira, Canary Islands) and the mountains of southern Arabia, Ethiopia, and tropical East Africa at high elevations, close to snow line. By contrast, there are few Recent and no fossil Nearctic species, and the genus has not been found south of the Mexican border. The American forms are all of the subgenus *Vitrina,* sensu stricto, and chiefly boreal, living to near sea level in the north, but only at higher elevations in the south (usually in the Canadian and Hudsonian, seldom in the Transition or Alpine Tundra Life-Zones).

Pilsbry (1946:499-505) recognized three Nearctic species: *Vitrina pellucida angelicae* H. H. Beck (1837) in Greenland and Iceland, nominate *V. p. pellucida* (O. F. Müller, 1774) being Palearctic; *V. limpida* A. A. Gould (1850) in the northeast; and *V. alaskana* W. H. Dall (1905) in the west. In a revision of the boreal ("nordic") species, based on the shell and anatomy, L. Forcart (1955) reduced them to two species and one subspecies: *V. angelicae angelicae* (restricted to Greenland), *V. angelicae limpida* (the eastern form), and *V. alaskana* (retained provisionally for the western form, as he could not study its anatomy). *V. angelicae limpida* is Recent from Labrador (north to Blanc Sablon) and Newfoundland south to Pennsylvania (southernmost Station near Pittsburgh, ca 40° 20' N), and as far west as Michigan, Manitoba, and Alberta (westernmost Station, Red Deer, 52° 17' N, 113° 46' W).

The junior author has now dissected the western *alaskana* from fresh Arizona snails (collected by R. H. Russell, Sept. 11, 1969, at Lockett Meadow, west slope of Sugarloaf Mtn, San Francisco Mtn, 10 miles north of Flagstaff, at 8,500 feet, Coconino County) and from preserved Montana snails (collected by R. H. Russell, Oct. 24, 1964, at the junction of Deep Creek and Clark Fork River, 5 miles west of Missoula, at 3,000 feet, Missoula County). In both lots the reproductive anatomy was that of the Eurasian *V. pellucida,* as described and figured by L. Forcart (1955:159, fig. 1). A large penial sheath is present and the vas deferens is bound within it; there is a prominent swelling at the base of the spermathecal duct. Accordingly, *alaskana* should be treated as a subspecies of *V. pellucida,* kept as such only for convenience as the American offshoot of a Pleistocene immigrant from Eurasia via the Bering Strait. It is Recent in Alaska (west to the Aleutian Islands, British Columbia, Washington, Oregon, Idaho, Montana, Wyoming, the Dakotas, Colorado, Utah, Nevada, California, Arizona (all Stations in the Check List; southernmost, Miller Peak, Huachuca Mts, 31° 23′ N, 110° 17′ W), and New Mexico (southeasternmost Station, Cloudcroft, Sacramento Mts, 8,300 feet, ca 33° N, 105° 50′ W, Otero County). The present American ranges of *V. angelicae limpida* and *V. pellucida alaskana* nowhere overlap, although *limpida* extends farther west in the north than *alaskana* extends east in the south. It is of interest that the western *alaskana* is much more closely related to the Eurasian *Vitrinae* than the eastern *limpida.* Probably both *pellucida alaskana* and *angelicae limpida* were derived from Eurasian immigrants via the Bering Strait, in which case the ancestor of *limpida* (and *angelicae*) must have reached America much earlier than that of *alaskana,* since *angelicae* had the time to evolve into a species anatomically distinct from its Old World ancestor.

Discus cronkhitei should perhaps also be regarded as circumpolar, if it could be shown definitely that it is only the New World subspecies of the Palearctic *Discus ruderatus.* Meanwhile we follow Pilsbry in retaining it as specifically distinct, and will discuss it later as a widespread Nearctic element.

5. A **Pan-American** element, autochthonous in the New World, and Recent both north and south of the Tropic of Cancer, comprises *Retinella indentata, Hawaiia minuscula,* and *Striatura meridionalis* in Zonitidae, and *Pupoides albilabris* and *Gastrocopta pellucida* in Pupillidae. These species have in common an extensive, strictly American distribution, covering much of the Nearctic and most of the northern half of the Neotropical Regions, raising in-

teresting and at present debatable problems. Did all five have essentially the same evolution, that is, originate at one or several nearby centers of dispersal, in either the north or south section of their present-day ranges? Or are some of northern (Nearctic) and others of southern (Neotropical) ancestry? Although four of the five are known fossil from the Late Cenozoic in the United States only, this evidence is one-sided and may be misleading, since very little is known of the Cenozoic land mollusks in Mexico, Central America, the Antilles, and northern South America. The problem is complicated by the fact that the present-day range of some of these species is undoubtedly in part artificial, due to post-Columbian transport by man, the importance of which is difficult to assess.

The general distribution of *Retinella* was discussed before for the circumpolar *Nesovitrea hammonis*. As a species, the strictly American *R. indentata* (including *R. i. paucilirata*) occurs in the east from Maine (north to Oxford and Hancock counties, ca 45° N) and south Ontario (Muskoka District, ca 45° N, 79° 20′ W) to Florida and the Gulf of Mexico, west to Michigan, Manitoba, east Kansas (Shawnee County, ca 39° N, 95° 40′ W), Utah, Texas, New Mexico, and Arizona (northwest to Arch Canyon, Ajo Mts, Pima County, 32° 2′ N, 112° 42′ W). As *Retinella i. paucilirata,* it is known in Mexico from Baja California, Sonora, and Chihuahua (Sierra de la Breña, at 7,000 feet, collected by the junior author), and reaches central Guatemala (south to Salama, 15° 6′ N, 90° 18′ W, the T.L. of *R. i. paucilirata*).[24] It is the only species of *Retinella* of the Neotropical Region, where, moreover, it lives only at high elevations in the "Tierra Templada." As it is unknown in northwestern America, it seems to be an autochthonous Nearctic snail that originated in southeastern Appalachia, where the genus has its present center of density and from where it may have spread southward during the Pleistocene to beyond the Tropic of Cancer. Nominate *R. i. indentata* or *R. i. paucilirata* have been reported from the Pleistocene of Indiana, Illinois, Missouri, Kentucky, Mississippi, Louisiana, Oklahoma, and Texas, but thus far not of Arizona.

Although clearly a single specific entity, *R. indentata* is of evolutionary interest as a possible example of intraspecific differentiation into two geographic clines. The northern (cool temperate) and southern (warm temperate to tropical) populations show a slight, but noticeable difference in the shells. The umbilicus of the northern, nominate *R. i. indentata* is rimate, that is, punctiform to nearly closed; whereas that of the southern *R. i. paucilirata* is

---

[24]A shell with animal of *R. i. paucilirata* was taken from the crop of a house wren (*Troglodytes aedon*) of the Sierra Huachinera, Sonora, by J. T. Marshall.

distinctly open, though narrow. The difference in the shells of the two forms does not appear to have any adaptive significance. Nominate *indentata* is the form of southeastern Canada (west to Manitoba) and the eastern United States (south to Florida), *R. i. paucilirata* that of the southwestern States (from Texas to Utah and Arizona, where the species reaches its northwestern limit), Mexico, and Guatemala. They meet and overlap in eastern Kansas and Oklahoma, where the populations may be of either form, often with transitional snails, showing that the two forms interbreed freely and are not specifically distinct.

Hawaiia, with the species *H. minuscula,* is one of very few truly monotypic genera of land mollusks. It is a pronounced synanthropic and eurytopic snail, adaptable to a wide range of environments, being now nearly cosmopolitan in temperate and subtropical countries. Its original home was at one time a puzzle, and is only partly cleared up. Its frequent and widespread occurrence in Pliocene, Pleistocene, and sub-Recent deposits in Ohio, Iowa, Indiana, Illinois, Missouri, Louisiana, Mississippi, Nebraska, Kansas, Oklahoma, Texas, New Mexico, Arizona, Colorado, Wyoming, and Idaho, and its absence from them in the Old World, seem conclusive proof that it is indigenous and autochthonous in the Nearctic Region, and not originally circumpolar.

In North America *H. minuscula* is Recent and presumably native from south British Columbia (Vancouver Island), Manitoba, Idaho, Montana, Wyoming, the Great Lakes area, south Ontario, south Quebec, Newfoundland, and Maine, to Florida, Alabama, Texas, Oklahoma, Kansas, Colorado, New Mexico, and Arizona. South of the Mexican border it is known from Baja California, Sonora (several Stations, B. A. Branson et al, 1964:103), Tamaulipas, San Luis Potosi, Vera Cruz, Puebla, Nayarit, and Yucatan;[25] and farther south, in Guatemala and Costa Rica. Thus far there are no or very few Recent records from Washington, Oregon, Utah, Nevada, north California, or continental Alaska (reported only from the Aleutian Islands, Shumagin Island, and Muir Inlet, probably introduced by man). In south California, according to Pilsbry (1946:423), "some of the records are probably owing to importation with plants"; the record from the Late Miocene of Barstow Hills, San Bernardino County, by Hibbard and Taylor (1960:148) is puzzling, and may be due to contamination with Recent shells.

---

[25]According to H. B. Baker (1930a:35), *Punctum pygmaeum albeolum* H. W. Dall (1926, *Proc. California Acad. Sci.* (4) 15 [15]:481, Pl. 35, figs. 18-19) from Maria Magdalena Island of the Tres Marias Group, off Nayarit, is *Hawaiia minuscula.*

Hawaiia minuscula was described in 1841 from Ohio and Vermont, without precise locality; but it may be assumed that the cotypes were indigenous, not adventive snails, since they could hardly have been imported there by man at that early date. There is, however, much uncertainty about the indigenous status of some later North American records, particularly from south of the Mexican border. Under the circumstances, it would serve no useful purpose to determine precise limits for its Recent distribution. It is found also, certainly as human introductions, in Bermuda, all of the Greater Antilles, many of the Lesser Antilles, and sporadically in South America, particularly in the western foothills of the Andes of Ecuador and Peru. It is a feral adventive (in the open) in Hawaii and other Pacific islands, and, in the Old World, in Japan, Korea, Formosa, and the southern Maritime Province of Siberia (introduced in southeast Siberia in our opinion, although C. H. Lindroth, 1957: 116, gives it as probably indigenous there). In Europe, it is known only from greenhouses in Great Britain, Ireland, the Netherlands, and Switzerland. Its wide, disjunct, and often artificial distribution accounts for the many generic and specific synonyms, only a few being given in the Check List. It has also been confused with other similar lesser snails, adding to the uncertainty of its distribution. In our experience, it retains its diagnostic features wherever it occurs. The published named subspecies or varieties seem to be based on geography rather than on consistent, observable differences. The shell often varies within one population in minor details, such as size for the same number of whorls, shape, and sculpture.

The genus *Striatura* is peculiar to America (five species, south to south Mexico) and Hawaii (three species), an anomalous type of distribution possibly due in part to avian transport from America across the Pacific in the geologic past. Mainly according to H. B. Baker (1930a:36), *S. meridionalis,* the only Arizona species, lives in the southern United States from southern New Jersey (north to ca 40° N) and southeastern Pennsylvania to Florida in the east, and to Kentucky, Arkansas, Oklahoma, Texas, New Mexico, southeastern Colorado, and Arizona (west to Kitt Peak, Pima County, 31° 58′ N, 111° 36′ W) in the west. South of the Mexican border, it is unknown from Sonora, but occurs in Chihuahua (Sierra de la Breña, 7,000 feet, collected by the junior author), Puebla, Nuevo Leon, and Vera Cruz (south to Orizaba, 18° 54′ N, 97° 5′ W). It may be more widely distributed there, as it often eludes collectors due to its minute size. It is recorded from Bermuda, possibly as an adventive, but has not spread to the Old World. It shares with *Retinella indentata* the distinction of being one of the very few land snails that presumably originated in the temperate Nearctic Region and later migrated by natural means to the subtropics. It has not been found fossil thus far.

Pupoides albilabris, a strictly American snail, has one of the widest, mostly natural distributions known for a land snail. In North America it is Recent from Maine (north to ca 45° N) and southern Ontario to Florida and the Gulf of Mexico, west to the Dakotas, eastern Colorado (near Boulder, ca 40° N, 105° 18' W; Trinidad, 37° 6' N, 104° 30' W), Utah, and Arizona (northwest to Mt Trumbull, Mohave County, 36° 20' N, 113° 5' W; southwest to Palm Canyon, Kofa Mts, Yuma County, 33° 20' N, 114° 5' W). South of the Mexican border it is known from Baja California Territorio Sur (San Ignacio, 27° 20' N, 112° 50' W, 500 feet; San José de Comondú, 26° 5' N, 111° 50' W, 1,500 feet; and some islands in the Gulf of California), Chihuahua (Presa Chihuahua, 7 miles south of the city of Chihuahua, 4,800 feet, R. H. Russell, 1971), Sonora (wash at west side of Pinacate Peak, 31° 50' N, 113° 30' W, collected by F. Werner, 1969; and elsewhere), Sinaloa (Rio Fuerte near San Blas, 26° 5' N, 108° 46' W), Nuevo Leon (Monterrey), and Tamaulipas (Tampico; drift of Rio Purificación, 24 miles west of Padilla, by the senior author, 1958). It has been recorded as *P. albilabris nitidulus* from Bermuda, the Bahamas, the Greater Antilles, Curaçao, and the coastal areas of Venezuela and Colombia.[26] The recently described *Pupoides albilabris peruvianus* W. K. Weyrauch (1960, *Archiv f. Molluskenk.* 89 [4-6]:117, Pl. 11, figs. 1-2) from the west slope of the Andes in central Peru, 50 km northeast of Lima, extends the range of the species far south of the Equator (if actually based on native snails and not on human transport, which could only be settled definitely by finding it as a Late Cenozoic fossil in the area). According to the author, it is a southern relict from a more extensive Tertiary natural range of the species, continuous once with that north of the Equator. At the T.L., *P. a peruvianus* was found fixed to the under side of stones in an arid rocky area at 3,750 feet, which is of interest to us because *P. a. albilabris* sometimes lives in Arizona also at similar, very arid Stations at about the same elevation.

---

[26]*Pupoides albilabris nitidulus* was originally described as *Bulimus nitidulus* L. Pfeiffer (1839, *Archiv f. Naturgesch.* 5 [1]:352; T.L.: Matanzas, Cuba), a specific name which, if acceptable, would take precedence over *Pupa albilabris* C. B. Adams (1841). Fortunately, L. Pfeiffer's *nitidulus* is antedated by *Bulimus (Bulimulus) nitidulus* H. H. Beck (1837, *Index Moll.*:67), which was validly proposed as a substitute for *Bulinus* [sic] *nitidus* W. J. Broderip (1832, *Proc. Comm. Soc. Zool. London* 16:31), itself based on a reference to a pre-publication copy of a figure of *Bulinus* sp. by J. B. Sowerby (published March 1833 in *Conchol. Illustrations,* Pt. 21). If the West Indian specimens of *P. albilabris* were worth separating subspecifically, which we doubt, they would require a new name; either *Pupa modica* A. A. Gould (1848) or *Pupa parraiana* A. d'Orbigny (1841) might perhaps be available for the purpose.

P. albilabris owes its wide range to its eurytopic tolerance to a wide variety of environments. In the Nearctic Region it lives from arid near-desert in west Arizona and Sonora to moist mesic forest in east Texas, and from the severe winters of the northeast United States to the subtropical heat of south Texas. A. B. Leonard and J. C. Frye (1962:26) noted that: "Although elsewhere it occupies various habitat situations, including some very humid ones, *P. albilabris* is one of the few common snails in the short-grass environment on the Great Plains, where it endures long periods of severe dessication and extremely high temperatures." In Arizona, it is often found in gardens, lawns and pastures, under logs, stones, or cowpads, or at the roots of grasses. Its great adaptability no doubt helps frequent transport by man and prompt feral settlement of new areas. For this reason it may be suspected that much of its present wide range is artificial, particularly in the West Indies. It is certainly indigenous (pre-Columbian) in Nearctic America, since it is a common Pliocene, Pleistocene, and sub-Recent fossil in Missouri, Illinois, Ohio, Iowa, Wisconsin, Nebraska, Kansas, Oklahoma, Texas, New Mexico, and Arizona. Unfortunately, as A. L. Metcalf (1967: 42) points out, "it is of little value as a paleoecological indicator because of its tolerance to a variety of habitats." Moreover, its Recent as well as fossil shells are highly variable, often within one population or at the same fossil horizon, in size (for same number of whorls), number of whorls in adults, shape (height, breadth, convexity of whorls), outline of mouth, reflection and strength of peristome, development of apertural callus, sculpture, etc—features sometimes trusted overly in defining subspecies. In referring shells from the Plomosa Mts, Yuma County, to *P. albilabris,* Pilsbry (1948:923) stated: "In Arizona the shell is often small and delicate, length 4 mm, hardly over five whorls, with scarcely any callous pad in the angle of the mouth (Fig. 499, 4), thus resembling *P. modicus;* but in the same lot there are also larger shells." A Pleistocene fossil from Murray Springs in the San Pedro Valley, Cochise County, agrees well with the description and figures of *P. modicus,* and even better with the original figure of *Pupa arizonensis* W. G. Gabb (1866; synonymized by Pilsbry with typical *P. albilabris*), which was based on a dead shell from the old Fort Grant in the San Pedro Valley, Pinal County, perhaps a fossil washed up in drift.

The peculiar Recent distribution of the strictly American *Gastrocopta pellucida* suggests that it originated as an autochthonous Neotropical snail that migrated northward, eventually reaching the southern, warm temperate sections of the Nearctic Region. It is at present widespread in the Bahamas and Greater Antilles and, on the Continent, over a continuous area in Mexico, possibly its original center of dispersal (recorded from Baja California, Sonora, Sinaloa, Morelos, Nuevo Leon, San Luis Potosi, Tamaulipas, Vera Cruz, Hidalgo [El Dedho, five miles west of Zimapan, by the senior author, 1958], Tabasco, and Yucatan), and Guatemala (Salama; Guatemala City; Antigua, at 4,500 feet). Reported occurrences in Panama and Ecuador (E. von Martens, 1898:328) may be based on misidentifications.

In the southern United States it is uniformly distributed and common in Texas, Oklahoma, Kansas, New Mexico, and Arizona, but rare and sporadic in southeast Colorado (Trinidad, Las Animas County, 37° 10′ N, 104° 31′ W), east Utah (northernmost Station of the species, Willow Springs, Grand County, 38° 50′ N, 109° 30′ W), and south California (Mohave Mts, San Bernardino County, ca 34° 35′ N, 114° 40′ W; Palm Canyon, San Jacinto Mts, Riverside County, ca 33° 45′ N, 117° W). East of this southwestern range, and separated from it by a wide hiatus, the species reappears sporadically on the Atlantic seaboard from south New Jersey (Cape May, 38° 56′ N, 74° 56′ W) to Florida, but never far inland and often on offshore islands. These eastern coastal Stations may be due to aerial transport from the Caribbean by the frequent and powerful hurricanes.

G. pellucida was originally described from Cuba, and early writers on North American mollusks (F. von Roemer; W. G. Binney; G. W. Tryon) used the specific name *pellucida* for the continental snails also. Pilsbry (1890c:44, Pl. 1, figs. G-K), however, described shells from Arizona (without precise locality), Texas (New Braunfels), and Florida (St. Augustine) as a distinct species, *Pupa hordeacella*. When he recognized later that the Antillean *pellucida* and the continental *hordeacella* were cospecific, he called the shells from the Southwest and the Atlantic seaboard *Gastrocopta pellucida hordeacella*, although he had to admit later (1948:914) that those from Florida were "scarcely if at all distinguishable from West Indian *pellucida*." He had written earlier (1916-1918:78-79) of *hordeacella*: "This continental race differs from *pellucida* by having a slight crest behind the outer lip, and a somewhat longer lower palatal plica. There is often no projection on the columellar side of the angulo-parietal lamella. Also by the pale brown color and the average larger size. In the original description [of *hordeacella*] a specimen of the minimum size was selected as type, but in the same lot the size is variable, from length 1.8, diam 0.76, to length 2.5, diam 1 mm. All of the characters distinguishing *hordeacella* from *pellucida* vary so much that in some individual cases, without a large series, there is little or no difference; but it is only the smallest individuals of any lot of *hordeacella* which would be taken for *pellucida*. The status of the subspecies is rather uncertain, and possibly it might be abandoned with advantage." We are unable to separate *pellucida* and *hordeacella,* except by locality, and follow B. A. Branson et al (1966:149) in synonymizing both *G. p. hordeacella* and *G. p. parvidens* with *G. pellucida.* The species is known as a Pleistocene fossil in Kansas, Texas, New Mexico, and Arizona, but not farther east, which adds to the possibility that the sporadic occurrences on the Atlantic coast are late arrivals from the Caribbean.

6. The **California** element in the Recent Southwestern land fauna consists primarily of the large snails derived from typical Helminthoglyptidae of the Pacific Coast area (California Molluscan Province), namely from *Micrarionta* and *Helminthoglypta.* At the outset, these became gradually adapted to increasing aridity in the desert mountains of southeast California, producing highly xerophilous species of *Eremarionta, Sonorelix,* and *Mohavelix,* as well as a few peculiar *Helminthoglypta* that retained the original generic anatomy. After migrating east to more propitious territory, some desert *Eremarionta* evolved further, eventually blossoming out into the profusion of *Sonorellae,* now the dominant element of much of the Southwestern Province. The putative history and Recent distribution of the four genera of helminthoglyptids were analyzed before. Although of western California ancestry, their Southwestern species are all precinctive in the Province.

Except possibly for two species of *Punctum* to be discussed presently, there seems to have been little or no expansion of range, with concomitant adaptive radiation, for the lesser Californian land mollusks. This may be due mainly to the fact that the North American families of lesser snails, such as Endodontidae, Zonitidae, Valloniidae, and Pupillidae, are at present poorly or not represented in the Pacific Coast fauna.

In America the genus *Punctum,* in Endodontidae, is mostly Nearctic. Of the three species of the Southwestern Province, only *P. californicum* and *P. conspectum* are known Recent in Arizona. Their general distributions are so similar that their evolutionary history was presumably much the same. Their main Recent ranges are near the Pacific Coast and in adjoining mountains, while occurrences elsewhere are spotty and discontinuous. It may, therefore, be presumed that both originated within the present-day boundaries of the California Molluscan Province. From there they migrated at first north, in the case of *P. conspectum* as far as the Aleutian Islands. They invaded the Rocky Mountain Province later from this northwest area, and eventually reached their few known

Stations in the Southwestern Province, possibly by way of the Rocky Mountains.

The Recent range of *Punctum californicum* covers mainly California west of the deserts, from 42° N to 33°N. The species is more scattered in the Rocky Mountain Province (Montana and Colorado), and known in North Dakota only from drift. Its actual distribution in the Rockies is imperfectly known. In Arizona this snail is widespread, west to Yavapai County (ca 34° 30′ N, 112° W) and Pima County (Santa Catalina Mts), and east to Apache County (Black River near Horseshoe Bend, 32° 43′ N, 109° 27′ W), Graham County (Pinaleno Mts), and Cochise County (southernmost Station, head of Cave Creek, Chiricahua Mts, 8,000 feet, ca 31° 51′ N, 109° 16′ W).[27] There are no records from New Mexico, Sonora, or elsewhere in Mexico. It occurs as a Late Pleistocene fossil in the San Pedro Valley of Arizona. If *P. californicum* were regarded as the southern subspecies or representative of the northwestern *P. randolphii* (W. H. Dall, 1895), as now only suspected, its Recent distribution might be much easier to understand, *P. randolphii* being known from British Columbia, Washington, Oregon, and Idaho.

Punctum conspectum extends farther north than *P. californicum*. It is known from the south coast of Alaska (Sitka, 57° 10′ N, 135° 30′ W), the Aleutian Islands (northwesternmost Station, Unalaska, 53° 30′ N, 166° 30′ W), British Columbia (Vancouver Island), Washington, Idaho, Montana (northeasternmost Station in Sweet Grass County, ca 46° N, 110° W), Washington, Oregon, and California (west of the deserts, south to San Diego County). It seems to bypass the Rocky Mountains, but reappears in east Arizona at two Stations in Apache County (mouth of Reservation Creek, ca 33° 42′ N, 109° 29′ W; two miles below Thomas Peak summit, ca 33° 55′ N, 110° W), and at one locality in Catron County, west New Mexico (southeasternmost Station, Willow Creek, Mogollon Mts, ca 33° 10′ N, 108° 30′ W), if these records are based on correct identifications. It has not been recognized as a fossil thus far. A subspecies, *P. c. jaliscoense* (H. A. Pilsbry, 1926), described from Mexico (Jalisco and Mexico, D.F.), is a zoogeographic riddle.

7. Twelve remaining species of lesser land snails, in seven genera, will be discussed together as a heterogeneous group of widespread, essentially autochthonous **Nearctic** elements. They

---

[27]Snails from the Black River were at first referred by Pilsbry and Ferriss (1919a:326) to *Punctum pygmaeum* (J. P. R. Draparnaud, 1801), a European species not now recognized in North America. The record was later transferred to *P. californicum* by Pilsbry (1948:648).

are *Discus cronkhitei, Punctum minutissimum, Helicodiscus eigen-manni,* and *H. singleyanus* in Endodontidae; *Zonitoides arboreus* in Zonitidae; *Gastrocopta pentodon, G. procera, G. contracta, Vertigo milium, V. ovata,* and *V. gouldii* in Pupillidae; and *Vallonia perspectiva* in Valloniidae. Most of them have their center of distribution and presumably of evolution outside the Southwestern Province, usually in the Eastern Molluscan Division of the Nearctic Region. They are, therefore, intruders in the arid Southwest. It is noteworthy that, in spite of the extensive, physiographically uniform area connecting the Eastern and Western Molluscan Divisions, few typically eastern land mollusks have succeeded in spreading by natural means west of the Mississippi River, and even fewer beyond the 100th Meridian.

In other respects these southwestern Nearctic elements are a mixture. The majority are at present strictly Nearctic species, but some (*Gastrocopta pentodon, G. contracta, Zonitoides arboreus,* and possibly others) seem to have extended their natural ranges to the northern Neotropics. Some are clearly of northeastern origin and are now widespread throughout North America; others probably evolved in the southeastern United States and cover now only a restricted area. It may be significant that, with some exceptions perhaps due to incomplete knowledge (*Punctum minutissimum, Helicodiscus eigenmanni, Gastrocopta procera, G. contracta*), they occur as Late Cenozoic fossils in Arizona, are of long standing in the Southwestern Province, and are true members of the original, native local fauna, not post-Columbian introductions by man. Since each species has its own pattern of distribution, both inside and outside the Province, they must have reached the Southwest at different times. Obviously also some have been more successful (adaptive) migrants than others.

By far the most interesting to the zoogeographer is *Discus cronkhitei.* The strictly Holarctic genus *Discus,* with many fossil and Recent species from the Upper Mesozoic onward in both the Old and New Worlds, must have evolved much earlier somewhere in the Holarctic. It may have originated in Eurasia, where more fossil and Recent species are known than in America. Pilsbry (1948:598-622) recognizes nine Recent, native Nearctic species, of which two only occur in the Southwestern Province.

Discus shimekii has a restricted range and was discussed before as a Rocky Mountain element. *Discus cronkhitei,* on the other hand, has perhaps the most extensive range of any strictly Nearctic land snail. It is known Recent from Alaska (north to Dyea Valley, 59° 29' N, 135° 21' W; Unalaska, Aleutian Islands, westernmost

Station of the species, 53° 30′ N, 166° 30′ W), Yukon, the Northwest
Territories (northernmost Station of the species, Fort Simpson, 61°
50′ N, 121° 31′ W; Great Slave Lake, 61° 30′ N, 114° 30′ W), Manitoba, Ontario, Labrador, Newfoundland (ca 48° 30′ N, 56° 20′ W),
and Nova Scotia, southeast to Maryland, Kentucky, Arkansas,
Missouri, and Illinois, west to Minnesota, South Dakota, Wyoming,
north Nebraska (alive in Hooker County, ca 42° N, 101° W, D. W.
Taylor, 1960:79; C. W. Hibbard and D. W. Taylor, 1960:143), southwest Colorado (Florida River, La Plata County, ca 37° 20′ N, 107°
40′ W, A. L. Metcalf, 1967:47), New Mexico (northeast to Beulah,
San Miguel County, 35° 45′ N, 105° 30′ W; southeast to Guadalupe
Mts, Otero County, one mile north of the Texas State line, A. L.
Metcalf, in litt., 1967), Arizona (south to Miller Canyon, Huachuca
Mts, Cochise County, 31° 25′ N, 110° 17′ W), and south California
(south to San Bernardino Mts, ca 34° 8′ N, 117° W, S. S. Berry,
1909:76). South of the Mexican border it is known from one Station
in North Chihuahua (southernmost record of the species, Sierra de
la Breña, 30° 10′ N, 108° 10′ W, collected by the junior author
at 7,000 feet). It is unknown Recent in the southeast United States,
Texas, and the High Plains (Kansas and Oklahoma, where it lived
in the Pleistocene, when the area was presumably more wooded). In
the northeast and northwest it now lives to near sea level, but in
the Southwest only at high elevations (from 2,750 feet on Tumamoc
Hill near Tucson to 12,000 feet on San Francisco Mtn, Coconino
County, Arizona). It is fossil from the Middle Pliocene or Pleistocene to the sub-Recent in Quebec, Ohio, Indiana, Iowa, Mississippi,
Kansas, Missouri, Illinois, South Dakota, Oklahoma, Texas, New
Mexico, Arizona, Wyoming, California, and no doubt elsewhere.

   D. cronkhitei is closely related to the Palearctic *Discus ruderatus* (S. Studer, 1820), a species occurring over most of Eurasia, from
the British Isles to the Bering Strait. W. H. Dall (1905, *Harriman
Alaska Exped.* 13:50; copied by Pilsbry, 1948:605) observed that
dead Alaska shells of *cronkhitei* were undistinguishable from those
of Bering Island and Kamchatka *ruderatus,* while living snails dif-
fered in the color of the mantle. If there is no other reliable difference between the two, it might be proper to regard them as only
subspecifically distinct, making *D. ruderatus* a circumpolar species
that originated in Eurasia, and presumably migrated to America in
the Late Tertiary or Pleistocene, after the genus *Discus* had already
settled otherwise in the New World.

   The Recent distribution of *Punctum minutissimum,* "a dwarf
among pygmies" (Pilsbry, 1948:645); 1 to 1.5 mm wide, 0.7 to 1 mm
high), is incompletely known. After careful study, H. B. Baker

(1930a:9) recognized it from Newfoundland (48° 30′ N, 56° 20′ W),
the Magdalen Islands, Prince Edward Island, Maine, Massachusetts,
New York, New Jersey, Pennsylvania, Delaware, Maryland, Virginia, North Carolina, Florida, Tennessee, Alabama, northeast Ohio,
Michigan, Illinois, Iowa, and South Dakota. Two shells he saw
from New Mexico (Cloudcroft, Sacramento Mts, Otero County, 32°
50′ N, 105° 40′ W) were very close to *minutissimum;* while those
from Arizona (without precise Station), Colorado, and Montana
"seemed closer to *californicum* or *conspectum.*" It is also reported
by others from south Ontario, West Virginia, Kentucky, Missouri,
Oregon, Idaho, Wyoming, and Colorado (Boulder County). It is not
clear that Pilsbry (1948:644-645) studied specimens from all these
States. S. S. Berry (1922, *Victoria Mem. Mus. Bull.* 36:15) reported
it from Kananaskis, Alberta, as *P. pygmaeum.* A. B. Leonard (1959,
Univ. Kansas Mus. Nat. Hist. Misc. Public. 29:136) marks it on a
map (fig. 55) for ten eastern Kansas counties (west to Phillips
County, ca 39° 50′ N, 99° 20′ W). Finally, H. B. Baker (1930a:5 and
9) collected it in Mexico (Necaxa, Puebla, 2,600 to 5,000 feet, 20° 14′
N, 97° 52′ W), the southernmost record of the species, implying that
it occurs elsewhere south of the Mexican border. It has not been
found in Sonora, Chihuahua, or Texas. There is no precise Recent
record thus far from Arizona; but the senior author refers to it as
a dead shell from drift of the Little Colorado River (two miles east
of Holbrook, Navajo County, 4,800 feet, 34° 54′ N, 110° 11′ W),
probably washed down from headwaters in the White Mts. It is recorded as a Pleistocene or sub-Recent fossil from Indiana, Illinois,
Missouri, Mississippi, Ohio, and Louisiana, but not from Texas,
New Mexico, or Arizona; whether these fossils were always correctly identified is uncertain. At present it can only be stated that
true *P. minutissimum* is basically an eastern Nearctic snail, that
extends, perhaps sporadically, west of the 100th Meridian.

    Helicodiscus eigenmanni H. A. Pilsbry (1890), with subspecies
arizonensis II. A. Pilsbry and J. II. Ferriss (1906:157), is the Southwestern representative of the basically Nearctic subgenus *Helicodiscus*, sensu stricto, in which Pilsbry (1948:625-635) includes *H. parallelus* (T. Say, 1821), *H. fimbriatus* A. G. Wetherby (1881), and *H.
salmonaceus* W. G. Binney (1890). These four species share a disk-
like shell, with flattened upper surface, a sculpture of spaced
spiral threads, a broad umbilicus, and occasional small internal
teeth in the last whorl. The presence and number of these teeth
vary, however, sometimes within one population; as the teeth are
often resorbed by the snail and their material redeposited later,
they provide no reliable specific characters. The spiral threads may

be strong and sometimes are fringed ridges, or they may be smooth and blunt, or very weak, differences perhaps to some extent specific. These threads are apt to wear off with age, mainly on the early whorls, which makes specific identification often uncertain, especially for fossils. All four forms are closely related, forming a complex that obviously had a common ancestor. At present it must be left undecided whether they are a monophyletic superspecies ("Artenkreis"), in which each form has progressed to a full biological (reproductively isolated) species, or a single polytypic species, the several forms having achieved only geographical, subspecific differentiation.[28] It remains in abeyance whether *H. parallelus* and *H. eigenmanni* are specifically or subspecifically distinct, the observable differences between them, either in shell or in anatomy, being extremely slight. Our discussion of the distribution of *H. eigenmanni* can, therefore, only be tentative. However, after studying much material from Texas and Arizona, the senior author is unable to find reliable differential characters between nominate *eigenmanni* and subspecies *arizonensis,* so that the latter is now relegated to the synonymy. *H. eigenmanni* is then essentially a Southwestern snail, with a continuous Recent range from Texas (easternmost Station of the species, Livingston, Polk County, ca 30° 40′ N, 95° W, collected by the senior author; south to Uvalde County, ca 29° N, 100° W, collected by the senior author; north to Dallas County, ca 32° 30′ N, 96° 40′ W, several records as *H. parallelus,* E. P. Cheatum and C. E. Burt, 1934, *Field and Laboratory* 2 [2]:51; and west to the trans-Pecos area) and Oklahoma (doubtfully in Kansas), to New Mexico (widespread) and Arizona (westernmost Station of the species, drift of Agua Fria, six miles northeast of Bumblebee, Yavapai County, 34° 10′ N, 112° 5′ W), with isolated records from Colorado (Trinidad, ca 37° 10′ N, 104° 31′ W) and South Dakota (northernmost Station of the species, Deuel County, ca 44° 50′ N, 96° 30′ W).

---

[28]The taxonomic history of the *H. parallelus* complex has been unfortunate from the start. T. Say (1817, *Journ. Acad. Nat. Sci. Philadelphia* 1:18) described the nominate species at first, without definite T.L., as *Helix lineata;* but this trivial name being a homonym of *Helix lineata* G. Olivi (1792), was changed by him (1821, *Journ. Acad. Sci. Philadelphia* 2:164) to *Planorbis parallelus* (originally printed "*arallelus,*" an obvious typographical error, corrected by him in the Index: 407). Recently much unnecessary confusion was created in the group by an indiscriminate avalanche of alleged new species, reminiscent of the obnoxious taxonomy of J. R. Bourguignat and A. Locard: *Helicodiscus triodus* L. Hubricht (1958), *H. multidens* L. Hubricht (1966), and *H. diadema* F. W. Grimm (1967). All three are apparently local variants of *H. parallelus,* not even entitled to subspecific status. L. Hubricht (1962) pointed out that *Gastrodonta (Clappiella) saludensis* J. P. E. Morrison (1937) is a *Helicodiscus;* it too appears to be a synonym of *H. parallelus.* To complete the record, *Helicodiscus lineatus sonorensis* J. G. Cooper (1893:343), of Sonora, was based on a freshwater snail and is a synonym of *Drepanotrema aeruginosum* (A. Morelet) of our Check List.

South of the Mexican border it is known from Sonora (San Bernardino, Pilsbry, 1948:630), Chihuahua (several Stations), and Puebla (southernmost Station of the species, Necaxa, 4,600 feet, 20° 14′ N, 97° 52′ W, H. B. Baker, 1930a:14). There are several fossil records from the Pleistocene in Oklahoma and Texas (either as *H. eigenmanni* or *H. parallelus*), but none thus far in Arizona.

Helicodiscus singleyanus (H. A. Pilsbry, 1890) is only remotely related to the *H. parallelus* complex and is placed in a distinct subgenus *Hebetodiscus* H. B. Baker (1929; perhaps better raised to generic status). Among other differences, it lacks spiral threads on the shell, and does not produce internal teeth so far as known. *H. singleyanus inermis* H. B. Baker (1929) cannot be distinguished consistently, even by geography, from *H. s. singleyanus* in our opinion. Pilsbry (1948:636) admitted that these two forms are "so similar that their separation is most difficult." *Helicodiscus intermedius* J. P. E. Morrison (1942) also is synonymized by us with *H. singleyanus.*

The strictly Nearctic distribution of *H. singleyanus* is probably incompletely known at present, particularly as it is often confused with *Hawaiia minuscula,* but it does not appear to be as widespread as *H. eigenmanni.* There are reliable Recent records from New Jersey (northeast to Camden County, ca 39° 50′ N, 75° W), Pennsylvania, Indiana, Illinois, and Iowa, south to Arkansas, Tennessee, Florida (to south tip, ca 25° 30′ N, 80° 30′ W), Alabama, Mississippi, and Louisiana, and west to southeast South Dakota (north to Brule County, ca 43° 45′ N, 99° W), Colorado (El Paso County, ca 38° 45′ N, 104° 30′ W), Kansas, Oklahoma, and Texas (widespread, but not in the east counties and perhaps only fossil in the Panhandle; south to Brownsville, ca 25° 54′ N, 97° 29′ W). A record from Berkeley, California, was no doubt based on an introduction by man. South of the Mexican border it is known thus far only from Tamaulipas (drift of Rio Purificación 24 miles west of Padilla, ca 24° N, 99° W) and Sonora (southwesternmost Station of the species, four miles north of Ciudad Obregón, ca 28° N, 108° 50′ W, B. A. Branson, 1964:104; Nacozari, ca 30° N, 110° W, by the junior author, 1964). It is a Late Pliocene to sub-Recent fossil in Illinois, Missouri, Nebraska, Kansas, Oklahoma, Texas, New Mexico, and Arizona. It appears to have been originally a snail of the warm temperate section of the Eastern Molluscan Province that migrated to the Southwest during some Pleistocene interglacial period.

Zonitoides arboreus is a ubiquitous, autochthonous, and mainly Nearctic snail, most likely "an old resident, present throughout the

Pleistocene, which has become adapted, within its great area, to both subarctic and tropical climates" (Pilsbry, 1946:475). Where it originated in North America is a moot question for lack of pertinent evidence. Its Recent range extends from the Northwest Territories (north to Great Slave Lake, 61° 30′ N, 114° 30′ W), British Columbia (north to Cameron Lake), Alberta, Manitoba, Ontario, Quebec, Labrador, Newfoundland (48° 30′ N, 56° 20′ W), New Brunswick, and Nova Scotia, southward throughout the United States (except Nevada) to south Florida (Polk County, in crowns of palms, D. Wilson, 1960, *Nautilus* 73:138), but far from uniformly. South of the Mexican border it is known in Chihuahua (Sierra de la Breña, 30° 10′ N, 108° 10′ W), Morelos, Nuevo Leon, Vera Cruz (Córdoba), and Puebla (Necaxa, 20° 14′ N, 97° 52′ W, 4,500 feet), but not from Sonora. There are also records from Guatemala, Costa Rica, and the Greater and Lesser Antilles; but how much of this southern range is natural is uncertain, as the snail is easily spread by man. It has been introduced to many Old World localities in Europe, South Africa, Australia, Hawaii, and Japan. C. H. Lindroth (1957:111) regards it as probably native in Kamchatka, but in the senior author's opinion it is rather an introduction there, as it is in Japan. In boreal America it may occur to near sea level, but in the South, particularly in Arizona, New Mexico, and Chihuahua, it is truly native mainly at high elevations (up to 12,000 feet), while below 5,000 feet it is often only a temporary adventive. For instance, in the cities (such as Tucson) it is sometimes imported from nearby mountains with logs, firewood, or even camping equipment, or brought in with nursery stock. It is a frequent and common Pliocene to sub-Recent fossil in the United States, reported from Iowa, Indiana, Missouri, Illinois, Mississippi, Ohio, Louisiana, Kansas, Oklahoma, Arizona, New Mexico, Texas, and California.

For the purpose of the present discussion, *Gastrocopta pentodon* (T. Say, 1822) will include *Gastrocopta tappaniana* (C. B. Adams, 1842) and cover a complex of variants "found over a greater range than any other North American *Gastrocopta*" (Pilsbry, 1918-1920: 31; 1948:888). The shell varies excessively, in size for the same number of whorls, in shape, and particularly in the armature of the aperture (five to nine teeth in various combinations and sizes). E. G. Vanatta and H. A. Pilsbry (1906:121-128; Pl. 6-7) attempted to put some order in the chaos of the resulting named forms, eventually recognizing two of them as distinct species, mainly on size and shape of shell: smaller (0.8 to 1.1 mm in diameter, 1.5 to 1.8 mm long) and "varying from conic to subcylindric" in *pentodon;* "larger (1.1 to 1.2 mm in diameter, 1.7 to 2.0 mm long) than *pentodon,* markedly conic though obtuse" in *tappaniana*. These differences are, however, far from clear-cut, as may be seen readily from the 53 figures the authors refer to their two "species." Moreover, they were well aware of this, since they wrote: "It is often a difficult question upon which no two experts may agree, whether to refer a certain specimen to *tappaniana* or to some form of *pentodon;* while the larger size and globose-conic shape are characteristic of *tappaniana*, yet intermediate types and shapes occur occasionally, so that we disclaim any intention of setting up a definite boundary between the forms. It can only be claimed that a great majority of lots are quite readily separable."

After studying many Recent and fossil lots of this complex from Texas and Arizona, the senior author reached the conclusion that it is impossible to recognize consistently two biological species. So-called *pentodon* and *tappaniana* are extremes of a continuum, with intermediate variants which obviously interbreed freely with the extremes. While it is easy to refer single shells or small lots to one or the other, large populations usually contain shells that must be placed arbitrarily. At any rate, within the Southwest there is no evidence of either the more slender *pentodon* or more obese *tappaniana* predominating in any section of the territory, so that there is no justification for treating them as geographical subspecies. All named forms are individual variants of a single species, a conclusion reached over a century ago by A. A. Gould (1844, *Boston Journ. Nat. Hist.* 4:354-355) when he synonymized his own *Pupa curvidens* of 1841.[29]

As here defined, Recent *G. pentodon* ranges from British Columbia, southwest Alberta (Laggan, 5,200 feet, ca 50° 30′ N, 116° 25′ W, northwesternmost Station, G. W. Taylor, 1893, *Nautilus* 7: 86), Manitoba, Ontario, Quebec, Prince Edward Island, Magdalen Islands (47° 30′ N, 61° 40′ W), and Maine, south throughout the eastern United States to central Florida, Alabama, Mississippi, Louisiana, and Texas, west to Nebraska, South Dakota, Montana, south Colorado (one Station, western Custer County, 8,400 feet, ca 38° N, 105° 30′ W), Kansas, Oklahoma, trans-Pecos Texas (living in South McKittrick Canyon, Guadalupe Mts, 5,300 feet, ca 31° 50′ N, 104° 26′ W, senior author, 1966), New Mexico, and

---

[29]It is unfortunate that Pilsbry did not study the holotype of *Pupa tappaniana*, which he says (1948:889) was then in the C. B. Adams collection at Amherst, Massachusetts. This collection was since transferred in permanent loan to the Museum of Comparative Zoology, Cambridge, Massachusetts, where the holotype of *tappaniana* was found to be a typical *pentodon* by W. J. Clench, as explained in the Check List.

Arizona (all Recent Stations in the Check List; the two western-most, Jerome, Yavapai County, 34° 45′ N, 112° 7′ W, perhaps based on drift shells; and Crittenden at the east foot of the Santa Rita Mts, Santa Cruz County, 31° 40′ N, 110° 43′ W). Its distribution is extremely spotty in the Rocky Mountains. It may be widespread in Mexico, although there are at present only records from Chihuahua (Rio Piedras Verdes near Pacheco, 30° 8′ N, 108° 20′ W, 5,900 feet, Pilsbry 1953:161), San Luis Potosi (north of San Diegueto, ca 22° N, 99° 14′ W, cited by Pilsbry, 1916-1918:31), Tamaulipas (four miles west of Ciudad Victoria, 23° 15′ N, 99° 22′ W, Pilsbry, 1916-1918:31; drift of Rio Purificación, 24 miles north of Padilla, 24° N, 99° W, senior author, 1958), and Puebla (Necaxa, 20° 14′ N, 97° 42′ W, H. B. Baker, 1930a:5); also reported from Guatemala (Jocolo, on the north side of Lake Izabal, 15° 30′ N, 89° 10′ W, A. A. Hinkley, 1920, *Nautilus* 34:48, southernmost record of the species, and two other Stations). It is a common Upper Pliocene to sub-Recent fossil in Quebec, Ohio, Illinois, Missouri, Tennessee, Iowa, Florida, Mississippi, Louisiana, Texas, Oklahoma, Kansas, Nebraska, New Mexico, Arizona, and Washington.

Gastrocopta procera (including *G. procera mcclungi* G. D. Hanna and E. C. Johnston, 1913, and *G. procera sterkiana* H. A. Pilsbry, 1917) is here accepted with reservations as an indigenous (pre-Columbian) Southwestern land snail. Although it is generally thought to be Recent and native in New Mexico and Arizona (for instance, by Pilsbry, 1948:907-911; 1916-1918:62-68), published records from there are actually few, sporadic, and mostly based on dead drift shells. We are inclined to regard these (all listed for Arizona in the Check List) as post-Columbian introductions by man. This is certainly the case for our two lots of living snails; our collections from drift are most probably also modern shells, washed down from cultivated areas (pastures, fields, gardens, etc), or perhaps sometimes brought in with hay or feed from elsewhere.

The spotty occurrences of *procera* in the Southwest contrast sharply with the continuous distribution of its sibling relative *G. cristata*. Moreover, while *cristata* is a common Pleistocene fossil in Arizona and New Mexico, no truly fossil *procera* has been found there thus far, the nearest approach to it being some *procera* recorded from modern alluvium in the Rio Grande Valley near Las Cruces and El Paso by A. L. Metcalf (1967:41), an area of intense cultivation for the past century.[30] Outside the Southwestern Province,

---

[30]Only full-grown shells in good condition should be used in identifying Recent or fossil *procera* and *cristata;* shells with damaged or incomplete last

Recent *G. procera* is widespread in the southeast Nearctic Region, presumably its original home, where it is also found fossil. It appears to be truly native from Maryland (T.L., Baltimore, 39° 18′ N, 76° 36′ W, northeasternmost record of the species), Washington, D.C., West Virginia, and South Carolina, south to northeast Alabama (Valley Head, De Kalb County, 34° 32′ N, 85° 35′ W) and central Texas (west to Ozona, Crockett County, 30° 42′ N, 101° 10′ W, living at roots of cacti, collected by Del Weninger, 1960; in trans-Pecos Texas and south of 28° N only as dead drift shells; not known from the eastern Austro-Riparian section), west to Kentucky, Arkansas, Missouri, Kansas (widespread; west to Cheyenne County, 39° 50′ N, 101° 50′ W), Oklahoma (widespread), southeast Minnesota (Winona County, 44° N, 91° 42′ W, northwesternmost record of the species), and southern South Dakota (north to Pennington County, ca 44° N, 103° 20′ W). There are sporadic records from Ontario, Ohio, Indiana, Illinois, and Iowa in the east and from Colorado (Trinidad, Las Animas County, 37° 10′ N, 104° 31′ W; Pikes Peak, El Paso County, 38° 50′ N, 105° 3′ W), and southeast Wyoming (Platte County, ca 42° N, 104° 30′ W) in the west; but it is doubtful whether these are truly native occurrences or post-Columbian introductions by man, as this snail is very synanthropic. It is not known from Mexico. It is definitely a Pliocene or Pleistocene fossil in Nebraska, Kansas, Oklahoma, and Texas, more doubtfully so in Iowa, Illinois, Missouri, and Minnesota.

Gastrocopta contracta (T. Say, 1822) is discussed mainly to complete the record, as it is known only from a few fringe Stations in the southeastern section of the Southwestern Province. One of the most distinctive and ubiquitous Recent elements of the Nearctic Eastern Molluscan Province, it occurs in southeastern Canada and the eastern United States, from Manitoba, Ontario, Quebec, and Maine south to Florida and the Gulf of Mexico, and west to the Dakotas, Nebraska, Kansas, Oklahoma, and much of Texas. At present its southwestern limits are in trans-Pecos Texas (Jeff Davis County, at Big Aguja Canyon of the Davis Mts, 12 miles northwest of Fort Davis, 4,800 feet. Culberson County, in McKittrick Canyon, Guadalupe Mts, 5,300 feet. Both by senior author), south-

---

whorl or aperture should be ignored. The most reliable specific character appears to be the shape of the angulo-parietal tooth: simple, heavy, straight, and peg-like in *cristata;* weaker, slightly oblique, and bifid at the tip in *procera.* The strength and position of the post-apertural crest of the body whorl, while often characteristic, are too variable to be fully trusted; this is true also for the outer margin or lip of the aperture, usually sharp with a slightly expanded edge in *cristata,* but thickened, flattened, and with a labial callus in most *procera* (when fully developed).

east New Mexico (Walnut Creek in the eastern foothills of the Guadalupe Mts, Eddy County, by the senior author), and southeast Sonora (Arroyo San Rafael at San Fernando, 27° 30′ N, 108° 54′ W, westernmost Station of the species, by Pilsbry, 1953:161). Farther south it is known in Mexico from Sinaloa, Morelos, Nuevo Leon, Tamaulipas, San Luis Potosi, and Vera Cruz. Records from Cuba and Jamaica may be based on human importations. It has not been found thus far either Recent or fossil in Arizona. It is a Pliocene to sub-Recent fossil in Quebec, Ohio, Illinois, Missouri, Nebraska, Kansas, Oklahoma, Texas, Tennessee, Louisiana, and Mississippi. Since natural extensions and regressions of molluscan ranges must be going on today as in the past, the present-day sporadic occurrences of *G. contracta* in trans-Pecos Texas and Sonora might well be due to natural post-Pleistocene or modern migrations.

The discussion of *Vertigo ovata* will cover also *V. ovata diaboli* H. A. Pilsbry (1919:88), *V. ovata mariposa* H. A. Pilsbry (1919:88), and *Vertigo teskeyae* L. Hubricht (1961, *Nautilus* 75:62, figs. 2A-C), none of which are subspecifically distinct in the senior author's opinion.[31] Pilsbry (1918-1920:82-89; 1948:952-955) stated that *V. ovata* is "the most widely distributed species of the genus." Although this is correct, the actual Recent distribution of *ovata* is at present confused, as many published records and labels in collections do not state whether the snails were found alive or dead in drift, the latter being often washed-up fossils, particularly in localities where *ovata* no longer lives. The true north, south, and west limits of the species are therefore at present uncertain. In the Nearctic Region, where *V. ovata* is probably precinctive, it is recorded in the north from the Labrador Peninsula (Ungava Bay, ca 58° 30′ N, 67° W, northeasternmost Station of the species), Prince Edward Island, Quebec, Ontario (widespread; west to Bortwick Lake,

---

[31]*Vertigo teskeyae* was based on unusually large *V. ovata* (2.1, 2.6, and 2.9 mm long), darker than usual (when fresh), always without infraparietal and suprapalatal teeth, with a horizontal columellar lamella, and the basal tooth smaller and placed lower than in *ovata*. It is said to be "a species of the southern Atlantic and Gulf Coastal Plains" (reported from Maryland, North Carolina, South Carolina, Georgia, Florida, Alabama, and Mississippi), and to be often associated with *ovata*. However, abnormally large *ovata*, often with some or all of the features of *teskeyae*, are by no means restricted to the southeastern States. For instance, Pilsbry (1948: 953) mentions shells 2.45, 2.5, and 2.6 mm long from Ontario. The senior author saw similar large specimens in Texas (Brazos, Fayette, Burleson, Refugio, Nueces, Galveston counties, etc). The number, size, and shape of the apertural teeth vary so much in most large populations throughout the range of *ovata*, that they are wholly unreliable even for subspecific purposes; the holotype of *teskeyae* (fig. 2A) actually has more teeth than Pilsbry's (1948:954, fig. 513, 1) *ovata* from New York, which also lacks infraparietal and suprapalatal teeth.

Kenora District, ca 50° N, 93° W), Manitoba (Winnipeg, 49° 47' N, 97° 15' W), Alberta (Laggan, ca 51° 30' N, 116° 25' W), British Columbia (Vancouver Island), Washington (Puget Sound and elsewhere), north Oregon, and Alaska (Kodiak [or Kadiak] Island, 57° 20' N, 153° 30' W; Tigalda Island, 54° N, 165° W). It is believed to range southward over most of the United States (except Nevada) to the Florida Keys, the Gulf of Mexico, and the Mexican border, but that it actually lives uniformly at present over this vast area is questionable. In California it is known from one Station only (Wanona, Sequoia National Park, Mariposa County, 37° 30' N, 120° W, T.L. of *V. o. mariposa*). In Texas the senior author knows it living from only three east counties (collected alive by him in Liberty and Harris counties; presumably alive also in Harrison County). After four years' collecting he has found no reliable evidence that it now lives in the Panhandle, or in the central, trans-Pecos, and southeast counties (south of 28° N); moreover, the majority of published Texas records are of fossils collected in situ.

From all known occurrences, *V. ovata* is sporadic in the Western Molluscan Division and particularly in the Rocky Mountain States, a type of haphazard distribution often characteristic of relict species. In this particular case, however, the senior author is inclined to attribute it more to chance transport by waders or other shore birds of snails or their eggs in mud adhering to feet or plumage, since *ovata* often lives in muddy litter and on damp logs or semi-aquatic plants, close to the water's edge of ponds and shallow creeks. This natural zoochorous mode of dispersal, well known for succineids with similar habits, must have occurred frequently in the past, as it does at present, for other lesser land and freshwater mollusks.

There are Recent (presumably live) records for Montana (widespread), Wyoming (Grand Teton National Park, etc), Utah (Fruita, location dubious), South Dakota (Deuel and Clay counties), Nebraska (Monroe County, B. Walker, *Nautilus* 20:81; Brown County, D. W. Taylor, 1960:44), Colorado (Trinidad, Las Animas County, 37° 10' N, 104° 31' W), Kansas (widespread, but spotty), and Oklahoma (widespread). It is not clear whether any of the few records from New Mexico (Mesilla; Silver City; Albuquerque; Las Vegas; Rincon; Grant) were based on Recent shells. A. L. Metcalf (1967:44) found it fossil in "Recent alluvium" of Dona Ana County. All known Stations in Arizona are listed in the Check List, only six being accepted as based on live snails (west to Jerome, 34° 45' N, 112° 7' W; southeast to Sylvania Spring, Huachuca Mts, 31° 27' N, 110° 24' W); but it is common in river drift, mostly as washed-up fossils (west to Alamo Crossing of Bill Williams River, 30 miles north of Wenden, Yuma County, 34° 25′ N, 113° 25′ W; northeast to Adamana, Apache County, 34° 55′ N, 109° 31′ W). The only trust-worthy Mexican record is from Sonora (drift of Rio Sonoyta at Sonoyta, 31° 50′ N, 112° 50′ W, by R. H. Russell, Sept. 1968). E. von Martens' (1898:327) record from Vera Cruz was based on beach drift material, no doubt floated down from Texas, where *ovata* is common in beach debris. The reputed occurrence in the Greater Antilles (Cuba, Jamaica, Hispaniola, and Puerto Rico), accepted by Pilsbry (1918-1920:86-87;1948:953) with reservations, defies a logical explanation; most probably it was based on misidentifications. *V. ovata* is a frequent fossil from the Early Pliocene to the Pleistocene and sub-Recent in Ontario, Quebec, Iowa, Ohio, Colorado, Kansas, Oklahoma, Arizona, New Mexico, Texas, Louisiana, and Nebraska.

Vertigo milium, the smallest American member of the genus (1.5 to 1.8 mm high), is mainly a snail of the temperate eastern Nearctic Region. It is far from being as generally distributed in North America as sometimes stated.[32] It occurs Recent from south Ontario (north to Ottawa, 45° 30′ N, 75° 44′ W), south Quebec (near Quebec, ca 46° 53′ N, 71° 20′ W), and Maine, south to the Florida Keys and Alabama; westward it ranges to Minnesota (Clearwater, Wright County, 45° 20′ N, 94° 5′ W), South Dakota (Chamber-lain, Brule County, 43° 49′ N, 99° 20′ W), north Colorado (Kremm-ling, Grand County, 40° N, 106° 25′ W, northwesternmost Station of species), east Kansas (west to Reno County, ca 38° N, 98′ W), Okla-homa (fairly widespread near permanent water or springs, B.A. Branson, 1961, *Proc. Oklahoma Acad. Sci.* 41 for 1960:49-50), east-central Texas (west to Kerr and Bandera counties, ca 30° N, 99° 30′ W; only fossil, either in situ or in drift, in the west, northeast, and southeast counties), and southeast Arizona (all records in the Check List; westernmost Recent Station of the species, Sylvania Spring in Scotia Canyon, Huachuca Mts, 31° 27′ N, 110° 24′ W); no Recent or fossil records from New Mexico. Recent occurrences west of the 100th Meridian are sporadic and possibly relictual, since the species is widespread there in Pleistocene deposits; some of them may, however, be due to modern transport by water or shore birds, as in the case of *V. ovata.* It is definitely known from the Late Pliocene to Pleistocene in Illinois, Iowa, Ohio, Missouri, Mississippi, Nebraska, Kansas, Oklahoma, Texas, and Arizona. There

------

[32]*V. milium* was isolated in a monotypic subgenus *Angustula* by V. Sterki (1888). D. W. Taylor (1960:76) pointed out that the few differences between this species and those placed in the subgenus *Vertigo* (sensu stricto) are of minor impor-tance. We follow him in including the species in *Vertigo* (sensu stricto).

is no reliable evidence that it now lives or ever lived in Mexico, the record from Tamaulipas (Tampico, A. A. Hinkley, 1907, *Nautilus* 21:77) being based on two drift shells washed down from the Texas coast, where the snail is common in beach debris. The reported occurrences in the Greater Antilles (Jamaica and Hispaniola, accepted by Pilsbry, 1918-1920:147; 1948:945), if correct, must be due to accidental transport by man, perhaps in ship's ballast, from the southeastern United States.

The confused taxonomy of *Vertigo gouldii* is discussed in the Check List, which gives all known occurrences in Arizona of the named variants (four supposed subspecies) of the Western Molluscan Division.[33] The combined general distribution of the seven Recent subspecies recognized by Pilsbry (1918-1920:98-100, for northeastern "*V. gouldii*," and 115-118, for western "*V. coloradensis*"; 1948:971-976), is strictly Nearctic and similar to that of *V. milium,* though more boreal in the north, more extensive but still sporadic in the west, and entering northwest Mexico in the southwest. It ranges from British Columbia (Field, 51° 25' N, 116° 30' W, northwesternmost Station of the species), Alberta, Montana (Ward, 8 miles south of Hamilton, Ravalli County, ca 46° N, 114° 10' W; and elsewhere), Ontario (widespread), Quebec (widespread), Anticosti Island, Prince Edward Island, Magdalen Islands, Newfoundland (48° 30' N, 56° 20' W, northeasternmost Station of the species), and Maine, south to Maryland, West Virginia, North Carolina, east Kentucky (Harlan County, ca 36° 45' N, 83° 20' W), east Tennessee (west to Marion County, ca 35° 10' N, 85° 30' W), and northeast Alabama (Valleyhead, De Kalb County, 34° 33' N, 85° 35' W), and west to Colorado (Swift Creek, Custer County; Rio Blanco), Utah (Box Elder Canyon), west New Mexico (east to Eagle Creek, Lincoln County, ca 33° 30' N, 105° 30' W), and Arizona (west to Bill Williams Mtn, 35° 10' N, 112° 15' W, southwesternmost Station of the species). It is not known Recent in Texas, Oklahoma, Kansas, and Sonora. In Mexico it occurs at two Stations in Chihuahua, the southernmost for the species being Rio Piedras Verdes near Colonia Juárez, ca 30° 15' N, 108° W. Published records from the Antilles (Cuba, Jamaica) are highly dubious. It is reported from the Late Pliocene to Pleistocene in Idaho, Wyoming, Tennessee, Ohio, Colorado, Indiana, Texas, Illinois, Missouri, Mississippi, Louisiana, and Kentucky.

Mention should be made of two additional species of *Vertigo* reported as Recent from New Mexico. Although they are not

---

[33]In 1927-1935:95, Pilsbry had to admit that the "Rocky Mountain and northeastern races, [of *V. gouldii*], though not absolutely identical, seem to be barely distinguishable."

known at present in Arizona, or elsewhere in the Southwestern Province, they may possibly occur there; at least one of them is a common Pleistocene fossil in Arizona. *Vertigo elatior* V. Sterki (1894; described as *Vertigo ventricosa* variety *elatior*) is recorded, presumably as a Recent snail, by Pilsbry (1918-1920:95; 1948:956) from Newfoundland, Maine, and Ontario (north to James Bay and Hudson Bay), south to New York, Ohio, and Michigan, and west to Montana, Alberta (Laggan), British Columbia (Field), and at one Station in New Mexico (Oscura Mts, Socorro County, 33° 40′ N, 106° 31′ W, a stray record perhaps based on washed-up fossils). It is frequent in the Pleistocene in Indiana, Illinois, Ohio, and Arizona. It is questionable, moreover, whether *elatior* is specifically separable from *Vertigo ventricosa* (E. S. Morse, 1865), which occupies much the same territory. Pilsbry (1948:957) states that in J. Oughton's experience in Ontario, typical and nearly typical *ventricosa* "are to be found mixed in large lots of *V. ventricosa elatior.*" *Vertigo binneyana* V. Sterki (1890) is closely related to *elatior,* from which it seems to differ in little more than the shape of the shell. It is reported by Pilsbry (1918-1920:90; 1948:956) from Montana (Helena, T.L., 46° 30′ N, 112° W; also elsewhere, mostly in drift), Washington (Seattle), British Columbia (Vancouver Island), Manitoba (Winnipeg), Iowa (drift of Missouri River), and at one Station in New Mexico (Albuquerque, after V. Sterki; Station questioned by Pilsbry). J. Oughton reported it also from Ontario (Fort Severn on Hudson Bay), perhaps by error. All these records require reappraisal, since some of them were no doubt based on fossils, or even on misidentified *V. gouldii,* a common snail of the same general area.

The most interesting feature of the genus *Vertigo* to us is that in the Late Pleistocene (only some 10,000 years ago) it was much better represented in individuals and species in Arizona than nowadays. Of six fossil species recognizable in the deposits of the San Pedro Valley, associated with mammoth remains, only three are now living in the State, and even these are scarce; the others are extinct there, though still living elsewhere in North America. The local decline of the genus was no doubt due to general increasing aridity and to intense deforestation of the mountains by man.

Vallonia perspectiva is a closely related sibling of *V. cyclophorella,* a species discussed before as a Rocky Mountain element, both being clearly derived from a common ancestor. Possibly the larger, more robust *cyclophorella* is the scarcely modified primitive stock, and *perspectiva* its derivative, with a smaller, thinner shell with more spaced riblets. In spite of their being so similar, there is no evidence that they ever interbreed. Since their Recent and Pleistocene ranges overlap to some extent in the Southwest, and since they are even sometimes sympatric, they must be fully isolated reproductively. If *perspectiva* originated in the Southwest, as seems plausible, it may be a case of a species autochthonous there, but no longer precinctive, having spread far east of its original home. Its main continuous range is even now in Arizona, New Mexico, and trans-Pecos Texas, where it lives at 3,500 to 8,700 feet elevation. Elsewhere the records are erratic and sometimes uncertain, as shown below.

In Arizona it is known alive in Coconino County (northwesternmost Station of the species in the Grand Canyon of the Colorado River, ca 36° N, 112° 30′ W), Pima County (southwest to Arch Canyon, Ajo Mts, 3,500 ft, 32° 2′ N, 112° 42′ W), Yavapai County, Greenlee County (northeast to Eagle Creek, ca 33° 10′ N, 109° 30′ W), Gila County, Pinal County, Santa Cruz County, Cochise County, and Graham County. Records from farther northeast in Navajo County (Holbrook, 34° 54′ N, 110° 11′ W) and Apache County (Buell Park, 35° 50′ N, 109° 5′ W) are for drift shells, probably washed-up fossils. In southwest New Mexico, the Recent range is fairly continuous from Hidalgo County (Big Hatchet Mts, 31° 30′ N, 108° 20′ W) north to Catron County (Mogollon Mts, 33° 20′ N, 108° 40′ W) and Socorro County (ca 34° N, 107° 30′ W), east to Otero County (Sacramento Mts, 32° 50′ N, 105° 40′ W) and Dona Ana County (Organ Mts, at 7,000 feet, ca 32° 20′ N, 106° 40′ W, A. L. Metcalf, 1967:45). In trans-Pecos Texas, it occurs alive in El Paso County (east side of North Franklin Mts, at 7,000 feet, ca 31° 40′ N, 106° 28′ W, A. L. Metcalf and W. E. Johnson, 1971:90) and Culberson County (South McKittrick Canyon of Guadalupe Mts, 5,400 feet, ca 31° 50′ N, 104° 40′ W, A. R. Mead et al, 1969). Published records from east of the Pecos River in Texas are of Pleistocene fossils in situ or washed up in drift. The senior author saw live adult and young snails collected on the Mary Davis Coupe Ranch, eight miles southeast of Eldorado, Schleicher County, by Mrs. Dorothea Caskey Mangun, Jan. 1964, evidently a human introduction with plants. In Mexico, it occurs in Sonora (mountains five miles south of Magdalena, at 4,250 feet, 30° 55′ N, 111° W, by the junior author. Banks of Rio Nacozari, ca 30° N, 110° W, and drift of Yaqui River, four miles northwest of Ciudad Obregón, southernmost Station of the species, both B. A. Branson et al, 1964:104) and Chihuahua (Sierra de la Breña, 11 miles above Mata Ortiz, ca 30° 10′ N, 108° 10′ W, 7,000 feet, Pilsbry, 1953:165, and the junior author. Rio Piedras Verdes, 5.5 miles above Colonia Juárez, ca 30° 15′ N, 108° W, Pilsbry, 1953:165).

Occurrences outside the Southwestern Province, here listed so far as known, are often of drift shells. They are, moreover, so sporadic that some appear to be based either on washed-up fossils or on human adventives. One record, the northeasternmost of the species, at Ventnor near Atlantic City, New Jersey, 39° 21′ N, 74° 27′ W, in a salt marsh, dates from 1893, but the Station is now destroyed (Pilsbry, 1948:1033). Others are from southeast Maryland (Elliott Island, Dorchester County, R. W. Jackson, 1960, *Nautilus* 73:160; Washington and Allegany counties, W. Grimm, 1959, *Nautilus* 73:21-22); Virginia (Shenandoah National Park, Warren County, W. Grimm, 1959, *Nautilus* 73:22); northeast West Virginia (Franklin, Pendleton County, 38° 40′ N, 79° 20′ W); northwest Illinois (one record, 4 miles north of Savanna, Carroll County, 42° 5′ N, 90° 9′ W); Iowa (one record, in drift of Missouri River); central Missouri (one record, Sedalia, Pettis County, 38° 43′ N, 93° 12′ W); east Tennessee (one record, Knoxville, Knox County, 35° 58′ N, 83° 57′ W); northeast Alabama (one record, T.L., Woodville, Jackson County, 34° 40′ N, 86° 20′ W); south Minnesota (two records in Wright County: Clearwater, 45° 30′ N, 94° W; Rockford, 45° 10′ N, 93° 40′ W, H. E. Sergent, 1895, *Nautilus* 9:89); southwest North Dakota (one record, northernmost of the species, in drift, Medora, Billings County, 46° 50′ N, 103° 30′ W); southeast South Dakota (one record, Chamberlain, Brule County, 43° 49′ N, 99° 20′ W); and southwest Utah (one record, Zion National Park, Washington County, 37° 10′ N, 113° W). It is definitely known as a Pleistocene or sub-Recent fossil in Arizona, New Mexico, Texas, and Kansas, and reported from Missouri, Illinois, and Mississippi, but whether always from correctly named specimens is uncertain.

Terrestrial Mollusks and the Arid Environment

From the foregoing analysis emerges a picture of a diversified, yet highly distinctive, Southwestern mollusk fauna derived from several sources, though predominantly autochthonous. Furthermore, the wealth of strictly precinctive genera and species proves that extreme aridity, combined with prevailing high temperature, conditions supposedly adverse to molluscan life, seem to have enhanced rather than deterred evolutionary activity in these animals. This may be surprising, particularly to malacologists accustomed to the faunas of lusher, more temperate sections of North America. It is, however, by no means a unique or even exceptional situation. On the contrary, it appears to be a fairly general rule for the molluscan faunas of arid and semiarid lands throughout the world.

The Old World has the Earth's largest belt of deserts and near-deserts, stretching some 800 to 1,000 miles wide north and south of the Tropic of Cancer, from the Atlantic to the Indian Ocean, over the Sahara, most of Egypt, and Arabia, and beyond the confines of Iran into Central Asia. Of course this territory of nearly 4,000,000 square miles has vast sandy or rocky areas wholly barren of mollusks; but it also harbors at selected, often widely separated, places terrestrial faunas of much diversity, including precinctive genera and subgenera (*Eremina, Sphincterochila, Deserticola, Obeliscella, Calaxis,* etc), as well as precinctive species of more ubiquitous genera. For instance, P. Pallary (1924) listed 22 native, mostly precinctive, species in six genera from the very arid, mountainous, and rocky Sinai Peninsula, an area of about 23,000 square miles (ca one-fifth of Arizona), and several species have since been added to his list.

In arid South-West Africa (Kalahari, Ovampoland, Damaraland, Namaqualand), an area of some 6,000,000 square miles north and south of the Tropic of Capricorn, M. Connolly (1931) recognized 42 Recent terrestrial species in 14 genera, 31 being precinctive and the majority belonging to five precinctive genera (*Sculptaria, Trigonephrus, Dorcasia, Microstele, Xerocerastus; Namibiella,* a recently described, precinctive monotypic genus, should be added to his list).

The extensive deserts of central and west Australia, north and south of the Tropic of Capricorn, which cover about two-thirds of this subcontinent (ca 2,000,000 square miles), have a relatively poor terrestrial snail fauna, but at least two precinctive genera (*Bothriembryon* and *Angasella*), each with a fair number of highly variable species.

Our partial synopsis of Recent desertic faunas shows that mollusks strictly adapted to a dry and hot environment evolved independently in widely separated areas and from several unrelated lines of descent. In the Old World, strictly peculiar xero philous genera occur in seven families: true Helicidae (*Eremina, Deserticola, Sphincterochila*), Achatinidae (sensu lato, including the Ferrussaciinae and Subulininae: *Xerocerastus, Calaxis, Obeliscella, Namibiella*), Endodontidae (*Sculptaria*), Bulimulidae (*Bothriembryon*), Acavidae (*Trigonephrus, Dorcasia*), Camaenidae (*Angasiella*), and Pupillidae (*Microstele*). Other more widespread genera produced additional species restricted to arid country: *Trochoidea, Helicopsis, Levantina,* etc in true Helicidae; *Petraeus* and *Buliminus* in Enidae; *Achatina, Zootecus,* and *Subulina* in Achatinidae; and *Pupoides* in Pupillidae. In the New

World, the strictly precinctive genera of the arid North American Southwest, discussed before in this Section, belong to three families: Helminthoglyptidae (*Eremarionta, Sonorella, Mohavelix*),[34] Polygyridae (*Ashmunella*), and Pupillidae (*Chaenaxis*). Additional Southwestern xerophilous species in more widespread genera are Bulimulidae (*Bulimulus* subgenus *Rabdotus*), Helminthoglyptidae (*Helminthoglypta*), Oreohelicidae (*Oreohelix*), and Pupillidae (*Gastrocopta, Pupilla, Vertigo*).

It may be assumed that xerophilous taxa evolved in these diverse lines of descent at different times in the geological past, but when and where they first appeared may never be known. Unlike marine and freshwater mollusks, land snails are as a rule poorly preserved as fossils, due to their specialized living habitats and the fragile texture of most shells, which are often thin and easily destroyed by chemicals in the soil. Moreover, the shell, the only part of the snail found fossil, is not always a trustworthy guide for determining the affinities of the animal. It is often necessary to examine the external features of the animal and its internal anatomy (radula, genitalia, etc), in order to place the snail correctly in genus, species, or even family, and this cannot be done for fossils. Their taxonomic status is always more or less arbitrary.

The occurrence of so many xerophilous genera and species of land snails in the Recent fauna is remarkable. Aridity is usually regarded as the youngest major type of present-day environments, one of the final stages in the Earth's steady loss of free water since the beginning of geological time. The frequency and variety of strictly xerophilous Recent genera and species on all Continents would seem to have required repeatedly a fairly rapid pace of progressive evolution in a phylum generally believed to be more conservative than the Vertebrata.

Speciation and supraspecific evolution may have been actually stimulated and assisted in large land snails by the very hazards of an arid ambience. Recurring dry and hot periods favored prolonged spatial and reproductive isolation of suitable genetic mutations, preceded and followed by rapid and wide dispersal when the climate was milder (damper and cooler). Complete spatial isolation of populations is easily assured in arid country for the large land mollusks (*Sonorella, Eremarionta, Oreohelix,* etc) by their size, weight, and adaptive behavior (hiding under

---

[34]None of the so-called "desert helices" of western North America are true Helicidae, where they were placed formerly; all are Helminthoglyptidae, an anatomically distinct, strictly New World family.

rocks, burrowing in the soil, sealing off the aperture, etc), which normally minimize passive dispersal of living individuals by wind or floods. As pointed out before, this contrasts sharply with the ease with which living lesser snails may be spread far and wide by physical means, particularly wind, so that spatial isolation has hardly influenced their evolution in arid lands.

From an evolutionary viewpoint it is of interest that the extreme and often puzzling intraspecific variability of many large Recent desert snails suggests continuing active evolution at present in arid environments. S. S. Berry (1953) called attention also to the relative plasticity of shell characters in most large xerophilous snails, which makes specific recognition more troublesome than usual, and contrasts with the greater specific stability of their internal and especially reproductive organs.

It is regrettable that, with one notable exception presented below, no comprehensive and critical analysis of a foreign desert molluscan fauna has been attempted thus far that could be compared with our discussion of the mollusks of the North American Southwest. Most published information on such matters is scattered, fragmentary, and too uncritical to be useful.

M. Connolly's (1931) revision of the non-marine Mollusca of South-West Africa (the Namib-Kalahari Desert) is unique in this respect. It presents all reliable pertinent information available at that time, with precise sources. It distinguishes clearly between truly native or adventive species, precinctive or widespread taxa, and Recent or fossil specimens. It gives also the Recent general distribution of the ubiquitous genera and species outside this territory. With a few later accessions, the area now has 43 known Recent (living) terrestrial species in 15 genera; 32 of them are precinctive, and of these 25 belong to six precinctive genera. All 11 widespread species of the area occur also in neighboring subtropical South Africa, and a few of them extend north of the Equator or even to the deserts of North Africa or Arabia.

Since South-West Africa is slightly over four times the size of Arizona, its total molluscan fauna is much poorer, but has a greater proportion of precinctives (74.4 pct for 68.2 pct in Arizona). This relative poverty is due mainly to its location within the zone of trade winds at the Tropic of Cancer, one of the Earth's most arid belts, while the American arid Southwest lies far north of the zone of trade winds of the Tropic of Cancer. South-West Africa has a more extreme desertic climate, and therefore, a more generally distributed and more specialized xerophytic vegetation. The topography is also more uniform: it lacks cooler and damper highlands, and is isolated from most past and present-day dispersal routes. As we have seen, migrations from several sources played an important role in building up the complex Recent mollusk fauna of the American Southwest.

In discussing South-West Africa, M. Connolly (1931:330-331) concludes: "Its exclusiveness is quite remarkable, considering that there are no more formidable obstacles north and south than the Kunene and Orange Rivers, and that its eastern boundary is, roughly speaking, merely the 20th degree of E longitude. . . . This account of the climate and country which offers no violent lines of regional demarcation, provides but a meagre solution to the peculiarity of much of its molluscan fauna, of which the salient features are the extremely restricted distribution of certain species and even genera." Obviously the molluscan faunas of South-West Africa and of the American arid Southwest are basically similar. Both were evidently produced by much the same evolutionary processes under selection pressure from arid ecological conditions, although derived from totally unrelated terrestrial ancestors.

II. Annotated Check List of Recent Arizona Mollusks

GASTROPODA

Pulmonata Geophila

Family Helicidae

* **Helix (Cryptomphalus) aspersa** O. F. Müller, 1774. T.L.: Italy. Introduced in New World from Europe; now fully established (feral) and widespread in warm temperate United States, Mexico, and Central and South America. Not yet feral in Arizona, but often common in irrigated gardens and nurseries in Yuma, Prescott, Phoenix, Tucson, etc. First State record from Tucson by A. R. Mead (1951:21; 1952a:90; 1952b:30; 1953:11; 1963:27).

* **Otala (Otala) lactea** (O. F. Müller, 1774, as *Helix*). T.L.: not given. Native in southern Europe and North Africa; frequently imported in United States in quantity as an edible snail; established in the open in some southern States (e.g. in east Texas). Reported in Arizona from Phoenix (NE section, Apr. 1970) and Tucson (living snails bought at a store on Speedway in 1964 by D. B. Sayner; living adults and immatures repeatedly found in the open by Floyd G. Werner and family in N Jackson Ave area, in irrigated gardens and adjoining vacant lots from 1965 to 1970). Whether it will become feral in the State away from human settlements is doubtful. Often confused with *Otala (Eobania) vermiculata* (O. F. Müller, 1774), also introduced to North America from southern Europe, but thus far not in Arizona. Shape and color of the shell and color of aperture are not reliable for distinguishing the two species, only the sculpture being fully trustworthy: in *lactea* the shell has many fine, irregular, incised spiral lines, which are lacking in *vermiculata;* the species also differ in the anatomy of the genitalia.

Family Helminthoglyptidae (Xanthonychidae)

Eremarionta H. A. Pilsbry, 1913

Eremarionta, proposed as subgenus of *Micrarionta* C. F. Ancey, 1880, agrees with it in having a dart sac seated on the vagina and two mucus glands, both features lacking in *Sonorella.* However, the species now left in *Eremarionta,* after removing some others to *Sonorelix,* differ from subgenus *Micrarionta,* sensu stricto, in having a prominent spermathecal diverticulum and a shell which resembles *Sonorella* in size (10 to 24 mm in diameter), shape, smoothness, and color pattern (light with a single dark peripheral band). It seemed therefore appropriate to raise *Eremarionta* to generic rank, since *Micrarionta,* sensu stricto, does not occur in Southwestern Molluscan Province. The shells of some *Eremarionta* and *Sonorella* are so similar that dissection of genitalia is needed for positive generic distinction. The general distribution of *Eremarionta* is also discussed in the section on Zoogeography. It contains the most pronounced desert snails of North America, maintaining flourishing populations under extreme arid conditions. It is restricted to low mountain ranges in the Mohave, Colorado, and Yuma deserts, northwest Sonora, and northeast Baja California, approximately from 28° 30′ to 36° N and from 112° 30′ to 116° 30′ W (map, Fig. 2).

1a. **Eremarionta r. rowelli** (W. Newcomb, 1865:181, as *Helix*); *Helix (Ampelita) rowelli* W. G. Binney and T. Bland, 1869:185, only the upper fig. 326 of holotype. T.L.: merely given as "Arizona," collected by Frick. Believed to be the first land snail *described* from the State. Pilsbry and Ferriss (1923:99-101, Pl. 3, 4 figs. 7; copied by Pilsbry, 1939:228, 4 figs. 114a) figured the holotype, No. 27517 in Cornell Univ. Coll. In 1923 they also selected as precise T.L. Tinajas Altas, in Tinajas Altas Mts, 32° 18′ N, 114° 3′ W, at 1,300 ft, 26 mi S of Wellton and ca 45 mi E of Colorado Riv, on the strength of 2 dead shells found there by E. A. Mearns, Feb. 21, 1894 (U.S. Nat. Mus. No. 187478, one shown in 1923 on Pl. 3, 3 figs. 6), which agree fully with Newcomb's holotype. Live snails were found there by W. O. Gregg and junior author, Jan. 30, 1958, and A. Ross, Jan. 23, 1965. Pilsbry (1939:229) mentions a dead shell found by E. C. Jaeger 8 mi N of Tinajas Altas. *E. r. rowelli* is known also in Yuma Co. at Heart Tank in Sierra Pinta, 32° 16′ N, 113° 33′ W, at 1,400 ft (N. M. Simmons, alive on Dec. 23, 1962, and Feb. 16, 1964, some hibernating, lying free under rocks, the aperture completely closed with a white, thin, stiff epiphragm), and in Sonora from Sierra del Tuseral, just S of Arizona border, 50 mi SE of Wellton, ca 32° N, 114° W (at Mus. Comp. Zool., Cambridge, Massachusetts), and along Mex. Hwy 2 in S extension of Tinajas Altas Mts, 8.9 mi W of El Puerto (junior author, March 30, 1963, banded snails). Genitalia of snails from T.L. and Heart Tank, dissected by junior author, are essentially as described for *E. r. hutsoni* by H. A. Pilsbry (1907).

Due to the vague T.L., *E. rowelli* was misunderstood by malacologists until Pilsbry and Ferriss (1923) studied the original types. H. A. Pilsbry and H. N. Lowe (1934:67) concluded that *E. rowelli, E. hutsoni, E. hutsoni desertorum,* and *E. rowelli mexicana* were "local races of one species," and Pilsbry (1939:228-239) treated these and other forms from the California deserts as sub-species of *rowelli.* To this complex may be added *E. r. bechteli,* mentioned below, from Isla de San Esteban, and an undescribed form recently found at Rampart Cave in N Mohave Co., ca 36° 2' N, 113° 55' W, by R. H. Russell and junior author. *E. rowelli* is a widespread snail (from ca 28° to 36° N and from ca 114° to 117° W) "with an astonishingly discontinuous range" (H. A. Pilsbry and H. N. Lowe, 1934:67).

The shells of two subspecies described, but not dissected, from Sonora differ little from nominate *E. r. rowelli. Eremarionta rowelli mexicana* (H. A. Pilsbry and H. N. Lowe, 1934:67, as *Micrarionta*); figured by R. J. Drake, 1957:76, Pl. 14, 3 figs. 1 (para-types) and 2 figs. 2 (topotypes), and Pl. 15, 5 figs. 1 (shells from rocky hills near Puerto Penasco). T.L.: Sierra de San Francisco, on road to Puerto Penasco, 12 mi S of Sonoyta, ca 31° 40' N, 113° 15' W, elevation not given (H. N. Lowe, Feb. 27, 1934); also near La Libertad (= Punta Libertad of H. N. Lowe, in Pilsbry, 1939:230), ca 30° N, 112° 40' W (R. J. Drake, 1957:76, Pl. 15, 4 figs. 2; and R. R. Humphrey, 1971). *Eremarionta rowelli bechteli* (W. K. Emer-son and M. K. Jacobson, 1964:327, 3 figs. 5, as *Micrarionta*); T.L.: Isla de San Esteban near Tiburon I, Sonora side of Gulf of Cali-fornia, 28° 40' N, 112° 35' W.

1b. **Eremarionta rowelli acus** (H. A. Pilsbry, 1939:234, 4 figs. 116a, as *Micrarionta*). T.L.: Needles Peak ("The Needles"), 4 mi S of Topock, ca 4 mi E of Colorado Riv, ca 34° 40' N, 114° 40' W, Mohave Co., elevation not given. Found close to or at T.L. and dis-sected by junior author, in Mohave Mts, 3 mi from U.S. Hwy 66, along dirt road starting S at 3 mi E of California end of Colorado Riv bridge, ca 1,100 ft, Jan. 26, 1958. Known also in California in Mohave Mts opposite Topock and in mountains NE of Essex, both San Bernardino Co.

1c. **Eremarionta rowelli desertorum** (H. A. Pilsbry and J. H. Ferriss, 1908:134, Pl. 11, figs. 6-8 of holotype and 9-10 of geni-talia, as *Micrarionta desertorum*). T.L.: "Small range of mountains

12 mi S of Parker," ca 10 mi E of Colorado Riv, ca 34° 5' N, 114° 30' W, Yuma Co. (W. J. Gilchrist), elevation not given. Found in the same area by J. H. Ferriss (1922, 10 mi from Parker, in Pilsbry, 1939:234), G. Willett (1930:4-5, who made it subsp. of *E. hutsoni*), M.L. Walton (1946, at 8.8 mi S of Parker), and W. O. Gregg and junior author (on State Hwy 72 at a point 8.4 mi S of Parker RR station, at 700 ft, Jan. 27, 1958). Dissected by Pilsbry (1908) and junior author (1958).

1d. **Eremarionta rowelli hutsoni** (G. H. Clapp, 1907:136, Pl. 9, figs. 1-4 of holotype, Carnegie Mus., Pittsburgh, No. 5659, as *Epiphragmophora [Micrarionta] hutsoni*; Pilsbry, 1907:138, Pl. 9, figs. 5-8, radula, jaw, and genitalia of paratypes). T.L.: originally "about 8 mi from Quartzsite, in the foothills at an altitude of about 1,600 ft," without other directions, Yuma Co., G. G. Hutson. The only suitable sites of *Eremarionta* within an 8 mi radius of Quartzsite are in the N foothills of Dome Rock Mts, ca 33° 35' N, 114° 30' W, where snails agreeing in shell and anatomy with *hutsoni* were found in some numbers by G. Willett (1930:4-5, at 4 mi from Quartzsite), J. H. Ferriss (1922, cited by Pilsbry, 1939:231), and W. O. Gregg and junior author (in rock slide on W facing slope of ravine at N side of U.S. Hwy 60-70, 9.2 mi W of junction with State Hwy 95 in Quartzsite, at 1,100 ft, Jan. 29, 1958; dissected by junior author from this lot). Unfortunately Pilsbry later (1939:230) chose to ignore G. H. Clapp's original T.L., substituting for it: "About 20 miles south of Quartzsite, 12 miles north of Kofa, Yuma County in foothills of the Short Horn Range, at an alt. of about 1,600 feet"; with a footnote: "After publication of Dr. Clapp's description the locality given by him was corrected by Mr. Hutson as above. It is in the north end of the Eagle Tail Mountains." The Eagle Tail Mts are some 45 mi SE of Quartzsite and no snails have ever been found there. The Short Horn Mts are not shown on any published map; but the name has sometimes been used locally for the Kofa Range, as explained by B. H. Granger (1960:385). There is no definite record of snails having been found thus far in Kofa Mts. H. C. Rawls (1969), speculating at length on the Short Horn Mts being the T.L. of *hutsoni*, was led to place it "about 20 mi SE of Quartzsite"; he fails to mention, however, that he found the snail there. Since *E. r. hutsoni* is definitely known from the N end of Dome Rock Mts, within the area originally given as T.L., Hutson's supposed correction can only be due to the collector's faulty memory. Pilsbry did not collect the snail and never visited the area.

Eremarionta (including Section *Chamaearionta* S. S. Berry, 1930a:75) of SE California, from Death Valley to Mexican border (references and distribution in H. A. Pilsbry, 1939:226-266): *Eremarionta rowelli acus* (H. A. Pilsbry, 1939; also in Arizona). *E. rowelli unifasciata* (G. Willett, 1930), synonym: *Micrarionta hutsoni hilli* G. Willett, 1930. *E. rowelli mccoiana* (G. Willett, 1935). *E. rowelli bakerensis* (H. A. Pilsbry and H. N. Lowe, 1934). *E. rowelli amboiana* (G. Willett, 1931). *E. rowelli granitensis* (G. Willett, 1935). *E. rowelli chuckwallana* (G. Willett, 1935). *E. rowelli chocolata* (G. Willett, 1935). *E. immaculata* (G. Willett, 1937). *E. millepalmarum* (S. S. Berry, 1930). *E. brunnea* (G. Willett, 1935). *E. orocopia* (G. Willett, 1939). *E. argus* (H. M. Edson, 1912). *E. indioensis* (L. G. Yates, 1890), synonym: *Micrarionta callinepius* S. S. Berry, 1930. *E. indioensis wolcottiana* (P. Bartsch, 1903). *E. indioensis cathedralis* (G. Willett, 1930). *E. indioensis xerophila* (S. S. Berry, 1922). *E. indioensis remota* (G. Willett, 1937). *E. morongoana* (S. S. Berry, 1929). *E. aquae-albae* (S. S. Berry, 1922).

At least one undescribed species of *Eremarionta* is known from northern Baja California (N of 28° N).

Sonorelix of SE California deserts (formerly placed in *Micrarionta* or *Eremarionta*) (references and distribution in H. A. Pilsbry, 1939:226-266): *Sonorelix melanopylon* (S. S. Berry, 1930). *S. harperi* (F. W. Bryant, 1900), synonym: *Micrarionta orcuttiana* P. Bartsch, 1904. *S. borregoensis* (S. S. Berry, 1929). *S. borregoensis ora* (G. Willett, 1929). *S. borregoensis carrizoensis* (G. Willett, 1937). *S. rixfordi* (H. A. Pilsbry, 1919), synonyms: *Micrarionta aetotis* S. S. Berry, 1928, and *Micrarionta depressispira* S. S. Berry, 1928. *S. avawatzica* (S. S. Berry, 1930). *S. avawatzica eremita* (H. A. Pilsbry, 1939). *S. baileyi* (P. Bartsch, 1904). *S. angelus* W. O. Gregg, 1949a: 100, Pl. 23.

Mohavelix micrometalleus (S. S. Berry, 1930), of SE California desert.

Helminthoglypta of SE California deserts (references and distribution in H. A. Pilsbry, 1939:159-170): *Helminthoglypta mohaveana* S. S. Berry, 1926. *H. graniticola* S. S. Berry, 1926. *H. crotalina* S. S. Berry, 1928. *H. jaegeri* S. S. Berry, 1928. *H. fontiphila* W. O. Gregg, 1931. *H. caruthersi* G. Willett, 1934. *H. greggi* G. Willett, 1931. *H. isabella* S. S. Berry, 1938. *H. fisheri* (P. Bartsch, 1904). *H. micrometalleoides* W. B. Miller, 1970.

Sonorelix of northern Baja California (N of 28° N): *Sonorelix peninsularis* (H. A. Pilsbry, 1916a:100, Pl. 2, 3 figs. 4, as *Sonorella*). *S. inglesiana* (S. S. Berry, 1928:76, Pl. 2, figs. 10-18, as *Micrarionta* subgenus *Eremarionta*). *S. chacei* (G. Willett, 1940:81, Pl. 20, as *Micrarionta* subgenus ? *Eremarionta*).

Helminthoglyptidae of northern Baja California (N of 28° N) not placed at present in the correct genus: *Sonorella merrilli* P. Bartsch, 1904:192. *Micrarionta evermanni* H. A. Pilsbry, 1927a:182, Pl. 12, figs. 4-6.

The generic identity and true locality of the following species, described from dead Arizona shells, are unknown or uncertain and may never be established.

Micrarionta praesidii H. A. Pilsbry and J. H. Ferriss, 1919a: 312, Pl. 6, figs. 8 and 8a-b; *Sonorella* (?) *praesidii* Pilsbry, 1939:334, 3 figs. 203. Based on one dead shell, Acad. Nat. Sci. Philadelphia No. 58121, collected by G. H. Horn about 1863-1864, possibly at old Fort Grant, at confluence of Arivaipa Creek and San Pedro Riv, Pinal Co. (not Graham Co., where the fort was moved in 1872-1873 after G. H. Horn had left Arizona); it was admitted by the authors (1919a:314) that the shell might have come from somewhere else in western Arizona. For the vicissitudes of Fort Grant see W. C. Barnes, 1935:188, B. H. Granger, 1960:128, and R. J. Drake, 1962b.

Micrarionta newcombi H. A. Pilsbry and J. H. Ferriss, 1923: 101, Pl. 3, 3 figs. 8; Pilsbry, 1939:239, 3 figs. 119. Based on one dead shell from "Arizona," found in Cornell Univ. Coll. with the holotype of *Eremarionta rowelli* (W. Newcomb), "presumably from the same locality."

Sonorella H. A. Pilsbry, 1900

Sonorella comprises helminthoglyptid snails with a depressed globose, helicoid shell, 12 to 30 mm in diameter, umbilicate or perforate, with a wide, unobstructed mouth and a thin, barely expanded peristome, smoothish or slightly sculptured with growth-lines, occasionally with fine oblique or spiral granulation and short hairs (particularly on the early whorls), lightly colored, and normally with a dark peripheral band. Its most characteristic features are, however, in the genitalia, which lack a dart sac and mucus glands. The essential specific differences also are in the genitalia, the shells of only a few species showing reliable diagnostic characters. Unless otherwise stated, genitalia of species and subspecies here recognized, as well as of their synonyms, were dissected by junior author on sexually mature, unpreserved snails, in most cases from populations collected at T.L. General distribution discussed in section on Zoo-geography.

In 1939 Pilsbry divided *Sonorella* into 4 subgenera, based primarily on differences in genitalia, although the sculpture of the embryonic whorls was also mentioned. These were: *Sonorella,* sensu stricto (type: *Epiphragmophora hachitana* W. H. Dall, 1896); *Masculus* (type: *Sonorella virilis* H. A. Pilsbry, 1905); *Sonoranax* (type: *Sonorella dalli* P. Bartsch, 1904); and *Myotophallus* (type: *Sonorella fragilis* H. A. Pilsbry, 1939). Most species were kept in *Sonorella,* sensu stricto, and distributed among 8 species-groups. Junior author's anatomical revision (W. B. Miller, 1967c) led him to discard Pilsbry's system and to arrange all species in four complexes, none of which should, in his opinion, be given subgeneric rank or names, since they are connected by transitional forms. The complexes do not correspond to Pilsbry's subgenera or groups. Thus, the species of the former "*S. coloradoensis* Group" are now distributed among three of the complexes.

A. *Sonorella hachitana* Complex. Penis with a usually long, slender, narrowly pointed verge; in extreme forms, verge thick and club-shaped. Shell relatively large and capacious, with smooth, silky-lustrous periostracum, usually with apical spirally descending threads. Contains 26 species, mostly placed by Pilsbry in his *S. hachitana* and *S. ambigua* Groups of *Sonorella,* sensu stricto, but some also elsewhere, and even one in his subgenus *Masculus.*

B. *Sonorella binneyi* Complex. Penis usually with a short, more or less thick, bluntly rounded verge. Shell relatively small, globose, occasionally depressed, with smooth, silky-lustrous periostracum, usually with apical spirally descending threads. Contains 15 species, placed by Pilsbry in his *S. binneyi* Group of *Sonorella,* sensu stricto, with one exception.

C. *Sonorella granulatissima* Complex. Penis with a usually stout and truncate verge, reaching extremes of diminution in some species or gigantism in others. Shell minutely granulose or wrinkly-granulose, with readily peeling periostracum, mostly without apical spiral threads. Contains 23 species, mostly placed by Pilsbry in his *S. tumamocensis, S. dragoonensis, S. clappi,* and *S. granulatissima* Groups of *Sonorella,* sensu stricto, but including also the types and other species of his subgenus *Masculus* and *Sonoranax.*

D. *Sonorella rooseveltiana* Complex. Inner tube of penis slender, but enveloped in an extremely thick, muscular sheath, tapering posteriorly; length of the whole about one-fifth of the diameter of shell or less; verge lacking or vestigial. Vagina very short. Shell relatively small, depressed-globose, with smooth, silky-lustrous periostracum, with spirally ascending and descending threads on embryonic whorls, prominent on young, and fainter or worn on older shells. Contains 2 species. *S. fragilis,* the type of Pilsbry's subgenus *Myotophallus,* is only a subspecies of *S. roosevel-*

tiana, placed by him in *Sonorella,* sensu stricto, but he saw only immature genitalia; both *fragilis* and *rooseveltiana* agree in lacking a verge. The closely related *S. allynsmithi,* however, has a minute, vestigial verge, and connects this complex with the others.

A. *Sonorella hachitana* Complex

2. **Sonorella delicata** H. A. Pilsbry and J. H. Ferriss, 1919. Known only from T.L.:"N end of Peloncillo Mts, ca 6 mi S of Gila Riv, on the old toll road from Solomonsville to Clifton, at ca 4,800 ft," Greenlee Co. (not Graham Co.); on U.S. Geol. Surv. map this Station is now at the S side of Hwy 666, in Tollhouse Canyon, 3.7 mi SW of Gila Riv bridge, at 4,350 ft, where the species was collected again in 1967.

3. **Sonorella waltoni** W. B. Miller, 1968:61, figs. 1A-C and 3D. Known only from T.L.: Peloncillo Mts, NE facing side of West Doubtful Canyon, in N part of Range, at 4,800 ft, Cochise Co. (collected in 1967). A dead, immature shell from this area was tentatively referred by Pilsbry (1939:277) to *Sonorella hachitana peloncillensis* H. A. Pilsbry and J. H. Ferriss (1915), which does not occur in Arizona. No form of *S. hachitana* (W. H. Dall, 1896) is known at present from Arizona.

4. **Sonorella caerulifluminis** H. A. Pilsbry and J. H. Ferriss, 1919. T.L.: originally given as "San Francisco Riv, 6 mi above its confluence with Blue River"; elevation not given; however, as implied by the specific name, this was an obvious error for Blue Riv, 6 mi above its confluence with San Francisco Riv, in Blue Mts, Greenlee Co., where the species was collected again in 1967 at 4,200 ft. Also known elsewhere in Greenlee Co. and in Graham Co. Syntypes are labeled in some collections by error as from San Francisco Mtn, Coconino Co., where *Sonorella* is unknown.

5. **Sonorella optata** H. A. Pilsbry and J. H. Ferriss, 1910. T.L.: Chiricahua Mts, at head of Big Emigrant Canyon, in the N section of Range, ca 6,000 ft. Cochise Co. Also at nearby Stations in same Range, at 4,500 to 5,500 ft.

6a. **Sonorella g. galiurensis** H. A. Pilsbry and J. H. Ferriss, 1919. T.L.: Galiuro Mts, 1.5 mi S of Copper Creek Mining Camp, Pinal Co. (not Graham Co.); elevation not given, but collected there recently at 4,200 ft, on a N facing slope above the Site of Sombrero Butte Camp. Other Stations mentioned in 1919 (pp. 299 and 307) are also in NW area of same Range, in Pinal Co.: E gate of John Rhodes Ranch; Rhodes Canyon; N slope on Whitlock (or Whitlow) Ranch, E of Sombrero Peak (Merced Ranch of present map); Copper Creek Mining Camp; trail halfway between Copper

Creek Camp and Table Mtn; NE slope of Table Mtn; 2 mi E of Table Mtn.

Pilsbry (1939:301) refers to *S. galiurensis* snails from Picket Post Mtn, ca 4 mi SW of Superior, at 4,370 ft, Pinal Co.; and from "Nantes" (probably misspelling of Nantanes) Mts, 10 mi N of Rice, Gila Co.; these were perhaps *S. g. superioris,* which was not recognized until 1939. In 1939:303, he also mentions under *S. galiurensis* a dead shell collected by George H. Horn, about 1863 to 1866, at old Fort Grant, located in Horn's time in Pinal Co. (not Graham Co.), as noted before for *Micrarionta praesidii.* There is no clear evidence that *S. galiurensis* occurs anywhere in Graham Co., although the E section of Galiuro Mts is there. The two species of *Sonorella* known from Pinaleno Mts in Graham Co. are restricted to that Range and not closely related to *galiurensis.*

6b. **Sonorella galiurensis superioris** H. A. Pilsbry, 1939. T.L.: near the highway tunnel, 4 mi E of Superior, on Queen Creek, Pinal Co.; elevation not given, but found there recently at 3,700 ft. Also at nearby Stations in Gila Co. and Maricopa Co.

7a. **Sonorella s. sabinoensis** H. A. Pilsbry and J. H. Ferriss, 1919. T.L.: Santa Catalina Mts, Station 16 (of 1913), W side of Sabino Canyon, near mouth at ca 5,000 ft, Pima Co. Elsewhere in same Range, at 3,000 to 6,000 ft in S section, up to 7,000 ft in N section; and NW slope of Tanque Verde Mts, 2 to 3 mi E of HQ of Saguaro National Monument, 3,600 to 5,000 ft (collected and dissected by junior author, Nov. 1969). Synonyms: *Sonorella marmorarius* H. A. Pilsbry and J. H. Ferriss, 1919 (T.L.: Santa Catalina Mts, S side of Marble Peak, Pima Co.; elevation not given, but paratypes labeled 6,000 to 7,000 ft; dissected by Pilsbry, 1919). *Sonorella marmorarius imula* H. A. Pilsbry and J. H. Ferriss, 1919 (T.L.: Santa Catalina Mts, N foothills, 6 mi NW of Brush Corrall Ranger Station, Pima Co.; elevation not given; not dissected from T.L.). *Sonorella marmorarius limifontis* H. A. Pilsbry and J. H. Ferriss, 1919 (T.L.: Santa Catalina Mts, bluffs near Mud Springs in Pine Canyon, ca 7,000 ft, Pima Co.; collected at T.L. and dissected by junior author, 1967).

7b. **Sonorella sabinoensis buehmanensis** H. A. Pilsbry and J. H. Ferriss, 1919. T.L.: Santa Catalina Mts, Buehman Canyon near Korn Kobb Mine, in SE foothills of Range, Pima Co.; elevation not given, but collected there recently at 3,450 ft. Also elsewhere in Buehman Canyon.

7c. **Sonorella sabinoensis dispar** H. A. Pilsbry, 1939, as new name for *Sonorella s. occidentalis* of 1919. T.L.: Santa Catalina Mts, near W end of Range, at E side of Pima Canyon, Pima Co.; elevation not given, but collected there recently at 3,400 ft. Also elsewhere in SW area of same Range. Synonym: *Sonorella sabinoensis occidentalis* H. A. Pilsbry and J. H. Ferriss, 1919 (same T.L. as *S. s. dispar*) (not *Sonorella granulatissima occidentalis* H. A. Pilsbry and J. H. Ferriss, 1915, which is *S. clappi* H. A. Pilsbry and J. H. Ferriss, 1915).

7d. **Sonorella sabinoensis tucsonica** H. A. Pilsbry and J. H. Ferriss, 1923. T.L.: Tucson Mts, "Station 81," described as "cliffs above Station 68," in the authors' "Wild Pig Amphitheater," Pima Co.; elevation not given (present location of Station 68 unknown). Also elsewhere in N section of same Range, where it was found recently at 2,300 ft. Synonym: *Sonorella sabinoensis deflecta* H. A. Pilsbry and J. H. Ferriss, 1923 (T.L.: Tucson Mts, ca 5 mi N of Mountain Sheep Camp, Pima Co.; elevation not given; dissected by Pilsbry, 1923).

8. **Sonorella bequaerti** W. B. Miller, 1967c:57, figs. E-F, Pl. 2, figs. G-H. T.L.: Rincon Mts, canyon running NW from Rincon Peak and just W of Rincon Peak-Happy Valley Saddle Trail, at 6,000 to 6,100 ft, Pima Co. Also in nearby Chimenea Canyon of Tanque Verde Mts, at ca 4,000 ft.

9. **Sonorella papagorum** H. A. Pilsbry and J. H. Ferriss, 1915. Known only from T.L.: Black Mountain, near San Xavier Mission, ca 9 mi S of Tucson, ca 3,200 ft, Pima Co.

10. **Sonorella vespertina** H. A. Pilsbry and J. H. Ferriss, 1915. Known only from T.L.: Baboquivari Mts, W side of main ridge, near summit, 0.5 mi S of Baboquivari Peak, Pima Co.; elevation not given, but found there recently at ca 6,000 ft.

11. **Sonorella eremita** H. A. Pilsbry and J. H. Ferriss, 1915. Known only from T.L.: NW end of San Xavier Hill, Mineral Hill Group, ca 20 mi S by SW of Tucson, Pima Co.; elevation not given, but found there recently at 3,850 ft; restricted to one rock slide.

12. **Sonorella meadi** W. B. Miller, 1966:50, Pl. 1, figs. G-I, Pl. 2, fig. E. T.L.: Agua Dulce Mts, in canyons both S and W of Agua Dulce Pass, E of Quitovaguita Peak, ca 1,600 ft, Pima Co. Dead shells, possibly of this species, were seen from nearby Bates Mts and Granite Mts, Pima Co.

13a. **Sonorella h. huachucana** H. A. Pilsbry, 1905. T.L.: Huachuca Mts, Browns Canyon, Cochise Co.; elevation not given; snails from T.L. not dissected thus far. Specimens dissected from elsewhere in Huachuca Mts, Cochise Co.; and in Patagonia Mts, Santa Rita Mts, and Saddle Mtn (7 mi SE of Patagonia), Santa Cruz Co. at 3,800 to 5,400 ft. Only subfossil shells agreeing with typical *huachucana* are common in Mustang Mts, on N slope of easternmost high Dome, at ca 5,200 ft. Synonyms: *Sonorella walkeri aguacalientensis* H. A.

Pilsbry and J. H. Ferriss, 1915 (T.L.: Santa Rita Mts, N base of bluffs along banks of stream flowing from Agua Caliente Canyon, 4,100 to 4,200 ft, Santa Cruz Co.). *Sonorella patagonica* H. A. Pilsbry and J. H. Ferriss, 1919 (T.L.: Patagonia Mts, main canyon of Mt Washington, running W, ca 4,500 ft, Santa Cruz Co.; snails from T.L. dissected by Pilsbry, 1919). *Sonorella parietalis* J. H. Ferriss, 1919 (nomen nudum).

? 13b. **Sonorella huachucana elizabethae** H. A. Pilsbry and J. H. Ferriss, 1919. Known only from T.L.: originally given as Canelo (misspelled "Canillo") Hills; more precisely in 1923 as "Station 276, a large peak E of the Huachuca-Duquesne road"; elevation not given, but some paratypes are labeled ca 5,000 ft; dissected by Pilsbry (1919). J. H. Ferriss' itinerary (1919b;41), studied with recent maps, shows that Station 276 was on Lookout Knob in Canelo Hills, ca 5 mi slightly W by S of Canelo, Santa Cruz Co., where dead specimens were found at 5,500 to 5,700 ft in 1965 and 1967. In 1939 Pilsbry gives the T.L. by error as Mt Hughes, which is far W of Canelo Hills and close to Sonoita-Patagonia Hwy.

14. **Sonorella mustang** H. A. Pilsbry and J. H. Ferriss, 1919. T.L.: given originally as Mustang Mts; more precisely in 1923:102 (for Pl. 1, fig. 2 of holotype), as Station 153, which is defined on p. 56 as "N side tower, East Peak" of Range, placing it in Cochise Co.; elevation not given, but collected there recently at 5,400 ft. Also elsewhere in same Range, in Cochise Co. and Santa Cruz Co., but not in Pima Co.

15a. **Sonorella w. walkeri** H. A. Pilsbry and J. H. Ferriss, 1915. T.L.: Santa Rita Mts, Walnut Branch of Agua Caliente Canyon, W slope of Range, at ca 6,000 ft, Santa Cruz Co. Also elsewhere in Santa Rita Mts, at 4,400 to 5,700 ft, and Atascosa Mts, both in Santa Cruz Co.; found also recently at two Stations in Sonora. Synonym: *Sonorella montana* H. A. Pilsbry and J. H. Ferriss, 1919 (T.L.: Atascosa Mts, Montana Peak, near Montana Mine, Santa Cruz Co.; elevation not given; dissected by Pilsbry, 1923).

15b. **Sonorella walkeri cotis** H. A. Pilsbry and J. H. Ferriss, 1919. T.L.: Whetstone Mts, Station 3, 1 mi up from Ranger Station, Cochise Co.; elevation not given (present location of Station 3 unknown). Also elsewhere in same Range, at 4,800 to 5,100 ft.

16. **Sonorella rosemontensis** H. A. Pilsbry, 1939. T.L.: Santa Rita Mts, N end at Station 49, near Rosemont, Pima Co.; elevation not given, but collected there recently at ca 5,500 ft. Also elsewhere in same area.

17. **Sonorella ashmuni** P. Bartsch, 1904. T.L.: Richinbar, ca 30 mi SE of Prescott, 3,500 ft, Site of abandoned mine on Agua Fria

Riv, 3 mi E of Bumblebee. Also known from Mingus Mtn, 6,500 ft,
Yavapai Co.; Fish Creek, ca 20 mi NW Superior, 3,200 ft, Maricopa
Co.; and Sierra Ancha, 7,000 to 7,200 ft, Gila Co. Synonym: *Sono-
rella strongiana* S. S. Berry, 1948:151, figs. 1-3 and 8 (T.L.: Sierra
Ancha, 3 mi N of Reynolds Creek, near Pueblo Mine, Gila Co.; 1
dead shell; elevation not given; collected alive and dissected from
T.L., at 7,200 ft, by junior author).

18. **Sonorella compar** H. A. Pilsbry, in H. A. Pilsbry and J. H.
Ferriss, 1919:296, footnote 3. T.L.: Purtyman's Ranch in Oak Creek
Canyon, Coconino Co.; elevation not given, but collected at a nearby station in 1966 at 5,000 ft. Jaw, radula, and genitalia of snails
from T.L., collected by E. H. Ashmun, were figured as those of *S.
hachitana* by Pilsbry (1900:557; Pl. 21, figs. 1-5); and their shells
and genitalia as those of *S. hachitana ashmuni* by Pilsbry (1905:259;
Pl. 17, figs. 9-14; Pl. 20, figs. 1-5). Junior author dissected snails
from a nearby Station in Oak Creek Canyon and found genitalia
agreeing with Pilsbry's accounts of 1900 and 1905.

19. **Sonorella coltoniana** H. A. Pilsbry, 1939. T.L.: cliffs on S side
of Walnut Canyon, inside National Monument, 13 mi E of Flagstaff,
Coconino Co.; elevation not given. Junior author collected it in 1966
and 1967 on N side of Walnut Canyon, just outside National Monument, 6,500 ft; the shells and their genitalia were as described
and figured from T.L. by Pilsbry (1939:336, figs. 205a-b and 206A).
Pilsbry also referred in 1939 to *coltoniana* snails from Oak Creek
Canyon, of which he figured shell (fig. 205c) and genitalia (fig. 206B);
the identity of these cannot now be established. Dead shells from
"around Oatman, in western Mohave Co.," also tentatively called
coltoniana by Pilsbry (1939), were probably *S. coloradoensis mohaveana.*[35]

20. **Sonorella bicipitis** H. A. Pilsbry and J. H. Ferriss, 1910. T.L.:
Dos Cabezas Mts, NW branch of Buckeye Canyon, Cochise Co.;
elevation not given, but found there recently at ca 4,600 ft. Also
elsewhere in same Range.

21a. **Sonorella a. ambigua** H. A. Pilsbry and J. H. Ferriss, 1915.
T.L.: originally given as "Cababi Mts, . . . ca 75 mi W of Tucson,"
Pima Co.; more precisely in 1923 as "Station 143, N side of eastern
hill of the group"; (Cababi Mts are Ko Vaya Hills of present map);
elevation not given, but found there recently at 2,500 ft. Also known

[35]The supposed *S. coltoniana* reported by D. S. and H. A. Dundee (1958:52)
from a dry wash in desert just W of Rimmy Jims, on U.S. Hwy 66, Coconino Co.,
were *Oreohelix,* as appears from their shells being "characterized by having a
double band of pigment on the body whorl." R. H. Russell and ourselves repeatedly (1968-1970) collected specimens of *Oreohelix houghi* at that Station, but
found no *Sonorella* there.

from Roskruge Mts, 2,650 to 2,900 ft; Quinlan Mts at Kitt Peak, 5,500 to 6,300 ft; Coyote Mts, 3,400 to 4,000 ft; Robles Hills; Comobabi Mts; hills 5 mi W of Kom Vo ("Comovo" of J. H. Ferriss), 2,000 ft; Santa Rosa spur of Nariz Mts near Mexican border, 2,280 ft, dead shells only. Inhabits some of the most arid and hottest stations of any *Sonorella*.

21b. **Sonorella ambigua verdensis** H. A. Pilsbry, 1939. T.L.: Mazatzal Mts, in Verde Riv Valley, at Gila Co.-Maricopa Co. border; elevation not given. Found recently at Tonto Natural Bridge of Pine Creek, 6 mi S of Pine, ca 13 mi E of T.L., at 4,500 ft, Gila Co.

22. **Sonorella imitator** W. O. Gregg and W. B. Miller, 1972, MS. T.L.: Pinaleño Mts, SW slope of Mt Graham, N of Swift Trail, ca 21 mi W of Hwy 666, 9,000 ft, Graham Co. Also on Heliograph Peak of same Range, 10,000 ft.

23. **Sonorella tortillita** H. A. Pilsbry and J. H. Ferriss, 1919. T.L.: Tortolita [misspelled "Tortillita"] Mts, Station 41, E side of Hog Canyon (locality not found on map or traced locally), on the mountain top, Pinal Co.; elevation not given. Found recently at S end of same Range in Ruelas Canyon, 3,050 ft, Pima Co. (main Range is in Pinal Co.).

24. **Sonorella simmonsi** W. B. Miller, 1966:48, Pl. 1, figs. D-E, Pl. 2, fig. C. T.L.: Picacho Mts, W side of Canyon running SE from Newman Peak, 5 mi E of Picacho, 2,500 ft, Pinal Co. Also at Picacho Peak, 2,000 ft, and in Silver Reef Mts, 12.5 mi S of Casa Grande, 1,600 ft, both in Pinal Co.

25a. **Sonorella r. rinconensis** H. A. Pilsbry and J. H. Ferriss, 1910. T.L.: Rincon Mts, originally given as "Rincon Peak," without precise Station or elevation; placed at first by error in Cochise Co., corrected to Pima Co. by Pilsbry (1939); in 1910 dead shells from Mt Mica and Wrong Mtn in the same Range were also referred to it; but only syntypes from Rincon Peak (lot 94313 Acad. N. S. Phila.) were dissected by Pilsbry. Junior author found snails, agreeing in shell and genitalia with Pilsbry's descriptions and figures of typical *rinconensis*, in SW foothills of Rincon Mts, in Posta Quemada Canyon, 2 to 3 mi NE of Colossal Cave, 3,700 ft, Pima Co. Snails from higher elevations in Rincon Peak area were by dissection either *S. bagnarai* (at 8,000 to 8,200 ft) or *S. bequaerti* (at 4,000 to 6,100 ft).

25b. **Sonorella rinconensis hesterna** H. A. Pilsbry and J. H. Ferriss, 1919. T.L.: dead snails from Rincon Mts, "at Station 148 in a rock slide on the S side of the Tucson-Benson Hwy [of 1917], near the cave on Shaw's Ranch [later Day's Ranch, now Posta Quemada Ranch], S foothills of the Rincons, at about 3,500 ft." In 1964 and 1966 M. L. Walton found two colonies of *S. r. hesterna* 1 to 2 mi SW of the former Day's Ranch buildings, in small rock slides, 2 to 3 mi N of present (new) Tucson-Benson Freeway, one on S bank of Pantano Wash, ca 3,200 ft, the other on E bank of Agua Verde Creek, ca 3,300 ft. In 1967 junior author found a third colony at base of N facing limestone cliffs, ca 0.25 mi S of present Posta Quemada Ranch. Snails of all 3 colonies agree in shell characters with *S. r. hesterna,* and, with only minor differences, in genitalia with typical *rinconensis.* The snails from N end of Santa Rita Mts, provisionally referred by Pilsbry and Ferriss (1923:60 and 90) to *S. hesterna,* were described by Pilsbry in 1939 as a distinct species, *S. rosemontensis.*

26. **Sonorella santaritana** H. A. Pilsbry and J. H. Ferriss, 1915. T.L.: Santa Rita Mts, Station 5 in Walnut Canyon, an upper branch of Agua Caliente Canyon, 6,000 ft, Santa Cruz Co. Also on W side of same Range, at 5,700 to 7,000 ft; collected recently by junior author just below Walnut Spring at 6,850 ft.

B. *Sonorella binneyi* Complex

27. **Sonorella binneyi** H. A. Pilsbry and J. H. Ferriss, 1910. T.L.: Chiricahua Mts, Station 1, 2 mi up mouth of Horseshoe Canyon, on SE side of Range, Cochise Co.; elevation not given, but collected recently by junior author near T.L. at 5,000 ft. Also elsewhere in same canyon at 5,000 to 5,500 ft.

28. **Sonorella tryoniana** H. A. Pilsbry and J. H. Ferriss, 1923. Known only from T.L.: Sanford's, "on the bank of Sonoita Creek," near Sanford Butte, in S foothills of Santa Rita Mts, Santa Cruz Co.; elevation not given, but found there recently at 3,850 ft. The supposed *tryoniana* from Empire Mts, recorded in 1923, was described by Pilsbry (1939) as a distinct species, *S. imperatrix.*

29. **Sonorella imperatrix** H. A. Pilsbry, 1939. T.L.: Empire Mts, W side of a large peak 1.5 mi N of Total Wreck Mine, Pima Co.; elevation not given, but collected recently at a nearby Station in N section of Range, at 4,900 ft.

30. **Sonorella imperialis** H. A. Pilsbry and J. H. Ferriss, 1923. T.L.: Empire Mts, Station 151, "gulch west, starting W then turning N," Pima Co.; elevation not given; exact location on present-day maps uncertain. Reported also from Station 150 in same Range, "road to Forty-Nine Mining Camp, N slope of Mountain, 1.5 mi S of Camp, in gulch running N"; collected in 1966 near or at this Station, in S section of Range, at 4,800 ft.

31a. **Sonorella s. superstitionis** H. A. Pilsbry, 1939. T.L.: Superstition Mts, in N Pinal Co., without more precise Station or elevation. Collected recently in N section of same Range, in Boulder Canyon near Canyon Lake, 2,000 ft, Maricopa Co.

31b. **Sonorella superstitionis taylori** W. B. Miller, 1969:87, fig. 1A, and 1968:59, figs. 2A-C. Known only from T.L.: Salt River Mts, in South Mountain Park, S side of canyon at W end of Guadalupe Rd, 2.6 mi by road W of 56th St, Phoenix, ca 1,500 ft, Maricopa Co., collected by D. W. Taylor. The dead shells reported as *Ampelita rowelli* by H. Prime in 1882 (reference not traced) from Salt River Mts, 7 mi S of Phoenix, recognized by Pilsbry and Ferriss (1923: 100) and by Pilsbry (1939:230) as a species of *Sonorella,* were evidently *S. superstitionis taylori.*

32. **Sonorella franciscana** H. A. Pilsbry and J. H. Ferriss, 1919. T.L.: E bank of San Francisco Riv, at Station 92, opposite Sardine Creek, in Blue Mts, Greenlee Co.; elevation not given. Also elsewhere on same river. All Stations said by Pilsbry (1939) to be ca 4,000 ft. Collected recently in the area by junior author at 3,600 ft.

33a. **Sonorella b. baboquivariensis** H. A. Pilsbry and J. H. Ferriss, 1915. T.L.: Baboquivari Mts, E side of Range at Station 25, on ridge at head of Thomas Canyon, ca 1.5 mi S of Baboquivari Peak, Pima Co.; elevation not given. Also elsewhere on E slope of Range and on N side of Saucito Mtn, where it was collected by junior author at 4,300 ft, 3.5 mi SE of Kitt Peak.

33b. **Sonorella baboquivariensis cossi** W. B. Miller, 1966:46, Pl. 1, figs. A-C, and Pl. 2, fig. A. T.L.: Ajo Mts, Organ Pipe Cactus National Monument, at S left side of Arch Canyon, ca 1 mi above the Arch, 2,900 ft, Pima Co. Also at nearby Stations in same Range, 3,500 to 4,000 ft.

33c. **Sonorella baboquivariensis berryi** H. A. Pilsbry and J. H. Ferris, 1923. Known only from T.L.: Roskruge Mts, "Station 103, on small hill N of road, on E side of Range," Pima Co.; elevation not given. The Station could not be traced in the field and we did not collect specimens. Genitalia incompletely dissected by Pilsbry (1923).

33d. **Sonorella baboquivariensis depressa** H. A. Pilsbry and J. H. Ferriss, 1915. T.L.: Baboquivari Mts, Station 29, "low in upper Sycamore [= Brown] Canyon, ca 1 mi above the forester's cabin and not far above the dam," Pima Co.; elevation not given; dissected from T.L. by H. A. Pilsbry (1923). Also elsewhere in same Range; in Sierrita Mts, where the subspecific identity is uncertain; and at several Stations in Tucson Mts (collected recently by junior author on NW side of Wasson Peak, 4,250 ft, and its identity with *depressa* verified by dissection).

34. **Sonorella micromphala** H. A. Pilsbry, 1939. T.L.: Rim of Mogollon Plateau, "on the furthest peak visible from Pine, at 6,000 to 7,000 ft," Gila Co.; more precisely on W slope of Milk Ranch Point, above Pine, where it was collected recently by junior author at 6,600 ft.

35. **Sonorella xanthenes** H. A. Pilsbry and J. H. Ferriss, 1923. Known only from Kitt Peak in Quinlan Mts, Pima Co.; T.L. originally given as Station 23, near summit, without elevation; later said by Pilsbry (1939:321) to be at ca 5,500 ft. Collected recently in several ravines of N slope of Peak at 6,200 to 6,700 ft.

36a. **Sonorella c. coloradoensis** (R. E. C. Stearns, 1890, as *Helix*). T.L.: Grand Canyon of Colorado River, on S Rim, near Hance Trail, opposite Kaibab Plateau, at 3,500 ft, Coconino Co. Widespread in Grand Canyon, on both N and S walls, up to 5,700 ft; an abnormal, scalariform shell was found by L. E. Daniels (1912:39) at E side of Powell Plateau near Mohave Creek. Also recorded by Pilsbry (1939:339) from cliffs 2 mi NW of Chino RR stop (5 mi NW of Seligman), placed by him in Yavapai Co., but more probably in Coconino Co. Northernmost species of genus.

36b. **Sonorella coloradoensis mohaveana** W. B. Miller, 1968: 51, figs. 1D-F and 3B. T.L.: Black Mts near Oatman, Mohave Co., in rock slide on N side of Rd to Goldroad, E of Sitgreaves Pass, 17 mi W of junction of Goldroad Rd and U.S. Hwy 66, 21 mi NE of Needles, in Arizona, ca 3,200 ft. A closely related, not yet described or dissected *Sonorella* was found in 1969 near Rampart Cave, 1,700 ft, 36° N, 113° 55′ W, Mohave Co., northwesternmost known Station of genus.

37a. **Sonorella s. sitiens** H. A. Pilsbry and J. H. Ferriss, 1915. T.L.: Las Guijas (misspelled "Las Gijas") Mts, NW end of Range, above Las Guijas Mine, Pima Co.; elevation not given, but collected there recently at 4,000 ft. Known also from Pajaritos Mts, Santa Cruz Co., where it was collected recently at 4,000 ft; and in Pima Co. from Quijotoa Mts, hills near Ventana, Comobabi Mts, Cababi Mts (= Ko Vaya Hills of U.S. Geol. Surv. map; found there recently at 2,700 ft), and Chutum Vaya Canyon of SW section of Baboquivari Mts at 3,500 ft. Synonym: *Sonorella sitiens comobabiensis* H. A. Pilsbry and J. H. Ferriss, 1915 (T.L.: Comobabi Mts, on N side of highest section of Range, ca 4,000 ft, Pima Co.; not dissected thus far from there).

37b. **Sonorella sitiens montezuma** H. A. Pilsbry and J. H. Ferriss, 1919. T.L.: Huachuca Mts, originally given as "S end of Range in Montezuma Canyon"; more precisely later (1923:79) as Station 327, described (1923:59) as "deep double gulch, W of last rock-fronted mountain on S side of Montezuma Canyon"; elevation not given, but collected recently at or near T.L. at 5,400 ft. Also known from other Stations at S end of Huachuca Mts, and from Patagonia Mts at 4,850 ft. Found by junior author also in Sonora at Sierra Pajaritos, 25 mi E of Ures, 29° 25′ N, 110° 10′ W, 3,300 ft, southernmost locality known for genus.

38. **Sonorella neglecta** W. O. Gregg, 1951:156, Pl. 52, 3 figs. T.L.: Chiricahua Mts, N of Portal-Paradise road, ca 3 mi W of Portal, at 5,300 ft, Cochise Co. Also at nearby Stations. Genitalia described from topotypes by junior author (1967c).

C. *Sonorella granulatissima* Complex

39a. **Sonorella g. granulatissima** H. A. Pilsbry, 1905. T.L.: Huachuca Mts, originally given as "Spring Canyon, near Fort Huachuca"; corrected by Pilsbry and Ferriss (1910a:500) to Ramsey Canyon, on E slope of Range, Cochise Co.; elevation not given, but found there recently at 5,750 to 6,000 ft. Also elsewhere in same Range, up to 9,000 ft.

39b. **Sonorella granulatissima latior** H. A. Pilsbry, 1905. T.L.: Huachuca Mts, Brown Canyon, Cochise Co.; elevation not given. Also elsewhere in same Range, near N end; collected recently in Ramsey Canyon, at 5,750 ft, where it intergrades with *S. g. granulatissima*.

40. **Sonorella danielsi** H. A. Pilsbry and J. H. Ferriss, 1910. T.L.: Huachuca Mts, at head of Bear Canyon, Cochise Co.; elevation not given. Also elsewhere in S part of same Range at 6,000 to over 7,500 ft.

41. **Sonorella bowiensis** H. A. Pilsbry, 1905. Known only from T.L.: originally given by error as "Bowie, Cochise Co."; corrected by Pilsbry and Ferriss (1910c:66) to "Quartzite Hill, back of Dixon's place, about 1 mi S of old Fort Bowie," at NW end of Chiricahua Mts, Cochise Co.; elevation not given, but found there recently at 6,200 ft.

42. **Sonorella parva** H. A. Pilsbry, 1905. T.L.: NW end of Huachuca Mts, between Fort Huachuca and former Manila Mine, Cochise Co.; elevation not given, but ca 5,400 ft; found recently in nearby Huachuca Canyon, at 6,000 ft; also elsewhere in N part of Range.

43. **Sonorella dalli** P. Bartsch, 1904. Known only from T.L.: Huachuca Mts, N end in Tanner (now Garden) Canyon, Cochise Co.; elevation not given; found there recently at 5,600 to 6,000 ft.

44. **Sonorella magdalenensis** (R. E. C. Stearns, 1890, as *Helix*). T.L.: Sonora, on top of a mountain, 1,000 ft above Magdalena; ele-

vation not given, but found in 1965 at the probable T.L. in Sierra Magdalena, ca 1 mi N of Magdalena, at 3,650 ft (which is ca 1,000 ft above the town). Also found recently elsewhere in Sonora, as far S as Sierra Pajaritos, ca 24 mi by road E of Ures, 29° 25′ N, at 3,000 ft, southernmost known locality of *Sonorella*. In Arizona in Santa Cruz Co. (Tumacacori Mts, at 3,720 ft; San Cayetano Mts, at 4,750 ft) and Pima Co. (Cerro Colorado, at 4,100 ft; Roskruge Mts; S part of Tucson Mts, at 3,300 ft; N foothills of Santa Rita Mts, 5,500 to 5,800 ft; Tumamoc Hill near Tucson, at 2,750 ft). Synonyms: *Sonorella tumamocensis* H. A. Pilsbry and J. H. Ferriss, 1915 (T.L.: Tumamoc Hill, at W city limits of Tucson, ca 2,750 ft, Pima Co.). *Sonorella sitiens arida* H. A. Pilsbry and J. H. Ferriss, 1915 (T.L.: S end of Cerro Colorado, ca 2 mi from Cerro Colorado Mine, Pima Co.; elevation not given, but found there in 1965 at 4,100 ft). *Sonorella hinkleyi* H. A. Pilsbry and J. H. Ferriss, 1919 (T.L.: San Cayetano Mts, on southernmost peak [Mt Shibell of present maps], 2 mi above Calabasas, Santa Cruz Co.; elevation given in print as 7,500 ft, but labels of paratypes give 6,500 ft; both wrong elevations, the highest peak of Range reaching 6,007 ft, and Mt Shibell only 5,146 ft; found in 1965 at T.L. at 4,750 ft). *Sonorella hinkleyi fraterna* H. A. Pilsbry and J. H. Ferriss, 1919 (nomen nudum). *Sonorella tumacacori* H. A. Pilsbry and J. H. Ferriss, 1919 (T.L.: Tumacacori Mts, NE side of Tumacacori Peak, Santa Cruz Co.; elevation not given, but found there in 1965 at 3,720 ft). *Sonorella cayetanensis* H. A. Pilsbry and J. H. Ferriss, 1919 (T.L.: "San Cayetano Mts," Santa Cruz Co., in 1919; more precisely later by Pilsbry, 1939:342, 343, as the highest peak of the Range, "from near the top to rather low" [the top is ca 6,000 ft, not 7,500 ft]). *S. linearis* H. A. Pilsbry and J. H. Ferriss, 1923 (T.L.: N end of Santa Rita Mts, on W side of Saddle, overlooking Helvetia, Pima Co., without elevation; found there in 1965 at ca 5,588 ft). Pilsbry (1939:341) synonymized *S. hinkleyi*, *S. tumacacori*, *S. fraterna*, and *S. cayetanensis* with *S. arida*, which he then raised to specific rank. In junior author's unpublished experiments (W. B. Miller, 1968a) snails from T.L. of *magdalenensis*, *arida*, *tumamocensis*, and *linearis* interbred freely, producing viable F_1 offspring, which in turn produced many F_2 snails; further, his dissections of snails of *magdalenensis*, *tumacacori*, *tumamocensis*, *arida*, and *linearis* from their T.L. showed no significant difference in their genitalia. *Sonorella arizonensis* (W. H. Dall, 1896, as *Epiphragmophora*), based on one dead shell from drift of Santa Cruz Riv at Tucson, is probably also a synonym of *S. magdalenensis*.

45. **Sonorella insignis** H. A. Pilsbry and J. H. Ferriss, 1919. T.L.:

"Whetstone Mts," Cochise [not "Pima"] Co. in 1919:21; specified in 1923:59 and 94, as their Station 304, "Peak N of Station 303, facing N, and the highest cleft peak"; elevation not given; found at or near T.L., March 1968, by junior author, A. R. Mead et al in limestone rock slide at 6,600 to 6,800 ft, on SE slope of French Joe Peak, NE facing slope of ravine above right bank of French Joe Creek, Cochise Co. (R19E, T19S, Sec. 5, SE ¼, of U.S. Geol. Surv. map); dissected by Pilsbry (1923) and junior author (1968). A dead shell, presumably washed down from Whetstone Mts, was found by Mrs. Edith Hipple in 1955 in drift of San Pedro Riv at Redington, Pima Co.

46a. **Sonorella v. virilis** H. A. Pilsbry, 1905. T.L.: "Chiricahua Mts, at 7,500 ft," Cochise Co.; specified by Pilsbry (1939:380) as Rucker Canyon, on W side of Range, at 5,900 to 6,000 ft. Also elsewhere over much of same Range at 5,400 ft (foot of Reeds Mtn in Cave Creek Canyon) to ca 8,400 ft (Barfoot and Buena Vista Peaks). Synonym: *Sonorella virilis circumstriata* H. A. Pilsbry, 1905 (T.L.: Chiricahua Mts, in Cave Creek Canyon, Cochise Co.; elevation not given; dissected by Pilsbry, 1905).

46b. **Sonorella virilis leucura** H. A. Pilsbry and J. H. Ferriss, 1910. T.L.: Chiricahua Mts, S side of Whitetail Canyon, Cochise Co.; elevation not given. Also elsewhere in N part of same Range; according to H. A. Pilsbry (1919b:70), it occurs in N branch of Pinery Canyon at 6,500 to 7,000 ft.

47. **Sonorella micra** H. A. Pilsbry and J. H. Ferriss, 1910. T.L.: Chiricahua Mts, N side of Whitetail Canyon, below the great cliffs along Indian Creek, Cochise Co.; elevation not given. In 1939 Pilsbry said it occurs at 5,500 to 7,000 ft. Also elsewhere in N part of same Range.

48. **Sonorella bartschi** H. A. Pilsbry and J. H. Ferriss, 1915. T.L.: Mule Mts, on Mt Ballard, in Escabrosa Ridge, ca 2 mi W of Bisbee, on a ledge of N side near summit, Cochise Co.; elevation not given, but found there recently at 6,900 to 7,600 ft. Also known from one other Station in same area.

49. **Sonorella dragoonensis** H. A. Pilsbry and J. H. Ferriss, 1915. T.L.: Dragoon Mts, Bear Canyon, central part of Range, halfway down, Cochise Co.; elevation not given, but found there recently at 6,100 ft; Bear Canyon, not named on present maps, the southwesternmost branch of Stronghold Canyon East. Also known from one other Station in same area.

50. **Sonorella apache** H. A. Pilsbry and J. H. Ferriss, 1915. T.L.: Dragoon Mts, E side of rocky bed of Cataract branch of Tweed Canyon [Stronghold Canyon West of present map], Cochise Co.;

elevation not given, but found there recently at 6,000 ft. Also elsewhere on W side of same Range.

51. **Sonorella ferrissi** H. A. Pilsbry, 1915. T.L.: Dragoon Mts, N side of N ridge of Tweed Canyon [Stronghold Canyon West of present map], in amphitheater, 0.25 mi W of Signal Peak [present Mt Glenn], Cochise Co.; elevation not given, but found there recently at 7,000 ft. Also elsewhere on slopes of N and NW canyons of same Range, down to 4,800 ft.

52. **Sonorella clappi** H. A. Pilsbry and J. H. Ferriss, 1915. T.L.: Santa Rita Mts, Station 8 in Madera Canyon, "an extensive rock pile in the bed of the Canyon ca 10 ft above the stream, opposite the saddle at head of Agua Caliente Canyon," Santa Cruz Co.; elevation not given, but found there recently at 5,700 to 6,000 ft. Also elsewhere in same Range at 5,600 to 6,500 ft. A dead shell, apparently of this species, labeled "Santa Rita Mts" at U.S. Nat. Mus., and sent to I. Lea in 1860 by H. C. Grosvenor, appears to be the first land snail *collected* in Arizona. Synonym: *Sonorella granulatissima occidentalis* H. A. Pilsbry and J. H. Ferriss, 1915 (T.L.: Santa Rita Mts, Camperel Canyon [= Gardner Canyon of present map], NE side of Old Baldy [Mt Wrightson of present map], Santa Cruz Co.; elevation not given, but probably ca 6,800 ft); (not *S. sabinoensis occidentalis* H. A. Pilsbry and J. H. Ferriss, 1919, renamed *S. sabinoensis dispar* by Pilsbry, 1939).

53. **Sonorella anchana** S. S. Berry, 1948:154, figs. 4-7, 9-10, and 12-16. T.L.: Sierra Ancha, Reynolds Creek, 3 mi above Ranger Station, on Pleasant Valley Rd, 6,000 ft, Gila Co. Also elsewhere in same Range, at 5,500 to 7,200 ft.

54a. **Sonorella o. odorata** H. A. Pilsbry and J. H. Ferriss, 1919. T.L.: Santa Catalina Mts, head of Alder Canyon, Pima Co., above 7.500 ft. Also elsewhere in same Range, up to 8,800 ft; probably restricted to this Range. Synonym: *Sonorella odorata* form *populna* H. A. Pilsbry, 1939:369 (T.L.: allegedly collected near Spud Rock Ranger Station of Rincon Mts, Pima Co.; elevation not given; not found again anywhere in Rincon Mts, and the occurrence of *S. odorata* there questionable).

54b. **Sonorella odorata marmoris** H. A. Pilsbry and J. H. Ferriss, 1919. T.L.: Santa Catalina Mts, Old Dan's Gulch, NW side of Marble Peak, near N end of Range, Pima Co.; elevation not given, but found there recently at 6,000 to 7,000 ft. Also at nearby Stations in same Range.

55. **Sonorella bagnarai** W. B. Miller, 1967c:55, figs. A-E. Known only from T.L.: Rincon Mts, NE side below summit of Rincon Peak, 8,000 to 8,200 ft, Pima Co.

56. **Sonorella grahamensis** H. A. Pilsbry and J. H. Ferriss, 1919. T.L.: Pinaleno Mts, Mud Spring on Mt Graham, "on the summit," Graham Co.; elevation not given. Found recently elsewhere on Mt Graham, at 6,300 to 9,000 ft.

D. *Sonorella rooseveltiana* Complex

57a. **Sonorella r. rooseveltiana** S. S. Berry, 1917. T.L.: Roosevelt, rock slides of N slopes, 2,200 ft, Gila Co. Also at nearby Stations, one at 2,250 ft.

57b. **Sonorella rooseveltiana fragilis** H. A. Pilsbry, 1939. T.L.: Cliff ruins at Roosevelt Lake, Gila Co.; elevation not given, but found there recently at 3,000 ft. Also at a nearby Station at 2,900 ft.

58. **Sonorella allynsmithi** W. O. Gregg and W. B. Miller, 1969: 90, figs. 1A-B, 1D and 2A-C. Known only from T.L.: Phoenix Mts, near Squaw Park, on E side of Squaw Peak Rd, in N area of Phoenix, ca 0.25 mi from Lincoln Drive, 1,100 ft, Maricopa Co. Also at other nearby Stations.

Sonorella betheli J. Henderson, 1914, described by error from Grand Canyon of Colorado Riv, was *Helminthoglypta traskii* (W. Newcomb) of California, according to H. A. Pilsbry (1939:174).

Sonorella of New Mexico (references and distribution in H. A. Pilsbry, 1939:273-278 and 322): *Sonorella hachitana* (W. H. Dall, 1896). *S. hachitana flora* H. A. Pilsbry and J. H. Ferriss, 1915. *S. hachitana peloncillensis* H. A. Pilsbry and J. H. Ferriss, 1915. *S. orientis* H. A. Pilsbry, 1936 (also in trans-Pecos Texas). *S. animasensis* H. A. Pilsbry, 1939.

Sonorella of trans-Pecos Texas: *Sonorella orientis* H. A. Pilsbry, 1936 (also in New Mexico).

Sonorella of Sonora: *Sonorella hachitana* (W. H. Dall, 1896) (also in New Mexico). *S. nixoni* W. B. Miller, 1967a:116, Pl. 6, figs. G-I, Pl. 7, figs. A-B. *S. greggi* W. B. Miller, 1967a:114, Pl. 6, figs. A-F, Pl. 7, figs. C-D. *S. sitiens montezuma* H. A. Pilsbry and J. H. Ferriss, 1919 (also in Arizona). *S. mugdalenensis* (R. E. C. Stearns, 1890) (also in Arizona) *S. mearnsi* P. Bartsch, 1904:194, Pl. 32, 3 figs. 2. *S. mormonum huasabasensis* W. B. Miller, 1967b:2, figs. A, B and F, Pl. 1, figs. A-C. *S. perhirsuta* W. B. Miller, 1967b:4, figs. D, E and H, Pl. 1, figs. D-F. *S. w. walkeri* H. A. Pilsbry and J. H. Ferriss, 1915 (also in Arizona).

Sonorella of Chihuahua: *Sonorella nelsoni* P. Bartsch, 1904: 191, Pl. 31, 3 figs. 3, synonym: *S. goldmani* P. Bartsch, 1904:192, Pl. 32, 3 figs. 6. *S. mormonum* H. A. Pilsbry, 1948b:196, figs. 5A-a, Pl. 13, 4 figs. 6. *S. pennelli* H. A. Pilsbry, 1948b:195, figs. 5B-b, Pl. 13, 3 figs. 5.

Family Oreohelicidae

Oreohelix H. A. Pilsbry, 1904

Oreohelix is the most distinctive and most widespread genus of Recent large land snails of western North America. Pilsbry (1939: 411) retained it in Camaenidae, but segregated it in a subfamily Oreohelicinae, suggesting that it might even be ranked as a family Oreohelicidae, a course adopted by C. B. Wurtz in 1955 (*Proc. Acad. Nat. Sci. Philadelphia* 107:101) and followed here. More recently H. B. Baker (1963:240) included it in his family Thysanophoridae. The Recent and past distribution of the genus is discussed in the section on Zoogeography. Pilsbry (1905) recognized two subgenera. Most Arizona species belong in *Oreohelix*, sensu stricto, having the early nepionic (embryonic) whorls of the shell finely, vertically striate and often with spiral engraved lines; these species are also viviparous, the eggs hatching in the uterus, where the young grow for a time before leaving the mother, so that the albumin gland is rudimentary in the adult snail. In subgenus *Radiocentrum* H. A. Pilsbry (1905:283), about the first one and one-half nepionic whorls are coarsely, radially ribbed; the snail is oviparous, the eggs leaving the uterus before hatching, so that the adult snail needs a large albumin gland. *O. barbata* is somewhat transitional, with a very weak ribbing on the early three-fourths or whole of the first nepionic whorl, becoming irregular on the second; but it is viviparous and it has a rudimentary albumin gland, which place it in *Oreohelix*, sensu stricto.

Correct identification of *Oreohelix* requires dissection of the genitalia, which has not been done thus far for all Arizona forms; so that the specific or subspecific status of some of them is as yet uncertain.

59a. **Oreohelix (Oreohelix) strigosa depressa** (T. D. A. Cockerell, 1890, as *Patula*). T.L.: canyon near Durango, La Plata Co., Colorado. In Arizona restricted to Coconino Co.: N and S sides of Grand Canyon of Colorado Riv, at 3,000 to 8,300 ft; S side of Navajo Mtn. Widespread in Colorado, Wyoming, Montana, Utah, Nevada, and New Mexico. No Arizona *depressa* dissected thus far. A sinistral shell from Powell Plateau, Coconino Co., figured by L. E. Daniels, 1912:41, Pl. 5, fig. 17, and a scalariform shell from Jacobs Canyon, Kaibab Plateau, Pl. 5, fig. 16. Nominate *O. s. strigosa* (A. A. Gould, 1846) does not occur in the State, or elsewhere in Southwestern Province.

59b. **Oreohelix (O.) strigosa meridionalis** H. A. Pilsbry and J. H. Ferriss, 1919. T.L.: Y Salt House Branch of Eagle Creek, Greenlee Co.; elevation not given. Also elsewhere in Greenlee Co. and

Apache Co. (Black Riv, opposite mouth of Fish Creek). Snails of *meridionalis* from Black Riv dissected by Pilsbry (1919a).

An undescribed subsp. of *O. strigosa* was found recently near the summit of Pinal Peak, Gila Co., at 7,600 ft (dissected by R. H. Russell, 1969).

60. **Oreohelix (O.) anchana** W. O. Gregg, 1953:71, Pl. 14, 3 top figs. T.L.: Sierra Ancha, NE slope of Center Mtn, below the cave, ca 3 mi N of Reynolds Creek, 7,200 ft, Gila Co. Also elsewhere in same Range, down to 6,500 ft. Dissected by W. O. Gregg (1953) from T.L. and by junior author (1967).

61. **Oreohelix (O.) subrudis** (L. Reeve, 1854 as *Helix*). T.L. unknown. Widespread in Rocky Mts area from SE British Columbia to Colorado and Utah; disjunctive in S New Mexico and SE Arizona at high elevations. Known in Arizona from Apache Co. (White Riv and Black Riv areas, at 7,500 to 8,000 ft) and Greenlee Co. (Blue Riv area). Black Riv snails dissected by Pilsbry (1919a). Synonym: *Oreohelix cooperi* form *apache* H. A. Pilsbry and J. H. Ferriss, 1919 (T.L.: Black Riv, 2 mi above Fish Creek, Apache Co.; elevation not given, but ca 7,000 ft).

62a. **Oreohelix (O.) c. concentrata** (W. H. Dall, 1896, as *Pyramidula*). T.L.: Huachuca Mts, at Fort Huachuca, 4,300 ft, Cochise Co. Also elsewhere in Huachuca Mts, 4,300 to 9,225 ft (top of Carr Peak, R. C. Ewell, Apr. 1969); also alive in Whetstone Mts; subfossil in Cochise Co. in Little Rincon Mts (25 mi NW of Benson), Mustang Mts, and Dragoon Mts. Alive in N Sonora (San Jose Mts). Smaller, typical *concentrata* from Miller Peak, Huachuca Mts, dissected by Pilsbry (1916c:Pl. 22, fig. 9). Scalariform shell from Limestone Mtn [= Ramsey Peak], Huachuca Mts, figured by Pilsbry (1905:274, Pl. 24, fig. 32). Synonym: *"Pyramidula" strigosa huachucana* H. A. Pilsbry, 1902 (T.L.: Conservatory [= Ramsey] Canyon, Huachuca Mts, Cochise Co.; elevation not given; found there recently at 5,600 ft); paratype of larger *huachucana* form dissected by H. A. Pilsbry (1939:505, fig. 328).

62b. **Oreohelix (O.) concentrata grahamensis** W. O. Gregg and W. B. Miller, 1972, MS. T.L.: Pinaleño Mts, SE slope of Mt Graham, at ca 8,500 ft, Graham Co.

63. **Oreohelix (O.) houghi** W. B. Marshall, 1929. T.L.: Heber, Navajo Co.; elevation not given, but ca 6,400 ft. Known from Coconino Co. (Meteor Crater [= Coon Mtn], 18 mi W of Winslow, 5,200 to 5,700 ft; Canyon Diablo at Two Guns, ca 4,800 ft; small canyon 4 mi E of Rimmy Jims, ca 4,800 ft; Mormon Crossing, 23 mi NW of Heber, 5,500 ft; at bridge below dam of Chevelon Creek, ca 7,000 ft, W. O. Gregg, 1941, R. H. Russell and senior author, 1968),

Navajo Co. (Holbrook; Wildcat Canyon at Heber-Winslow Rd, 15 mi NW of Heber, Sanda Truett, 1965; Clear Creek, 8 mi SW of Winslow; Show Low Creek, 1.5 mi NW of Shumway, A. Phillips, 1952), and Apache Co. (12 mi S of St Johns; Hardscrabble Wash, NW of St Johns; at Zuni Sacred Lake; Nelson Reservoir, 18 mi N of Alpine, 7,000 ft, M. L. Walton). Dissected by Pilsbry (1939) from Meteor Crater and Canyon Diablo. Sinistral specimens found by R. H. Russell and senior author in 1968: one immature dead in Canyon Diablo; one full-grown alive, with 275 dextral shells, at Mormon Crossing. The Late Pleistocene *Oreohelix* sp. reported by A. D. Reger and G. L. Batchelder (1971:192 and 193) from Meteor Crater Site, at depths of 8 and 15 ft, in Silica Hill Shaft, were probably *O. houghi,* now also living in Meteor Crater. Synonym: *Oreohelix houghi winslowensis* W. B. Marshall, 1929 (T.L.: Clear Creek, 8 mi SW of Winslow, Coconino Co.; elevation not given, but ca 4,800 ft); not dissected thus far.

64a. Oreohelix (O.) y. yavapai H. A. Pilsbry, 1905. T.L.: Purtyman's Ranch in Oak Creek Canyon, Coconino Co.; elevation not given, but ca 5,000 ft. Also recorded from Yavapai Co. (Mingus Mtn area; several collectors have failed to find it there recently). Dissected from T.L. by Pilsbry (1905).

64b. Oreohelix (O.) yavapai cummingsi H. A. Pilsbry, 1934. T.L.: near reservoir of Yellow House ruins, 2 mi S of Endische Springs, at S foot of Navajo Mtn, Coconino Co.; elevation not given. Also at other Stations of Navajo Mtn in Utah. Not dissected thus far. Synonym: *Oreohelix yavapai clutei* H. A. Pilsbry, 1934 (T.L.: Red Rock Springs, at S foot of Navajo Mtn, Coconino Co.; elevation not given); not dissected thus far.

64c. Oreohelix (O.) yavapai profundorum H. A. Pilsbry and J. H. Ferriss, 1911. Known only from T.L.: Grand Canyon of Colorado Riv, in head of recess where Mystic Spring or Bass Trail zigzags down, ca 5,700 ft, Coconino Co. Dissected from T.L. by Pilsbry (1911).

64d. Oreohelix (O.) yavapai extremitatis H. A. Pilsbry and J. H. Ferriss, 1911. T.L.: Grand Canyon of Colorado Riv, at and near head of Bass Trail, ca 200 ft below S Rim (here at ca 6,800 ft), Coconino Co. Also elsewhere in same area, and reported from Wyoming and Montana. Dissected from Shell, Wyoming, by Pilsbry (1916c:351, Pl. 22, figs. 6-6a). Synonym: *Oreohelix yavapai angelica* H. A. Pilsbry and J. H. Ferriss, 1911 (T.L.: Grand Canyon, on Bright Angel Trail, at 100 to 400 ft below S Rim, Coconino Co.; not dissected thus far).

Oreohelix yavapai fortis T. D. A. Cockerell, 1927, and its synonym *Oreohelix yavapai vauxae* W. B. Marshall, 1929, were based on Pleistocene or sub-Recent fossils from Grand Canyon of Colorado Riv, Coconino Co.

65. **Oreohelix (O.) barbata** H. A. Pilsbry, 1905. T.L.: Chiricahua Mts, in Cave Creek Canyon, near the Falls [= Winn Falls of U.S. Geol. Surv. map, on a short W affluent of Cave Creek], Cochise Co.; elevation not given, but found there recently at 7,200 ft. Also elsewhere in same Range; only species of subgenus *Oreohelix,* sensu stricto, in Chiricahua Mts. Recorded also from Mogollon Mts, New Mexico. Dissected from T.L. by Pilsbry (1905) and junior author (1971). Synonym: *Oreohelix barbata minima* H. A. Pilsbry ,and J. H. Ferriss, 1910 (T.L.: Chiricahua Mts, at head of Rucker Canyon, Cochise Co.; elevation not given); not dissected thus far.

66a. **Oreohelix (Radiocentrum) c. clappi** J. H. Ferriss, 1904. T.L.: Chiricahua Mts, S side of Cave Creek Canyon at 5,000 to 7,000 ft, Cochise Co. Also elsewhere in same Range. Dissected from T.L. by Pilsbry (1905). Synonym: ? *Oreohelix clappi cataracta* H. A. Pilsbry and J. H. Ferriss, 1910 (T.L.: Chiricahua Mts, near the Falls, Cochise Co.; elevation not given); not dissected thus far.

66b. **Oreohelix (R.) clappi emigrans** H. A. Pilsbry and J. H. Ferriss, 1910. Known only from T.L.: Chiricahua Mts, at "Big Emigrant Mtn" [= Rough Mtn of U.S. Geol. Surv. map], on S side of Big Emigrant Canyon, ca 7,000 ft, Cochise Co. Not dissected thus far.

67a. **Oreohelix (R.) c. chiricahuana** H. A. Pilsbry, 1905. T.L.: Chiricahua Mts in Cave Creek Canyon, "on the slope with S exposure below the Cave" [= Crystal Cave]; elevation not given, but found there recently at 5,600 to 6,200 ft. Also elsewhere in same Range. Dissected from T.L. by Pilsbry (1905).

67b. **Oreohelix (R.) chiricahuana obsoleta** H. A. Pilsbry and J. H. Ferriss, 1910. T.L.: Chiricahua Mts, originally without precise Station, Cochise Co.; given as Whitetail Canyon, 6,000 to 7,000 ft, by Pilsbry in 1939:551. Also elsewhere in N section of same Range. Not dissected thus far.

67c. **Oreohelix (R.) chiricahuana percarinata** H. A. Pilsbry and J. H. Ferriss, 1910. T.L.: Chiricahua Mts, summit of Cross J Mtn [Wood Mtn of U.S. Geol. Surv. map], near mouth of Big Emigrant Canyon, at ca 7,300 ft. Cochise Co. Also in N section of same Range. Not dissected thus far.

Oreohelix of New Mexico (references and distribution in H. A. Pilsbry, 1939:412-553): *Oreohelix strigosa depressa* (T. D. A. Cockerell, 1890) (also in Arizona and farther north). *O. strigosa nogalen-*

sis H. A. Pilsbry, 1939. *O. subrudis* (L. Reeve, 1854), *O. swopei* H.
A. Pilsbry and J. H. Ferriss, 1917. *O. metcalfei* T. D. A. Cockerell,
1905. *O. metcalfei concentrica* H. A. Pilsbry and J. H. Ferriss, 1917.
O. metcalfei acutidiscus H. A. Pilsbry and J. H. Ferriss, 1917. *O.
metcalfei radiata* H. A. Pilsbry and J. H. Ferriss, 1917. *O. metcalfei
hermosensis* H. A. Pilsbry and J. H. Ferriss, 1917. *O. metcalfei
cuchillensis* H. A. Pilsbry and J. H. Ferriss, 1917. *O. pilsbryi* J. H.
Ferriss, 1917. *O. yavapai neomexicana* H. A. Pilsbry, 1905. *O. barbata* H. A. Pilsbry, 1905 (also in Arizona). *O. (Radiocentrum) hachetana* H. A. Pilsbry, 1915. *O. (R.) ferrissi* H. A. Pilsbry, 1915.
Only dead, possibly fossil shells known of *O. metcalfei florida* H. A.
Pilsbry, 1939; *O. yavapai compactula* T. D. A. Cockerell, 1905
(perhaps identical with Recent *O. yavapai neomexicana*); *O.
socorroensis* H. A. Pilsbry, 1905; *O. socorroensis magdalenae* H. A.
Pilsbry, 1939; *O. hachetana cadaver* H. A. Pilsbry, 1915; *O. ferrissi
morticina* H. A. Pilsbry, 1915.

In trans-Pecos Texas *Oreohelix* is known only from dead, presumably fossil shells of two species in Guadalupe Mts, Hueco Mts,
and Franklin Mts. They have been referred to *O. f. ferrissi* H. A.
Pilsbry and *O. s. socorroensis* H. A. Pilsbry and are discussed in the
section on Zoogeography.

In Sonora *Oreohelix* has been found thus far only at a single
locality of *O. concentrata* (W. H. Dall, 1896), in San José Mts,
ca 5 mi S of Arizona border; the species occurs also in Arizona.

Oreohelix of N Chihuahua: *O. caenosa* H. A. Pilsbry, 1948:
198, Pl. 13, 4 figs. 7 and fig. 7a. *O. labrenana* H. A. Pilsbry, 1948:
199, Pl. 13, 5 figs. 8. *O. almoloya* R. J. Drake, 1949b:110, Pl. 8, 4
figs. 1.

Oreohelix of S Nevada (references and distribution in H. A.
Pilsbry, 1939:429-435 and 534-536): *Oreohelix strigosa depressa*
T. D. A. Cockerell, 1890); *O. handi* H. A. Pilsbry, 1917; *O. handi
jaegeri* H. A. Pilsbry, 1931.

Oreohelix barely enters SE California with a single species,
O. californica S. S. Berry, 1931a:73, 1931n:115, figs. 1-2 (T.L.: W.
side of Clark Mtn, 7,000 ft, NE corner of San Bernardino Co., ca
10 mi from Nevada border).

An undescribed *Oreohelix* of subgenus *Radiocentrum* was recently discovered by the junior author in Baja California Territorio
Sur.

Family Polygyridae

Ashmunella H. A. Pilsbry and T. D. A. Cockerell, 1899

Ashmunella is closely related to the large and widespread North

American *Polygyra,* differing only in genitalia: a short penis, without papilla or verge, but continued in a very long epiphallus (epiphallus lacking in *Polygyra*), with a free, but short flagellum (lacking in *Polygyra*); a cylindric spermatheca, not enlarged distally (spermatheca with a slender duct and a terminal sac in *Polygyra*); and a penis partially covered with an imperfect and tenuous sheath (no penial sheath in *Polygyra*). The depressed, helicoid shell is so much like *Polygyra* that it may be inadequate for generic diagnosis. In *Ashmunella* the aperture is without or with teeth, which are variable in number (one to four), shape and size, and either well developed, obsolete, or lacking, even within the same species, subspecies, or population. The validity and true relationships of several of the named forms remain to be worked out, particularly in Arizona. Distribution of genus discussed in section on Zoogeography.

Pilsbry (1940:577; 1948:xxi) placed *Ashmunella* in subfamily Triodopsinae, owing to the presence of a penial sheath. However, it differs from some Triodopsinae, such as *Vespericola* and *Triodopsis,* in other genitalic features, viz the presence of a flagellum and the cylindric shape of the spermatheca, suggesting that *Ashmunella* was perhaps derived from another polygyroid stock, not closely related to *Triodopsis*. It may therefore be advisable to place it in a distinct subfamily Ashmunellinae, as proposed by G. R. Webb (1954:18).

68a. **Ashmunella c. chiricahuana** (W. H. Dall, 1896, as *Polygyra*). T.L.: Chiricahua Mts, Flys Park, Cochise Co.; elevation, given as 10,000 ft. exaggerated, Flys Park being at ca 9,200 ft, on NE slope of Flys Peak, 0.25 mi E of Tubs Springs. Also elsewhere in same Range. Dissected by Pilsbry (1905).

68b. **Ashmunella chiricahuana esuritor** H. A. Pilsbry, 1905, as *A. esuritor*. T.L.: Chiricahua Mts, Barfoot [or Bearfoot] Park, ca 8,000 ft. Cochise Co. Also elsewhere in same Range, at 7,500 to 8,000 ft. Possibly not separable from *A. c. chiricahuana*. Dissected by Pilsbry (1905). Synonym: *Ashmunella metamorphosa* H. A. Pilsbry, 1905 (T.L.: Barfoot [or Bearfoot] Park, ca 8,000 ft, Cochise Co.; dissected by Pilsbry, 1905); synonymized with *A. c. esuritor* by Pilsbry, 1935a:67 and 1940:953.

69. **Ashmunella ferrissi** H. A. Pilsbry, 1905. Known only from T.L.: Chiricahua Mts, foot of Reeds Mtn, at S side of Cave Creek, 0.5 mi. E of Southwestern Research Station, Cochise Co; elevation not given, but found there recently at 5,400 ft. Dissected from T.L. by Pilsbry (1910).

70a. **Ashmunella p. proxima** H. A. Pilsbry, 1905, as *A. levettei proxima*. T.L.: originally given as "Chiricahua Mts, Sawmill Canyon," without elevation; according to Pilsbry and Ferriss (1910c: 100), Sawmill Canyon is also known as Rigg's or Pine Canyon [Pine Canyon of U.S. Geol. Surv. map]; in 1940:954 Pilsbry defined the T.L. as "head of Pine Canyon at Barfoot [or Bearfoot] Park, 8,500 ft." Also elsewhere in same Range. Dissected from T.L.: by Pilsbry (1905).

70b. **Ashmunella proxima angulata** H. A. Pilsbry, 1905, as *A. angulata.* T.L.: "Chiricahua Mts, in the South Fork of Cave Creek," Cochise Co.; elevation not given, but found there recently at 5,500 ft. Also elsewhere in same Range at 4,800 to 7,200 ft. Dissected from T.L. by Pilsbry (1905). Synonyms: *Ashmunella proxima harveyi* H. A. Pilsbry, 1940 (T.L.: Chiricahua Mts, Rucker Canyon, ca 6,000 ft, Cochise Co; not dissected from T.L. thus far). *Ashmunella lenticula* W. O. Gregg, 1953:74, Pl. 14, 3 bottom figs. (T.L.: Chiricahua Mts, ca 2 mi above mouth of Horseshoe Canyon, on E slope of Range, 4,800 ft, Cochise Co.; dissected from T.L. by W. O. Gregg, 1953).

70c. **Ashmunella proxima emigrans** H. A. Pilsbry, 1910. T.L.: Chiricahua Mts, "on Big Emigrant Mtn, Big Emigrant Canyon," elevation not given, but found there recently at 6,000 to 7,000 ft, Cochise Co. [Big Emigrant Mtn is Rough Mtn of U.S. Geol. Surv. map]. Also elsewhere in same Range at 5,500 to 8,800 ft. Dissected from T.L. by Pilsbry (1910). Synonyms: *Ashmunella fissidens albicauda* H. A. Pilsbry and J. H. Ferriss, 1910 (T.L.: described from 8 Stations on the south side of Whitetail Canyon, to about 7,500 ft elevation, Cochise Co., but none strictly defined as the T.L.; dissected from this area by Pilsbry, 1910). *Ashmunella fissidens pomeroyi* H. A. Pilsbry and J. H. Ferriss, 1910 (T.L.: Chiricahua Mts, "in Hands Pass, head of Jhu Canyon," without elevation; Pilsbry later, 1940:957, footnote, defined this as ca 2 mi NW of head of Jhus Canyon, at ca 6,600 ft; not dissected from T.L. thus far).

70d. **Ashmunella proxima fissidens** H. A. Pilsbry, 1905. T.L.: Chiricahua Mts, originally given as "Cave Creek Canyon" (possibly by error), Cochise Co.; elevation not given. Also elsewhere in same range at 6,000 to 9,500 ft. Not dissected thus far from T.L. Synonym: *Ashmunella duplicidens* H. A. Pilsbry, 1905 (T.L.: Chiricahua Mts, Barfoot [or Bearfoot] Park, 8,500 ft, Cochise Co.; dissected by Pilsbry, 1905).

70e. **Ashmunella proxima lepiderma** H. A. Pilsbry and J. H. Ferriss, 1910. T.L.: Chiricahua Mts, Whitetail Canyon, Cochise Co.; elevation not given, but one lot of syntypes labeled 7,000 ft. Also elsewhere in same Range at 5,500 to 7,200 ft. Dissected by Pilsbry (1910).

71a. **Ashmunella l. levettei** (T. Bland, 1881, as *Triodopsis;* not 1882 or 1880, as often cited). T.L. originally given as "near Santa Fe, New Mexico, where two living and one dead specimen were collected by my friend, Dr. G. M. Levette, who presented me with one of the former." Only one of the three original syntypes known at present; no other shell agreeing with it collected since in New Mexico. Original locality now believed to be erroneous; perhaps merely sent to T. Bland from Santa Fe, but not collected in New Mexico. In 1940 Pilsbry proposed Miller Canyon of Huachuca Mts, Cochise Co., as true T.L., since shells fully agreeing with the one remaining, figured holotype are occasionally, although rarely, found there; they seem to occur in populations consisting mainly of *Ashmunella levettei ursina* H. A. Pilsbry and J. H. Ferriss, 1910 (T.L.: Huachuca Mts, in head of Bear Canyon, 7,500 ft, Cochise Co.), which is here regarded as a synonym of nominate *A. l. levettei.* As here understood, typical *A. l. levettei* occurs in several canyons on E slope of Huachuca Mts, at 5,800 to 9,200 ft. W. G. Binney (1886:36, Pl. 1, 2 figs. 15) figured a snail from Huachuca Mts which he referred to *A. levettei.* Dissections in Pilsbry and Ferriss (1910a: 504, Pl. 20, figs. 1 and 4) are of *A. l. ursina.*

71b. **Ashmunella levettei angigyra** H. A. Pilsbry, 1905. T.L.: Huachuca Mts, in "Conservatory" (now Ramsey) Canyon, at 6,000 ft, Cochise Co. Also elsewhere in same Range, at 5,000 to 6,000 ft. Dissected by Pilsbry (1905).

71c. **Ashmunella levettei bifurca** H. A. Pilsbry and J. H. Ferriss, 1910. T.L.: Huachuca Mts, in "Tanner" (now Garden) Canyon, Cochise Co.; elevation not given. Perhaps not separable from *A. l. angigyra,* occurring sometimes with it in same population. Not dissected thus far.

71d. **Ashmunella levettei heterodonta** H. A. Pilsbry, 1905, as var. *heterodonta.* T.L.: Huachuca Mts, Ida Canyon, a tributary of Cave Creek Canyon, "on the southern slope of the E end of the Huachuca Range," selected by Pilsbry (1905:241-242), Cochise Co., elevation not given. Both Cave Creek Canyon and Ida Canyon are shown in Pilsbry's (1905:212) and Pilsbry and Ferriss' (1910a:495) sketch-maps of Huachuca Mts. Also known from nearby Cave Creek Canyon. Not dissected thus far from T.L. Synonym: *Ashmunella microdonta* H. A. Pilsbry and J. H. Ferriss, 1910 (T.L.: Huachuca Mts, at the head of Cave Creek Canyon; elevation not given; dissected by Pilsbry, 1910a). Pilsbry (1940:969) suggested that *S. l. heterodonta* might be a hybrid of *A. l. levettei* and *A. l. varicifera.*

71e. **Ashmunella levettei varicifera** C. F. Ancey, 1901, as *Ashmunella chiricahuana* variety *varicifera*. T.L.: originally described from "Tucson, Arizona," an area where *Ashmunella* is unknown; the syntypes, collected by Mr. Cox, and first recorded as *Polygyra chiricahuana* by W. H. Dall (1897a:341, Pl. 32, figs. 9, 10, and 12), were presumably mailed from Tucson, but actually came from Huachuca Mts, where *varicifera* is widespread at 6,000 to 9,800 ft. Pilsbry (1940:871) selected as T.L. Miller Canyon, Huachuca Mts, Cochise Co., where shells agreeing with syntypes of *varicifera* occur at high elevations.

It is not impossible, though improbable, that snails agreeing with the holotype of *A. levettei* may yet be found in some canyon near Santa Fe, New Mexico, in which case the Huachuca species of *Ashmunella* will have to be called *A. varicifera* C. F. Ancey, 1901, with *ursina, angigyra, bifurca,* and *heterodonta* subordinated to *varicifera* as subspecies.

? 72. **Ashmunella mogollonensis** H. A. Pilsbry, 1905. T.L.: Whitewater Creek, 1.5 mi E of Mogollon Peak, 9,000 ft, Catron Co., New Mexico. Recorded in Arizona by Pilsbry and Ferriss (1919) from Greenlee Co., at 3 Stations "on or near the rim of the Blue Mts at 5,500 to 12,000 ft" (according to J. H. Ferriss; elevation exaggerated), an area only 50 mi W of T.L.; not seen by us. Snails from New Mexico dissected by Pilsbry (in Pilsbry and Ferriss, 1917:93, Pl. 10, fig. 3).

73. **Ashmunella pilsbryana** J. H. Ferriss, 1914. T.L.: E bank of San Francisco Riv, ca 10 mi above Clifton, Greenlee Co.; elevation not given. Also elsewhere in this area, at ca 4,000 ft. Dissected from T.L. by junior author.

Ashmunella of New Mexico (references and distribution in H. A. Pilsbry, 1940:919-978): *Ashmunella thomsoniana* (C. F. Ancey, 1887, as *Helix levettei* var. *thomsoniana*), synonyms: *Helix levettei* var. *oroboena* C. F. Ancey, 1887, *Ashmunella thomsoniana cooperae* T. D. A. Cockerell, 1901, and *A. antiqua* T. D. A. Cockerell, 1901. *A. thomsoniana porterae* H. A. Pilsbry and T. D. A. Cockerell, 1899. *A. thomsoniana pecosensis* T. D. A. Cockerell, 1903. *A. ashmuni* (W. H. Dall, 1897, as *Polygyra*). *A. ashmuni robusta* H. A. Pilsbry, 1905. *A. pseudodonta* (W. H. Dall, 1897, as *Helix*). *A. pseudodonta capitanensis* E. H. Ashmun and T. D. A. Cockerell, 1899. *A. rhyssa* (W. H. Dall, 1897, as *Polygyra*). *A. rhyssa miorhyssa* (W. H. Dall, 1898, as *Polygyra*), synonym: *Ashmunella rhyssa hyporhyssa* T. D. A. Cockerell, 1898. *A. rhyssa trifluviorum* H. A. Pilsbry, 1940. *A. rhyssa edentata* T. D. A. Cockerell, 1900. *A. altissima* (T. D. A. Cockerell, 1898, as *Polygyra*). *A. townsendi*

P. Bartsch, 1904. *A. townsendi nogalensis* H. A. Pilsbry, 1940. *A. organensis* H. A. Pilsbry, 1936. *A. tetrodon* H. A. Pilsbry and J. H. Ferriss, 1915. *A. tetrodon mutator* H. A. Pilsbry and J. H. Ferriss, 1915. *A. tetrodon inermis* H. A. Pilsbry and J. H. Ferriss, 1915. *A. tetrodon fragilis* H. A. Pilsbry and J. H. Ferriss, 1917. *A. tetrodon animorum* H. A. Pilsbry and J. H. Ferriss, 1917. *A. danielsi* H. A. Pilsbry and J. H. Ferriss, 1915. *A. danielsi dispar* H. A. Pilsbry and J. H. Ferriss, 1915. *A. mogollonensis* H. A. Pilsbry, 1905 (also in Arizona). *A. binneyi* H. A. Pilsbry and J. H. Ferriss, 1917. *A. mendax* H. A. Pilsbry and J. H. Ferriss, 1917. *A. cockerelli* H. A. Pilsbry and J. H. Ferriss, 1917. *A. cockerelli perobtusa* H. A. Pilsbry and J. H. Ferriss, 1917. *A. cockerelli argenticola* H. A. Pilsbry and J. H. Ferriss, 1917. *A. mearnsi* (W. H. Dall, 1896, as *Helix*). *A. hebardi* H. A. Pilsbry and E. G. Vanatta, 1923. *A. walkeri* J. H. Ferriss, 1904. *A. kochii* G. H. Clapp, 1908. *A. kochii amblya* H. A. Pilsbry, 1940 (also in trans-Pecos Texas). *A. carlsbadensis* H. A. Pilsbry, 1932 (also in trans-Pecos Texas). *A. rileyensis* A. L. Metcalf and P. A. Hurley, 1971.

Ashmunella of trans-Pecos Texas: *Ashmunella kochii amblya* H. A. Pilsbry, 1940 (also in New Mexico). *A. pasonis* (R. J. Drake, 1951:44, figs. 1-4, as *Polygyra*). *A. edithae* H. A. Pilsbry and E. P. Cheatum, 1951:88, Pl. 4, fig. 10. *A. bequaerti* W. J. Clench and W. B. Miller, 1966:1, figs, la-b, 2, and 1-2. *A. carlsbadensis* H. A. Pilsbry, 1932 (also in New Mexico).

Ashmunella of Chihuahua: *Ashmunella meridionalis* H. A. Pilsbry, 1948b:199, figs. 6E-I, Pl. 13, 3 figs. 3. *A. juarezensis* H. A. Pilsbry, 1948b:200, figs. 6C-D, Pl. 13, 2 figs. 2. *A. montivaga* H. A. Pilsbry, 1948b:202, figs. 6A-B, Pl. 13, 3 figs. 1. *A. intricata* H. A. Pilsbry, 1948b:203, Pl. 13, 4 figs. 4. An undescribed species was collected by R. H. Russell. Aug. 1971.

Family Thysanophoridae

74. **Thysanophora hornii** (W. M. Gabb, 1866, as *Helix*). T.L.: old Fort Grant, at junction of Aravaipa (misspelled "Arivapa") Creek and San Pedro Riv, Pinal Co.; now a marked Site, 8 mi N by 5 mi W of Mammoth. Widespread in Arizona at 2,400 to 7,000 ft (at near sea level in some Gulf of California islands); a typical snail of Lower Sonoran Life-Zone of valleys, bajadas, and lower canyons; only at a few suitable Stations in Upper Sonoran and Transition Zones. Recent in Yuma Co. (Palm Canyon, Kofa Mts, 2,500 ft, senior author, 1969), Mohave Co. (Mt Logan), Coconino Co. (Shinumo Canyon, Pilsbry and Ferriss, 1911:191. Oak Creek

Canyon, 3,500 ft), Yavapai Co. (Jerome, E. H. Ashmun, 1899:14. Prescott, H. A. Pilsbry, 1900b:98), Gila Co. (Salt Riv Canyon, 4 mi N of Seneca, 3,000 ft. Pinto Creek, 6 mi SE of Roosevelt), Pinal Co. (Galiuro Mts, Copper Creek, 3,800 ft. Aravaipa Creek, 15 mi NE of Mammoth, 2,500 ft, senior author), Pima Co. (Ajo Mts, Arch Canyon, 3,500 ft. Baboquivari Mts: Brown Canyon, 4,000 ft; Sycamore Canyon, 3,700 to 4,000 ft; Baboquivari Canyon, 3,500 to 4,000 ft. Coyote Mts, Mendoza Canyon, 3,600 ft. Quijotoa Mts. Las Guijas Mts, 3,500 to 4,000 ft. Tucson Mts, 2,400 to 3,000 ft. Tumamoc Hill, at Tucson city limits, 2,750 ft. Santa Catalina Mts: many canyons, from 2,800 ft in Sabino Canyon to 6,400 ft on S slope of Marble Peak, and in S foothills close to Tucson city limits. N foothills of Santa Rita Mts at Rosemont, 4,800 ft. Posta Quemada Canyon near Colossal Cave, 3,300 to 3,800 ft. Empire Mts, 4,800 to 5,000 ft. S foothills and slopes of Rincon Mts, 3,500 to 3,800 ft. W slope of Tanque Verde Hills, in Saguaro National Monument, near HQ, 3,500 ft), Santa Cruz Co. (Nogales. Patagonia Mts. Sanford Butte at Sonoita Creek. Tumacacori Peak, 3,700 ft. San Cayetano Mts, Mt Shibell, 4,000 to 4,750 ft. Atascosa Mts, 2.5 mi W of Ruby, 4,000 ft. Pajaritos Mts, 2.5 mi W and also 1.5 mi S of Pena Blanca Lake), Cochise Co. (Johnny Lyon Hills, ca 15 mi NNE of Benson. Whetstone Mts, 4,700 to 5,400 ft. Mustang Mts, 5,200 to 5,400 ft. Huachuca Mts: near Manila Mine, Pilsbry and Ferriss, 1906:126; Sylvania Spring in Scotia Canyon, 6,150 ft; 1 mi N of Bear Creek, 5,600 ft; Huachuca Canyon, 5,250 ft; Copper Canyon, 6,200 ft. Dragoon Mts: canyons at 4,700 to 5,700 ft. Cave 1.5 mi W of Dos Cabezas, 4,800 ft. Chiricahua Mts: canyons at 4,500 ft and over; alive in Pinery Canyon at 6,500 to 7,000 ft, senior author. Mule Mts, 2 mi SE of Warren, at SE end of Range), Graham Co. (Pinaleno Mts, Noon Creek, SE slope of Mt Graham, 5,100 ft), and Greenlee Co. (6 mi up Blue Riv and on Little Blue Riv, Pilsbry and Ferriss, 1919a:301. Eagle Creek, 7 mi SW of Morenci, 3,550 ft. N end of Peloncillo Mts, 4 mi SW of Gila Riv, 4,400 ft). Dead shells common in drift of most rivers, creeks, and washes. Few trustworthy fossil records; not known thus far from Late Pleistocene of San Pedro Valley, although living there at present. General distribution discussed in section on Zoogeography; Recent range includes S New Mexico, S Texas, and N half of Mexico from Baja California, Sonora, and Chihuahua to Jalisco, Sinaloa, Nuevo Leon, San Luis Potosi, and Tamaulipas; not recorded from California.

75. **Microphysula ingersollii** (T. Bland, 1875 [not 1874, as sometimes cited], as *Helix*). T.L.: not selected thus far; described from Howardsville, San Juan Co., and other Stations in SW Colo-

rado, at 9,300 to 11,000 ft. Known in Arizona only at high elevations in Coconino Co. (Grand Canyon of Colorado Riv; Watershed Road, SE slope of San Francisco Mtn, 10 mi N of Flagstaff, 8,500 to 9,300 ft; Bill Williams Mtn; Navajo Mtn at Endische Spring; Two-Spring Gulch in Kaibab Saddle, N of Grand Canyon, 6,700 ft), Yavapai Co. (top of Mingus Mtn), Navajo Co. (Show Low Creek, 1 mi S of dam), Apache Co. (Reservation Creek, 6,800 ft; 12 mi N of Fort Defiance, 7,000 ft), Greenlee Co. (Cosper's pasture, top of Blue Mts, ca 10,000 ft; Fish Creek; Horseshoe Bend of Black Riv; K.P. Cienega, ca 4 mi S of Hannagan Meadows, 9,000 ft), Gila Co. (Sierra Ancha, 6,500 to 7,000 ft), Pima Co. (Santa Catalina Mts, several Stations on Mt Lemmon, 8,000 to 9,000 ft), Graham Co. (Pinaleno Mts, S slope of Mt Graham at Hospital Flat, 9,000 ft), and Cochise Co. (Huachuca Mts, Miller Peak, 7,000 ft; Chiricahua Mts, several Stations, 7,500 to 9,400 ft). General range covers also British Columbia, Washington, Oregon, Idaho, Montana, Wyoming, Colorado, Utah, and New Mexico; not known from Sonora. Not known fossil. Synonym: *Thysanophora ingersolli meridionalis* H. A. Pilsbry and J. H. Ferriss, 1910 (T.L.: Chiricahua Mts, Long Park, 8,000 ft, Cochise Co.).

Family Bulimulidae (Orthalicidae)

76. **Bulimulus (Rabdotus) nigromontanus** W. H. Dall, 1897. T.L.: summit of Black Mtn, 12 mi S of Monument 77 of International Boundary, right bank of San Bernardino Riv, Sonora; elevation not given. Known in Arizona only from Pena Blanca Canyon, Pajaritos Mts, ca 4,000 ft, Santa Cruz Co.

Bulimulus of New Mexico (references and distribution in H. A. Pilsbry, 1946:13-14 and 18-19): *Bulimulus dealbatus neomexicanus* H. A. Pilsbry, 1946 (also in trans-Pecos Texas). *B. pasonis* H. A. Pilsbry, 1902 (also in trans-Pecos Texas).

Bulimulus of trans-Pecos Texas (references and distribution in H. A. Pilsbry, 1946:11-21): *Bulimulus dealbatus ragsdalei* H. A. Pilsbry, 1890 (also elsewhere in Texas). *B. dealbatus neomexicanus* H. A. Pilsbry, 1946 (also in New Mexico). *B. pasonis* H. A. Pilsbry, 1902 (also in New Mexico). *B. alternatus schiedeanus* (L. Pfeiffer, 1841) (also in Chihuahua, Coahuila, Durango, Jalisco, San Luis Potosi, and Puebla). *B. pilsbryi* J. H. Ferriss, 1925.

Bulimulus of Sonora: *Bulimulus nigromontanus* W. H. Dall, 1897 (also in Arizona). *B. sonorensis* H. A. Pilsbry, 1928:115, fig. 1. *B. baileyi* W. H. Dall, 1893:640, Pl. 71, fig. 1 (also in Chihuahua and Sinaloa).

Bulimulus of Chihuahua: *Bulimulus alternatus schiedeanus*
(L. Pfeiffer, 1841) (also in trans-Pecos Texas and elsewhere in
Mexico). *B. baileyi* W. H. Dall, 1893:640, Pl. 71, fig. 1 (also in
Sonora and Sinaloa).

Family Urocoptidae

Holospira E. von Martens, 1860

Holospira is an American genus of unicolorous, medium-sized
snails (adults 8 to 20 mm long). The turret-shaped, cylindric shell
of many, closely coiled whorls, has a persistent, conic top, a rounded,
partly detached, thickened peristome, and the axis of the spire
hollow, but closed or narrowly open at umbilicus. The wall of the
hollow axis may bear spiral lamellae or slight protuberances,
mainly in the penultimate whorl. The shape, number, length, and
arrangement of the internal lamellae, as well as the sculpture,
size, and shape of the shell, are highly variable, not only for the
several species, but even within the same species or population,
which complicates the taxonomy of the genus, at present in a
state of flux. According to F. G. Thompson (1964), the true inter-
specific relationships may perhaps eventually be based on anatomical features. Meanwhile, however, the shells of the 13 species of
south Mexico to which this author restricts subgenus *Bostrichocentrum* all have a single axial lamella, with the exception of
H. anomala. On the other hand, the species of Arizona and
New Mexico, placed by Pilsbry (1946) in *Bostrichocentrum,* vary
greatly in the number and size of the internal lamellae. F. G.
Thompson opines that these may have evolved independently from
holospiroid ancestors, and, pending anatomical examination, should
be removed from that subgenus. In accordance with this view,
senior author has adopted the subgeneric name *Eudistemma* W. H.
Dall (1896:3) for the Arizona species, its type species, *Holospira
arizonensis* R. E. C. Stearns (1890), being the first species described
from the State. The exact relationship of subgenus *Eudistemma* to
subgenus *Allocoryphe* H. A. Pilsbry (1946; 1953:152) proposed for
most of the species of Sonora, listed by us later, must be left in
abeyance at present. The general distribution of the genus *Holospira* was discussed in the section on Zoogeography.

The local species of *Holospira* are excessively variable in most
shell characters, such as size, shape, and sculpture, often even within a limited area or a single, obviously homogeneous and freely
interbreeding population. In spite of this, however, they have not
speciated as intensively as *Sonorella* in the same area under similar

environmental conditions. Being highly calcicolous, as well as xerophilous, they usually live at restricted and widely separated, often
very hot and extremely arid, limestone Stations, partly exposed to
sun and drought, under small rocks, in loose litter, or under bunches
of bear grass (*Nolina microcarpa*). Poorly sheltered in such situations, they may be picked up easily by rain or wind. The small
and relatively light immature shells in particular may then be
transported far and wide by wind, as explained in the section on
Zoogeography, where we pointed out that in arid country frequent
air-borne dispersal of small snails may counteract speciation by
isolation.

In *Holospira* individual variability is such that one is often
at a loss to decide whether a named form should be considered as
only a collector's item rather than an acceptable biological taxon.
Although Pilsbry was fully aware of this, he kept some forms as
distinct species or subspecies that are unrecognizable except by
locality. Anatomical studies and a critical appraisal of all southwestern species, including those of Sonora, may perhaps clarify this
unsatisfactory situation. At present, the anatomy has not been
published for any of the local species. Senior author has nevertheless attempted a revision of the described Arizona forms, reducing
Pilsbry's eight species to six, and his seven subspecies to three. In
addition, *H. campestris cochisei* has been transferred, as a subspecies, to *H. danielsi;* in Pilsbry's fig. 64 (1946:136), figs. 7-7f and
8-8b of *campestris cochisei* scarcely differ from figs. 4-4b and 5-5a of
a variety of *H. danielsi* which appears to connect *cochisei* with nominate *danielsi*.

77.   **Holospira (Eudistemma) arizonensis** R. E. C. Stearns, 1890:
208, Pl. 15, figs, 2-3. T.L.: "Cave at Dos Cabezas," Cochise Co.;
elevation not given; two dead shells and fragments collected by V.
Bailey, 1889; first record of genus for Arizona. Pilsbry and Ferriss'
(1910c:118) detailed description was from a large lot collected for
them near the Cave by Mort Wien in 1909; they stated that it is
not a cave snail. T.L. located more precisely in 1948 by M. L. Walton,
who guided the authors in 1971 to the small, rugged limestone hill
of the cave, where the snail lives in the open at 4,800 ft, ca 4 mi SW
of Dos Cabezas Peaks and 1.5 mi due W of Dos Cabezas, 32° 11′ N,
109° 39′ W (R26E, T14S, Sec. 25, SW¼ of U.S. Geol. Surv. map).
Also in Chiricahua Mts, Mule Mts, and Mustang Mts, Cochise
Co., at 5,000 to 5,500 ft. Synonyms: *Holospira arizonensis emigrans*
H. A. Pilsbry and J. H. Ferriss, 1910 (T.L.: head of Big Emigrant
Canyon, Chiricahua Mts; no elevation given). *H. arizonensis mularis*
H. A. Pilsbry and J. H. Ferriss, 1915 (T.L.: Mule Mts, N slope of

E Escabrosa Ridge, at 6,000 to 6,500 ft, Cochise Co.). *H. arizonensis mustang* H. A. Pilsbry and J. H. Ferriss, 1923 (T.L.: Mustang Mts, N side of East Dome, Cochise Co.; elevation not given, but at 5,000 to 5,200 ft). *H. whetstonensis arata* H. A. Pilsbry and J. H. Ferriss, 1923 (T.L.: Mustang Mts, on top and NW side of East Dome, Cochise Co.; elevation not given, but at 5,000 to 5,200 ft).

78. **Holospira (E.) whetstonensis** H. A. Pilsbry and J. H. Ferriss, 1923. T.L.: Whetstone Mts, Station 304, highest cleft peak facing N, Cochise Co.; elevation not given; collected again by A. R. Mead, junior author, et al, March 1968, at or near T.L., on SE slope of French Joe Peak, above right bank of French Joe Creek (R19E, T19S, Sec. 5, SE ¼ of U.S. Geol. Surv. map), at 6,000 to 6,800 ft. Also elsewhere in same Range; and in Empire Mts, 1.5 mi N of Total Wreck Mine, Pima Co., where it was collected recently at 4,800 ft. Two dead shells found in drift of San Pedro Riv near Benson by senior author (1963). Probably at best a subspecies of *H. ferrissi* H. A. Pilsbry, 1915.

79a. **Holospira (E.) t. tantalus** P. Bartsch, 1906. T.L.: unknown. Collected later at several Stations in Dragoon Mts, Cochise Co., at 4,800 to 5,600 ft. Synonym (new here): *Holospira millestriata* H. A. Pilsbry and J. H. Ferriss, 1915 (T.L.: Dragoon Mts, S of Tweed Canyon [Cochise Stronghold Canyon West of present map], at summit of limestone ridge between head of Cataract Gulch and next canyon opening westward, Cochise Co.; elevation not given, but ca 5,400 ft).

79b. **Holospira (E.) tantalus campestris** H. A. Pilsbry and J. H. Ferriss, 1915, as *Holospira campestris.* T.L.: Dragoon Mts, Station 26, mesa at W foot of Range, at S fence of Fourr Ranch, in mouth of Fourr Canyon, Cochise Co.; elevation not given, but collected there recently at 5,500 to 5,600 ft. Also elsewhere in same Range, down to 4,800 ft.

80a. **Holospira (E.) f. ferrissi** H. A. Pilsbry, 1905. T.L.: Huachuca Mts, W foothills near site of Manila Mine, NW end of Range, Cochise Co.; elevation not given, but found there recently at 5,200 ft. Also at nearby stations in same Range; in Mule Mts, Cochise Co.; and in western Canelo Hills and W section of Mustang Mts, Santa Cruz Co. Found dead by senior author in drift of Santa Cruz Riv and Rillito Creek in Tucson, Pima Co., and of Sonoita Creek at Patagonia, Santa Cruz Co. Synonyms: *Holospira ferrissi fossor* H. A. Pilsbry and J. H. Ferriss, 1915 (T.L.: S end of Mule Mts, ca 2 mi SE of Warren, Cochise Co.; elevation not given, but ca 5,000 ft). *Holospira ferrissi sanctaecrucis* H. A. Pilsbry and J. H. Ferriss, 1915 (T.L.: Tucson, Pima Co., in drift of Santa Cruz

Riv at Congress St). *Holospira ferrissi caneloensis* H. A. Pilsbry and J. H. Ferriss, 1923 (T.L.: western Canelo Hills, W of Duquesne Rd, Santa Cruz Co.; elevation not given). *Holospira ferrissi monoptyx* H. A. Pilsbry and J. H. Ferriss, 1923 (T.L.: Mustang Mts, near Site of Dan Mathew's Ranch, 2 mi W of East Dome of Range, Santa Cruz Co.; elevation not given, but ca 5,000 ft). *Holospira ferrissi fluctivaga* H. A. Pilsbry and J. H. Ferriss, 1923 (T.L.: Mammoth, Pinal Co., in drift of San Pedro Riv; also in drift of same river at Benson, Cochise Co.).

    H. f. ferrissi was found in April 1971 by V. Roth in N Sonora at Canyon de Evans, Sierra de los Ajos, 17 mi SE of Cananea, at 5,500 ft (ca 30° 50′ N, 109° 50′ W; southernmost Station of species). The shells agree in size (up to 8.2 mm long, 3.5 mm wide), shape, and sculpture with paratypes of *H. f. caneloensis.*

80b. **Holospira (E.) ferrissi cionella** H. A. Pilsbry, 1905, as *Holospira cionella.* T.L.: originally as "Fort Bowie, Cochise Co.," Site of old Fort in NW foothills of Chiricahua Mts, the "original types" without more precise Station; elevation not given, but topotypes collected recently by junior author at ca 6,000 ft; more precise locality given by Pilsbry and Ferriss (1910c:122) as lower slope of Quartzite Peak, toward Dixon's house [Knape Ranch of present map], shown in a photograph at (2) of fig. 5, p. 67. Shells from this Station said to "agree most closely with original types of *cionella.*" Two cotypes from Quartzite Peak were shown on Pl. 12, figs. 1-2, part of Lot No. 87,117, Acad. Nat. Sci. Phila.; the shell of fig. 1 (in 1910) was selected as holotype by Pilsbry (1946:145), and then shown enlarged in fig. 68 (30). Also elsewhere in Chiricahua Mts and Dos Cabezas Mts, at 5,500 to 6,500 ft. *H. cionella* is here made a subspecies of *H. ferrissi,* as first suggested by Pilsbry and Ferriss (1915a:388). Synonyms: *Holospira cionella intermedia* H. A. Pilsbry and J. H. Ferriss, 1910 (T.L.: Chiricahua Mts, at box of Whitetail Canyon, Cochise Co.; elevation not given, but found there recently by junior author at 5,500 ft). *Holospira cionella capillacea* H. A. Pilsbry and J. H. Ferriss, 1910 (T.L.: Chiricahua Mts, Whitetail Canyon, about halfway to summit of ridge, Cochise Co.; elevation not given, but found there recently by junior author at 5,600 ft).

81a. **Holospira (E.) d. danielsi** H. A. Pilsbry and J. H. Ferriss, 1915:373. T.L.: Dragoon Mts, NW side, in Tweed Canyon [Cochise Stronghold Canyon West of present map], Cochise Co.; elevation not given, but found there recently at 5,300 ft. Also elsewhere in same Range.

81b. **Holospira (E.) danielsi cochisei** H. A. Pilsbry and J. H. Ferriss, 1915:379, as *H. campestris cochisei.* T.L.: Dragoon Mts,

mesa in mouth of Tweed Canyon [Cochise Stronghold Canyon West
of present map], Cochise Co.; elevation not given, but ca 5,000 ft.
Also at other nearby Stations in same Range.

82. **Holospira (E.) chiricahuana** H. A. Pilsbry, 1905. T.L., as
given precisely by Pilsbry, 1946: Chiricahua Mts, "on the steep
slopes of a small dry ravine or wash tributary to Cave Creek, below
and near the entrance of the cave" [viz Crystal Cave, ca 1 mi W of
Southwestern Research Station], Cochise Co.; elevation not given,
but found there recently at 5,500 ft. Also elsewhere in same Range,
at 5,500 to 5,800 ft. Synonyms: *Holospira chiricahuana ternaria*
H. A. Pilsbry and J. H. Ferriss, 1910 (T.L.: Chiricahua Mts, halfway
up N slope of a ridge projecting in Cave Creek Valley from W
border, Cochise Co.; elevation not given). *Holospira chiricahuana
optima* H. A. Pilsbry and J. H. Ferriss, 1910 (T.L.: Chiricahua
Mts, "at base of N slope of ridge where *H. c. gracilis* was found"
[viz in Ash Fork of Cave Creek Canyon], Cochise Co.; elevation not
given). *Holospira chiricahuana gracilis* H. A. Pilsbry and J. H.
Ferriss, 1910 (T.L.: Chiricahua Mts, crest of a narrow spur from the
S wall of Cave Creek Valley ending in a high conical hill not far
from the creek, Cochise Co.; elevation not given [this hill now
called locally "Rattlesnake Hill," at 5,600 ft, ca 1 mi SW of Southwestern Research Station and ca 5 mi SW of Portal P.O.]).

    Holospira of New Mexico (references and distribution in Pils-
bry, 1946:114-135): *Holospira roemeri* (L. Pfeiffer, 1848) (also in
trans-Pecos Texas). *H. regis* H. A. Pilsbry and T. D. A. Cockerell,
1905. *H. cockerelli* W. H. Dall, 1897. *H. crossei* W. H. Dall, 1895.
H. bilamellata W. H. Dall, 1895, synonym: *H. mearnsi* W. H. Dall,
1895.

    Holospira of trans-Pecos Texas (references and distribution in
Pilsbry, 1946: 114-135): *Holospira roemeri* (L. Pfeiffer, 1848) (also
in New Mexico). *H. hamiltoni* W. H. Dall, 1897 (also in Chihuahua),
synonym: *H. riograndensis* H. A. Pilsbry, 1946. *H. mesolia* H. A.
Pilsbry, 1912. *H. pasonis* W. H. Dall, 1895 (also in Chihuahua).
H. montivaga H. A. Pilsbry, 1946 (T.L. in Texas, not in New Mexi-
co), synonym: *H. pityis* H. A. Pilsbry and E. P. Cheatum, 1951:89,
Pl. 4, figs. 3-3a. *H. oritis* H. A. Pilsbry and E. P. Cheatum, 1951:89,
Pl. 4, figs. 5-5a. *H. yucatanensis* P. Bartsch, 1906:143, Pl. 3, figs. 2
(T.L. not in Yucatan).

    Holospira of Sonora: *Holospira minima* E. von Martens, 1897:
280, Pl. 16, fig. 18, as *H. pfeifferi* var. *minima*, synonym: *H. minima*
var. *percostata* H. A. Pilsbry, 1902:95, Pl. 24, fig. 7. *H. dentaxis* H.
A. Pilsbry, 1953:154, Pl. 9, figs. 3-3a, synonyms: *H. dentaxis striatella* H. A. Pilsbry, 1953:155, Pl. 10, fig. 3, *H. dentaxis lamellaxis*

H. A. Pilsbry, 1953:155, Pl. 9, fig. 6, and *H. dentaxis potamia* H. A. Pilsbry, 1953:155, Pl. 9, fig. 2. *H. kinonis* J. L. and R. I. Baily, 1940:94, Pl. 12, fig. 1. *H. remondii* (W. M. Gabb, 1865:208, Pl. 19, figs. 10-13, as *Cylindrella*), synonyms: *Haplocion guaymasensis* P. Bartsch, 1943:56, fig. 1, *H. remondii laevior* H. A. Pilsbry, 1953: 153, Pl. 10, fig. 2, *H. remondii yaquensis* H. A. Pilsbry, 1953:154, Pl. 10, figs. 4-5, and *H. remondii forticostata* H. A. Pilsbry, 1953: 154, Pl. 9, figs. 5-5a. *H. cyclostoma* H. A. Pilsbry, 1953:148, Pl. 8, figs. 1 and 1a-b. *H. ferrissi* H. A. Pilsbry, 1905 (also in Arizona).

Holospira of Chihuahua: *Holospira wilmoti* (P. Bartsch, 1947: 288, fig. 3, as *Haplocion*). *H. hamiltoni* W. H. Dall, 1897 (also in trans-Pecos Texas). *H. semisculpta* R. E. C. Stearns, 1890:208, Pl. 15, figs. 1 and 4. *H. pasonis* W. H. Dall, 1895 (also in trans-Pecos Texas). *H. bryantwalkeri* H. A. Pilsbry, 1917b:124, Pl. 4, fig. 6. *H. freytagi* (P. Bartsch, 1950:265, 2 figs. 1, as *Coelostemma*). *H. coahuilensis* (W. G. Binney, 1865b:50, Pl. 7, figs. 4-5, as *Cylindrella*) (also in Coahuila).

Family Achatinidae (including Subulinidae)

* **Rumina decollata** (C. von Linné, 1758, as *Helix*). Originally described from S and E Europe, without precise T.L. Introduced from the Mediterranean area; the first North American record from North Carolina in 1813. Now fully naturalized in SE United States, as far W as trans-Pecos Texas (Longfellow, 16 mi W of Sanderson; Alpine; El Paso). Often common in gardens, but not feral thus far, in Arizona; reported from Yuma Co. (Yuma), Maricopa Co. (Phoenix; Tempe; Mesa), Pinal Co. (Florence), and Pima Co. (Tucson, first noticed in the State by A. R. Mead in 1952).

* **Lamellaxis (Allopeas) gracilis** (T. Hutton, 1834, as *Bulimus*). T.L.: Mirzapur, India; true original home uncertain, but probably tropical America. Introduced in North America and fully established in South Carolina, Georgia, Florida, Alabama, Louisiana, and Texas. Known thus far in Arizona only from Santa Cruz Co. (two incomplete dead shells in drift of Santa Cruz Riv, 1.5 mi E of Amado, senior author, 1967) and Pima Co. (Ajo, a living colony in a garden, collected by R. D. Cross, Aug. 1971). Recorded from Sonora, without precise locality, by R. J. Drake (1953a:154).

Family Euconulidae

83. **Euconulus fulvus** (O. F. Müller, 1774, as *Helix*). T.L.: Fridrichsdal, Denmark. Recent and native in Arizona in all counties, except Yuma, at 5,000 to 10,500 ft, mainly in Transition and Canadian Life-Zones, but only above ca 6,000 ft in S half of State; in

Mohave Co. in litter of Hualapai Peak at 8,000 ft, by M. D. Robinson and R. L. Bezy, June 1967. Circumboreal over most of Holarctic Realm; in North America in Alaska, much of Canada, Greenland, and most of United States (except Gulf and S Atlantic States from North Carolina to central Texas); Recent in New Mexico, trans-Pecos Texas, and NW Chihuahua; not known from Sonora, where it may be expected in some mountains; general distribution discussed in section on Zoogeography. Fossil in Pliocene, Pleistocene, and Holocene of Europe and North America; in Arizona in Late Pleistocene of Coconino Co. (Winona Site, 13 mi E of Flagstaff, R. D. Reger and G. L. Batchelder, 1971: 191 and 193, as *E. fulvus alaskensis*) and Cochise Co. (San Pedro Valley: Lehner Mammoth and Murray Springs Sites, associated with mammoth remains 10,000 to 11,000 yrs old). Synonyms: *Conulus fulvus alaskensis* H. A. Pilsbry, 1899 (T.L.: Dyea Valley, Alaska, 59° 29′ N, 135° 21′ W). *Helix trochiformis* G. Montagu, 1803, *Test. Britann.* 2:427, Pl. 11, fig. 9 (England).

Family Zonitidae

* **Oxychilus draparnaldi** (H. Beck, 1837, *Index Moll.*:6, as *Helicella;* substitute name for *Helix lucida* J. F. R. Draparnaud, 1801, with France as T.L.; not *Helix lucida* H. Pulteney, 1799). Native in W Europe; introduced in North America ca 1850 or earlier. Alive in a nursery greenhouse in Phoenix, Maricopa Co. (C. A. Westerfelt, Jr., Oct. 1961), and in the open in a nursery on Oracle Rd in Tucson, Pima Co. (J. R. Hershey and R. H. Russell, March 1970). Reported also from several eastern States and from Oregon, Washington, California, and Colorado.

84. **Retinella (Glyphyalinia) indentata paucilirata** (A. Morelet, 1851, as *Helix paucilirata*). T.L.: Salama, Guatemala. Recent in most Arizona highlands at 3,500 ft (in Ajo Mts) to 8,500 ft (on San Francisco Mtn); known alive from the following counties: Coconino, Navajo, Apache, Yavapai, Pinal, Gila, Pima, Santa Cruz, Cochise, Graham, and Greenlee; dead shells common in riparian drift in the same counties; found fossil in Late Pleistocene or sub-Recent of trans-Pecos Texas, but not thus far in New Mexico or Arizona; general distribution discussed in section on Zoogeography. According to Pilsbry (1946:291; mostly after H. B. Baker), Recent in S United States from Indiana, Kentucky, and North Carolina to Florida and Gulf States, W to Oklahoma, Texas, New Mexico, Utah, and Arizona; not known from California; also in Mexico (Baja California, Sonora, Chihuahua, etc) and S to Guatemala. In the mid-Atlantic States it intergrades with nominate *R. i. inden-*

tata (T. Say, 1823). Synonym: *Vitrea indentata* var. *umbilicata* T. D. A. Cockerell, 1899:120 (T.L.: Texas, without more precise locality).

85. **Nesovitrea hammonis electrina** (A. A. Gould, 1841, as *Helix electrina*). T.L.: shores of Fresh Pond, Cambridge, Massachusetts. Known in Arizona, presumably Recent, at very few Stations of NE section of Apache Co. (dead, bleached shells, without more precise locality, H. B. Baker, 1930:196; one shell from Black Riv, 2 mi above Fish Creek, H. A. Pilsbry and J. H. Ferriss, 1919a:326) and Greenlee Co. (drift of Eagle Creek, 6 mi SW of Morenci, senior author, 1962). Formerly more widespread in the State, being common in Late Pleistocene of San Pedro Valley at Lehner Mammoth and Murray Springs Sites, associated with mammoth remains 10,000 to 11,000 yrs old. Distribution of Holarctic *N. hammonis* discussed in section on Zoogeography; Palearctic *N. h. hammonis* (I. A. af Stroem, 1767) from Iceland and British Isles through N and central Eurasia to Kamchatka and Japan; Nearctic *N. h. electrina* from Alaska through Washington, Ontario, Nova Scotia, and Newfoundland, S to Virginia, Kansas, Utah, Oregon, New Mexico, and NE Arizona; not known from Mexico.

86. **Hawaiia minuscula** (A. Binney, 1841, after April 3, as *Helix;* not 1840 or 1845 as sometimes given). T.L.: "Ohio and Vermont"; Pilsbry (1946:421) chose Ohio as T.L., but without more precise locality. Truly native (pre-Columbian) in Arizona, as shown by local Pleistocene fossils; widespread Recent in the State, at 2,500 to 8,500 ft; perhaps its most common snail, also in irrigated, artificial habitats of towns (Tucson, Phoenix, Nogales, etc). Known alive and from riparian drift in all counties except Greenlee; fossil in Late Pleistocene of Coconino Co. (Winona Site, 13 mi E of Flagstaff, R. D. Reger and G. L. Batchelder, 1971:191 and 193) and Cochise Co. (San Pedro Valley: El Paso Pipeline Site, 7 mi S of Benson, R. S. Gray, 1965, D. W. Taylor, 1966b:94; Lehner Mammoth and Murray Springs Sites, associated with mammoth remains 10,000 to 11,000 yrs old). General distribution discussed in section on Zoogeography. Recent in New World from Alaska to Quebec and Newfoundland, S to Florida, Mexico (Baja California, Sonora, etc), Central America, and Antilles; carried by man after Columbus' time to Old World and Pacific Islands and now nearly cosmopolitan. Occurrence in North American Pleistocene and absence from it in Eurasia prove that this snail was originally an autochthonous native of New World. Synonym: *Zonites minusculus* var. *alachuana* W. H. Dall, 1885 (T.L.: Alachua Co., Florida); although first based

on Florida snails from near sea level, the name sometimes has been applied to Arizona specimens from high elevations.

87. **Zonitoides arboreus** (T. Say, 1817, as *Helix*). Described from North America without precise T.L.; none selected thus far. Common in most Arizona mountains, usually above 5,000 ft and up to 12,000 ft; below 5,000 ft only as introductions by man under artificial conditions of moisture and shelter (e. g. in Tucson). Known Recent from all counties except Yuma, Maricopa, and Pinal. Fossil in Late Pleistocene or sub-Recent in Coconino Co. (Winona Site, 13 mi E of Flagstaff, R. D. Reger and G. L. Batchelder, 1971: 191-193), Apache Co. (Fort Defiance, B. Walker, 1915:2), and Cochise Co. (San Pedro Valley at Murray Springs Site, associated with mammoth remains 10,000 to 11,000 yrs old). Definitely indigenous in North America only, in S Canada (N to 61° N), most of United States, and much of Mexico, but no Sonora record thus far; farther south, in Central America and Antilles, perhaps only a human introduction gone feral. Widespread as an adventive in Old World; doubtfully native in Kamchatka. Recent and past distribution discussed in section on Zoogeography.

88. **Striatura meridionalis** (H. A. Pilsbry and J. H. Ferriss, 1906, as *Vitrea*). T.L.: drift of Guadalupe Riv above New Braunfels, Comal Co., Texas. Widespread in Arizona at 4,800 to 9,000 ft, mainly in Upper Sonoran Life-Zone. Known Recent in Coconino, Yavapai, Gila, Maricopa, Pima, Santa Cruz, Cochise, Greenlee, and Graham counties; probably overlooked elsewhere because of minute size. Not known fossil. Recent and native in most southeastern States, N to New Jersey, W to SE Colorado, New Mexico, Arizona, and Texas; in Mexico in Chihuahua (Sierra de la Breña, 7,000 ft, junior author, 1966), Puebla, Nuevo Leon, and Vera Cruz; no Sonora record; probably introduced by man in Bermuda. Perhaps only the southern subspecies of *Striatura milium* (E. S. Morse) of N United States.

Family Vitrinidae

89. **Vitrina pellucida alaskana** W. H. Dall, 1905, as *Vitrina alaskana,* substitute name for *Vitrina pfeifferi* W. Newcomb, 1861, with same T.L.: Carson Valley, Churchill Co., Nevada. Widespread, but disjunct, in Arizona, at 6,000 to 10,000 ft, mainly in Canadian and Hudsonian Life-Zones. Known Recent in Mohave Co. (Mt Trumbull, Pilsbry and Ferriss, 1911:191), Coconino Co. (4 Stations by Pilsbry and Ferriss, 1911:191: Bill Williams Mtn; Grand Canyon of Colorado Riv at Seep Spring, 2 mi W of Mystic Spring [or Bass] Trail; Kaibab Plateau at Warm Spring Canyon and Riggs Spring.

Navajo Mtn, Pilsbry, 1946:503. Drift of Oak Creek Canyon, 7 mi N of Sedona, 3,500 ft, senior author. San Francisco Mtn, 4 Stations by R. H. Russell, 1969: SE slope on Watershed Rd, 10 mi N of Flagstaff, 10,000 ft; NW slope near Snow Bowl, 8,500 to 9,500 ft; Lockett Meadow, W side of Sugarloaf Mtn, 10 mi N of Flagstaff, 8,000 ft; Orion Spring below Schultz Pass, 5 mi N of Flagstaff, 8,500 ft), Navajo Co. (TaBiko Canyon near Betatakin Ruin, D. T. Jones, 1940:37), Apache Co. (Reservation Creek, 6,800 ft, Pilsbry and Ferriss, 1919:327; East Fork of Black Riv, 3 mi N of Buffalo Crossing, 7,700 ft, senior author), Greenlee Co. (Rim of Blue Mts, ca 10,000 ft [not in Graham Co. or at 12,000 ft as given], Pilsbry and Ferriss, 1919:327), Graham Co. (Pinaleno Mts, S slope of Mt Graham at Hospital Flat, 9,000 ft, senior author), Gila Co. (Sierra Ancha, Workman Creek below falls, 6,500 ft, senior author), Pima Co. (Santa Catalina Mts: Marshall Gulch, 7,600 ft, and Bear Wallow, 7,800 to 8,000 ft, senior author; N side of Mt Lemmon, 9,000 ft, Pilsbry and Ferriss, 1919:302; ski area of Mt Lemmon, ca 9,000 ft, W. B. Miller, R. H. Russell, et al, Apr. 1971), Cochise Co. (Huachuca Mts: Wickersham Rock near Miller Peak, ca 9,200 ft, Pilsbry and Ferriss, 1910a:514; Carr Canyon, ca 6,000 ft, T. R. Van Devender, 1971. Chiricahua Mts: 3 Stations by Pilsbry and Ferriss, 1910c:130, at Long Park, 8,000 ft, Barfoot Park, 8,000 ft, and head of Cave Creek, 8,000 ft; 1 mi S of Tubs Spring on N slope of Flys Peak, 9,400 ft, J. R. Hershey, 1969; Winn Falls on W tributary of Cave Creek, 2.5 mi SW of Southwestern Research Station, 7,200 ft, R. L. Reeder and junior author, Apr. 1971). Dissections of San Francisco Mtn snails by junior author (1970) show that *V. alaskana* agrees in shell and genitalia with Old World *V. pellucida* (O. F. Müller, 1774), of which it is the American subspecies. Taxonomy and general distribution discussed in section on Zoogeography. Recent and native in western North America from Alaska, Montana, and Wyoming E to South Dakota and S to California, Arizona, and New Mexico; not recorded from Mexico. Synonym: *Vitrina pfeifferi* W. Newcomb, 1861 (T.L.: Carson Valley, Churchill Co., Nevada).

Family Limacidae

* **Limax (Limax) maximus** C. von Linné, 1758. Described from Europe without precise T.L., but presumably from Sweden, as the *Fauna Suecica* is cited. Adventive in Arizona; known thus far at one locality in Coconino Co.: 3 mi NW of Flagstaff, near Museum and Research Center of Northern Arizona, 7,100 ft, collected by Barton Wright, Aug. 1969, but was perhaps there for the past 10

years. Native of Europe, carried by man to most temperate regions. First noticed in North America in 1867 by G. W. Tryon, Jr., at Philadelphia; now feral, but strictly synanthropic, from British Columbia, Ontario, and Newfoundland over most northern States, S to S California, N Arizona, Utah, Colorado, Texas, Kansas, Missouri, and Maryland.

* **Limax (L.) flavus** C. von Linné, 1758. Described from Europe without precise T.L., but presumably from Sweden, as the *Fauna Suecica* is cited. Adventive in Arizona in Maricopa Co. (Phoenix, C. A. Westerfelt, Jr., 1960), Pima Co. (Tucson, first State record, A. R. Mead, 1952:30, 1963:27), and Cochise Co. (cienega of Babocomari Creek at Babocomari [now Brophy] Ranch, A. R. Mead et al, 1969). Native in temperate Europe, widely spread by man and now nearly cosmopolitan. First noticed in North America at Philadelphia by T. Say, perhaps before 1825, but it may have reached America earlier; now feral over most of the United States and N to British Columbia.

* **Limax (Lehmannia) valentianus** A. E. J. d'A. de Férussac, 1821, *Tabl. Syst. Anim. Moll., Tabl. Syst. Limaces:*21; Atlas, Pl. BA, figs. 5-6. T.L.: Valencia, Spain. A fairly common adventive in Arizona, perhaps in the process of becoming feral. Known from Maricopa Co. (Phoenix, H. W. Waldén, 1961:90, as collected by C. A. Westerfelt, Jr., 1960), Pima Co. (Tucson, first State record, A. R. Mead, 1953:11; 1963:27, as *Limax poirieri*), Santa Cruz Co. (Oak Ridge section of Nogales, 4,000 ft, C. Spitzer, 1965. Santa Rita Mts, Madera Canyon, 5,000 ft, J. F. Burger, 1968), and Cochise Co. (Huachuca Mts, Ramsey Canyon, 5,000 ft, 1967. Chiricahua Mts, Cave Creek at Southwestern Research Station, 4,400 ft, 1966-1970); all records based on dissections by junior author. Native in Old World in Mediterranean area, especially the Iberian Peninsula, and Atlantic islands; carried by man to many parts of the Earth. First North American record by T. D. A. Cockerell (1917) from a greenhouse at Boulder, Colorado, as *Limax marginatus;* reported also, often as *marginatus,* from Newfoundland, New York, Massachusetts, New Jersey, Maryland, Michigan, Missouri, Ohio, Oklahoma, Kansas, and California. Synonyms: *Limax poirieri* J. Mabille, 1883, *Bull. Soc. Philom. Paris* (7) 7 (1):52 (T.L.: Gran Canaria, Canary Is). *Limax marginatus* H. A. Pilsbry, 1948:529 (not of O. F. Müller, 1774).

* **Deroceras reticulatum** (O. F. Müller, 1774, as *Limax*). T.L.: Rosenburg and Fridrichsdal, Denmark. Adventive in Arizona; known thus far from one slug in Phoenix, D. B. Carver, May 9, 1969, on nursery plants shipped from California (sent to A. R. Mead by P. F. Min of Arizona Comm. Agric. Hort.). A. R. Mead informs us that it was found previously in a garden in Tucson, but not reported in print. Native in temperate Europe; introduced by man to New World and elsewhere. First record in North America not traced; now frequent in cultivated areas from British Columbia, Ontario, and Newfoundland S to north Alabama, Colorado, and south California. Whether all published American records are of true *D. reticulatum,* or of the closely related *Deroceras agreste* (C. von Linné, 1758, as *Limax*), is uncertain.

90. **Deroceras laeve** (O. F. Müller, 1774, as *Limax*). Originally described without T.L.; Fridrichsdal, Denmark, selected as such by H. E. Quick (1960, *Bull. Brit. Mus. Nat. Hist. Zool.* 6 [3]:172). Native in Arizona, normally at 4,500 to 8,000 ft; at lower elevations in cultivated areas. Recent at natural Stations in Coconino Co. (Snake Gulch at Big Spring near Grand Canyon of Colorado Riv; Rio de Flag Canyon near Museum of Northern Arizona, 3 mi NW of Flagstaff, 6,600 ft, R. H. Russell, 1970) and Cochise Co. (Huachuca Mts: Miller, Brown, Tanner [now Garden], and Huachuca canyons; Sylvania Springs in Scotia Canyon, 6,150 ft. Chiricahua Mts: Barfoot [or Bearfoot] Park; Long Park. Cienega of Babocomari Creek near Babocomari [now Brophy] Ranch, 4,500 ft); no doubt elsewhere also, but often overlooked by collectors. Introduced in gardens at Tucson (at 1249 E Mabel St, A. R. Mead, 1953) and Nogales. Internal fossil shells similar to those of Recent *D. laeve* occur in Late Pliocene or Late Pleistocene of Cochise Co. (San Pedro Valley Sites: Post Ranch, 6 mi, El Paso Pipeline, 7 mi, and California Wash, 9.5 mi S of Benson, R. S. Gray, 1965; and D. W. Taylor, 1966b:94, who refers them to the extinct *Deroceras aenigma* A. B. Leonard, 1950, described from Kansas. Lehner Mammoth and Murray Springs Sites, associated with mammoth remains 10,000 to 11,000 yrs old). A circumboreal slug, widespread also in Eurasia; Recent in North America, native from Arctic Region to Florida and Central America, but probably only introduced by man farther south. Synonyms: *Limax campestris* A Binney, 1842, not 1844 as sometimes cited (described from New England, New York, Ohio, and Missouri; no T.L. selected thus far). *Agriolimax hemphilli ashmuni* H. A. Pilsbry and E. G. Vanatta, 1910 (T.L.: Huachuca Mts, Miller Canyon, Cochise Co., elevation not given. We follow Pilsbry (1948: 532) and A. Zilch (1959:267) in preferring the generic name *Deroceras* C. S. Rafinesque (1820) to *Agriolimax* O. A. L. Mörch (1865).

* **Milax gagates** (J. P. R. Draparnaud, 1801, as *Limax*). T.L.: France, without more precise locality, but probably from near

Montpellier. Introduced; known in Arizona only from Phoenix, Maricopa Co., where it was found by J. N. Roney, Feb. 1957, in 700 block of W Cambridge St, and again by M. M. Evans, March 1970, at N 17th St; identified in both cases by A. R. Mead. Indigenous in Mediterranean Subregion and W Europe; spread by man elsewhere in Old and New Worlds and now nearly cosmopolitan. First American record in 1872 from gardens in California (San Francisco), as *Limax hewstoni* J. G. Cooper, 1872.

Family Endodontidae (Punctidae)

91. **Discus (Discus) cronkhitei** (W. Newcomb, 1865:180, as *Helix*). T.L.: Klamath Valley, Klamath Co., Oregon. Native and Recent in Arizona at 2,750 to 12,000 ft; usually above 6,000 ft in SE mountains; rare in drift at lower elevations, mostly as washed-up fossils. Known living in Mohave Co. (Mt Trumbull. Kaibab Plateau, in Snake Gulch at Cattle Springs and Riggs Spring, Pilsbry and Ferriss, 1911:191. Hualapai Peak, 8,000 ft, M. D. Robinson and R. L. Bezy, 1967), Coconino Co. (Bill Williams Mtn, Pilsbry and Ferriss, 1911:191. Rio de Flag Canyon, 2 mi NW of Flagstaff, 6,800 ft, senior author. Woody Mtn, 10 mi SW of Flagstaff, 7,500 ft, M. Noller, 1964. San Francisco Mtn, widespread, 8,500 to 12,000 ft, R. H. Russell, 1969-1970), Yavapai Co. (Mingus Mtn, Pilsbry, 1948:600), Navajo Co. (Show Low Creek Canyon, 0.75 mi S of dam, A. R. Mead, 1953), Apache Co. (Reservation Creek, 6,800 ft [not 9,500 ft], Pilsbry, 1948:600. Baldy Peak [= Mt Thomas], summit at 11,470 ft [not 13,500 ft], Pilsbry and Ferriss, 1919a:326), Greenlee Co. (Cosper's pasture, in Blue Mts, 10,000 ft; Mogollon Rim of Blue Mts; both Pilsbry and Ferriss, 1919a: 326; 4 mi S of Hannagan Meadow, 9,000 ft, senior author, 1964), Gila Co. (Pinal Peak, 10 mi SW of Globe, 7,300 to 7,500 ft, Pilsbry, 1948:600, also J. Bagnara and junior author, 1967. Sierra Ancha: upper Workman Creek at falls, 6,500 ft; Reynolds Creek, 5,600 ft; both S. S. Berry, 1948:151, and senior author, 1964. Pine Creek at Tonto Natural Bridge, 4,500 ft, senior author, 1966), Pima Co. (Santa Catalina Mts, widespread, 7,000 to 8,500 ft, Pilsbry and Ferriss, 1919a:301. Tumamoc Hill, 2,750 ft, Pilsbry and Ferriss, 1923:55. Rincon Mts, Manning Camp, 1.5 mi S of Spud Rock, 8,000 ft, senior author, 1963), Cochise Co. (Huachuca Mts, widespread, 5,000 to over 8,000 ft [Wickersham Rock], Pilsbry, 1948:603. Chiricahua Mts: Flys Park, ca 9,200 ft, W. H. Dall, 1897a:342 and 366, as *Pyramidula striatella;* widespread, 5,400 to over 9,000 ft, Pilsbry and Ferriss, 1906:153 and 1910c:133. Dragoon Mts, Bear Gulch off Stronghold Canyon East, near Huzzar Mine, Pilsbry and Ferriss, 1915a:383), and Graham Co. (Pinaleno

Mts: Mt Graham, 6,300 to 9,000 ft, senior author, 1964). Strictly Nearctic; Recent from Alaska, Northwest Territories, and Newfoundland south to S California, Arizona, and New Mexico in the west, and to Kentucky and N Illinois in the east; in Mexico thus far only in N Chihuahua (Sierra de la Breña, at 7,000 ft, junior author, 1966); not known from Sonora. General Recent and fossil distribution in section on Zoogeography. Fossil in Late Pleistocene of Coconino Co. (Winona Site, 13 mi E of Flagstaff, R. D. Reger and G. L. Batchelder, 1971:191 and 193) and Cochise Co. (San Pedro Valley: Murray Springs Site, associated with mammoth remains 10,000 to 11,000 yrs old). Synonyms: *Helix striatella* J. G. Anthony, 1840 (not of P. S. Rang, 1831) (T.L.: Cincinnati, Ohio). *Pyramidula anthonyi* H. A. Pilsbry and J. H. Ferriss, 1906 (T.L.: Cincinnati, Ohio).

92. **Discus (Gonyodiscus) shimekii** (H. A. Pilsbry, May 1890a:3, as *Zonites;* Oct. 1890b:297, Pl. 5, figs. 9-11). T.L.: Iowa City, Iowa, as a Pleistocene fossil in loess. As shown in the section on Zoogeography, *Pyramidula cockerelli,* based on Recent shells, is not separable as a subspecies from nominate fossil *shimekii.* Known Recent in Arizona only at three Stations, in Coconino Co. (Big Pine Creek, W of Endische Springs, SW foot of Navajo Mtn, elevation not given [probably ca 10,000 ft], Pilsbry, 1948:619, as *D. shimekii cockerelli* [other Stations on Navajo Mtn appear to be in Utah]; Agassiz Peak of San Francisco Mtn, above timberline in Arctic-Alpine Life-Zone, 11,300 to 12,100 ft, R. H. Russell, Aug. 1970) and Cochise Co. (Reef of Carr Canyon, Huachuca Mts, 7,200 ft, Pilsbry and Ferriss, 1910a:514, as *D. shimekii cockerelli*). Recent also in Yukon, Alberta, Oregon, Montana, Wyoming, north California, Utah, Colorado, South Dakota, and north New Mexico; in the southern mountains only at high elevations; not known from Mexico. Recorded as a Pleistocene fossil from Illinois, Iowa, Missouri, and Nebraska, but not thus far from Arizona. Synonyms: *Pyramidula cockerelli* H. A. Pilsbry, 1898:85 (T.L.: Saguache Co., Colorado, selected by Pilsbry, 1948:619). *Zonitoides randolphi* H. A. Pilsbry, 1898:87 (T.L.: Lake Lindeman, Yukon [not Alaska]; synonymized with *D. shimekii cockerelli* by Pilsbry, 1948:618).

93. **Helicodiscus (Helicodiscus) eigenmanni** H. A. Pilsbry, 1890. T.L.: Beaver Cave near San Marcos, Comal Co., Texas. In Arizona Recent in SE and S central sections, mainly in Upper Sonoran Life-Zone, at 2,400 to 8,600 ft; known from Pima Co. (Santa Catalina Mts: Sabino Canyon, 4,500 ft; Marble Peak, 7,000 ft; etc. Tucson Mts, 0.25 mi NW of Picture Rocks, 2,400 ft. Rincon Mts, Spud Rock,

8,600 ft). Santa Cruz Co. (Santa Rita Mts: Gardner Canyon, 6,000
ft; etc), Cochise Co. (Huachuca Mts: Miller Canyon, 6,000 to 6,500
ft; etc. Dragoon Mts, 5,000 to 5,600 ft. Chiricahua Mts: Winn Falls
on NW tributary of Cave Creek, 7,200 ft; Bonita Canyon, 5,600
ft; etc), Graham Co. (Pinaleno Mts, Wet Canyon on SE slope of
Mt Graham, 6,300 ft), and Greenlee Co. (Cosper's Ranch on Blue
Riv, 5,060 ft); also in riparian drift in Pima Co. (Madrona Canyon,
SW foot of Rincon Mts), Graham Co. (Turkey Creek, Galiuro Mts,
3,400 ft), Greenlee Co. (Eagle Creek, 6 mi SW of Morenci, 3,550
ft), Gila Co. (Salt Riv at U.S. Hwy 60, 4 mi N of Seneca, 3,000
ft), and Yavapai Co. (Agua Fria Riv, NE of Bumblebee, 3,500 ft).
Not known fossil in the State. Recent also in South Dakota, Utah,
Colorado, New Mexico, and Texas; in Mexico in Sonora, Chihuahua,
and Puebla. Synonyms: *Helicodiscus eigenmanni arizonensis* H. A.
Pilsbry and J. H. Ferriss, 1906. T.L.: "Fort Bowie," defined by the
authors later (1910c:133) as "Quartzite Peak," a hill ca 0.5 mi
due S of Site of the fort, in NW foothills of Chiricahua Mts, Cochise Co.; no elevation given, but at ca 6,000 ft. *Helicodiscus lineatus* H. C. Yarrow, 1875:929 (record of Gila Riv, Arizona); not *Helix
lineata* T. Say, 1817, an early, invalid name for *Helicodiscus parallelus* (T. Say, 1821), widespread in eastern North America. However, *H. eigenmanni* is perhaps only the southwestern subspecies
of *H. parallelus*.

94. **Helicodiscus (Hebetodiscus) singleyanus** (H. A. Pilsbry,
1889, as *Zonites*). T.L.: New Braunfels, Comal Co., Texas. Widespread in Arizona; Recent at 2,400 to 8,700 ft, mainly in Lower
and Upper Sonoran Life-Zones; common also in drift; known from
all counties. A Pleistocene fossil in San Pedro Valley (Lehner Mammoth and Murray Springs Sites, associated with mammoth remains 10,000 to 11,000 yrs old). Recent in SE United States, N
to New Jersey and Indiana, W to South Dakota, Colorado, New
Mexico, and Arizona; only adventive in California; native also in
North Mexico (Sonora; Tamaulipas).

95. **Punctum californicum** H. A. Pilsbry, 1898. T.L.: Fish Camp,
Fresno Co., California. Probably widespread in Arizona, but overlooked due to small size (1.35 to 1.8 mm); living at 4,000 to 9,500 ft;
below this in drift, probably as washed-up fossils; known Recent
from Coconino Co. (NW slope of San Francisco Mtn, at Snow Bowl,
9,500 ft, 10 mi NW of Flagstaff), Yavapai Co. (Montezuma Castle
National Monument), Apache Co. (Black Riv, near Horseshoe
Bend), Pima Co. (Santa Catalina Mts, widespread, at 7,000 to 7,500
ft. Rincon Mts, Rincon Peak, 6,100 ft), Santa Cruz Co. (Santa Rita
Mts, widespread, 4,000 to 6,300 ft. Pajaritos Mts, 4,000 ft. E Canelo

Hills, 5,500 ft), Cochise Co. (Chiricahua Mts, widespread, 4,800 to 9,400 ft. Dragoon Mts. Huachuca Mts: widespread, 5,500 to 6,000 ft; Sylvania Springs in Scotia Canyon, 6,150 ft), and Graham Co. (Pinaleno Mts, S slope of Mt Graham, 6,300 to 9,000 ft). Fossil in Late Pleistocene of San Pedro Valley at Murray Springs Site, associated with mammoth remains 10,000 to 11,000 yrs old. Recent also in California, Colorado, Montana, and South Dakota (Pilsbry, 1948:648). Perhaps only a southwestern subspecies of northwestern *Punctum randolphii* (W. H. Dall, 1895).

96. **Punctum conspectum** (T. Bland, 1865, as *Helix*). T.L.: San Francisco, California. Recorded Recent in Arizona by Pilsbry and Ferriss (1919a:326) and Pilsbry (1948:651) at only two Stations in White Mts, Apache Co., at over 6,800 ft: 2 mi below summit of "Baldy Peak" [Mt Thomas of present map], at head of Black Riv; Reservation Creek, tributary of Black Riv, 6,800 ft; not seen by us. Known also from Alaska, British Columbia, Idaho, Montana, Washington, Oregon, California, and W New Mexico (Mogollon Mts); recorded from Kamchatka; a subspecies in Mexico, but not known in Sonora. Not known fossil.

A dead shell, probably a washed-up Pleistocene fossil, of *Punctum minutissimum* (I. Lea, 1841) was found in drift of Little Colorado Riv, 2 mi E of Winslow, Navajo Co. Recent distribution imperfectly known, discussed in section on Zoogeography.

97. **Radiodiscus millecostatus** H. A. Pilsbry and J. H. Ferriss, 1906:154, 3 figs. 10 of immature shell; adult topotype shown by H. A. Pilsbry, 1926c:132, 2 figs. 4A. T.L.: Huachuca Mts, Carr Canyon, Cochise Co., selected by Pilsbry (1948:656), without elevation, but seen by us from Carr Canyon at ca 6,000 ft. In Arizona above 5,000 and up to 11,000 ft; known from Coconino Co. (NW slope of San Francisco Mtn, at Snow Bowl, 9,500 ft, 10 mi NW of Flagstaff), Apache Co. (Reservation Creek, 6,800 ft), Greenlee Co. (Mogollon Rim of Blue Mts, ca 11,000 ft. Cosper's pasture, Blue Mts, ca 10,000 ft. Little Blue Riv), Pima Co. (Santa Catalina Mts: Mt Lemmon, 7,000 to 9,500 ft, at Congden Camp, Aspen Gulch, Cold Spring ca 1 mi ENE of summit, and Westfall's Mine; also Bear Wallow, 7,800 to 8,000 ft, and Kellogg Peak, 8,400 ft. Rincon Mts, NE slope of Rincon Peak, 8,200 ft), and Cochise Co. (Huachuca Mts: Carr Canyon, 7,200 ft; Miller Peak at ca 9,400 ft; Wickersham Gulch and Rock, ca 9,200 ft; Miller Canyon, 7,800 ft. Dragoon Mts: Tweed Canyon [= Cochise Stronghold Canyon West of present map]. Chiricahua Mts: head of Cave Creek, 8,000 ft; Barfoot [or Bearfoot] Park, 8,000 ft; Pine Canyon, 7,500 ft; Long Park, 8,000 ft; Box of Rucker Canyon, 6,800 ft; Rustler Park, 8,400 ft). Also

in SW New Mexico (Mogollon Mts, Socorro Co.), Mexican States of Chihuahua, Michoacan, and Tamaulipas, and Costa Rica. Not known as a fossil.

Family Philomycidae

98a. Pallifera (Pancalyptus) a. arizonensis (H. A. Pilsbry, 1917, as *Philomycus*). T.L.: "N of Mt Baldy" [= Mt Wrightson of U.S. Geol. Surv. map], E slope of Santa Rita Mts, at 6,800 ft, Santa Cruz Co. [this Station was in Camperel Canyon, the upper extension of Gardner Canyon]. Also elsewhere in same Range, down to 6,400 ft. Known from Sonora (Sierra Purica, ca 6,300 ft, W. B. Miller, 1967a: 116). Synonym: *Pallifera (Pancalyptus) pilsbryi santaritana* C. D. Miles and A. R. Mead, 1960:78, Pl. 5, fig. 1 (T.L.: Santa Rita Mts, Madera Canyon, 6,400 ft, Santa Cruz Co.).

98b. Pallifera (P.) arizonensis pilsbryi C. D. Miles and A. R. Mead, 1960:75, Pl. 5, figs. 2-3, as *Pallifera pilsbryi*. T.L.: Santa Catalina Mts, Bear Wallow, 7,600 ft, Pima Co. Also elsewhere in same Range, at 7,800 ft.

Family Succineidae

Most genera and species of Succineidae can be distinguished only by the genitalia. Dead shells, without dissections from the same population, cannot be definitely named. In Arizona dead succineids are common, either in natural habitats or washed up in drift, but live snails are difficult to obtain. Since most specimens seen from the State were collected dead, generic and specific names used here are only tentative, unless stated to be based on dissections. No attempt was made, therefore, to complete local or general specific distributions.

On the shells only, four native and one adventive Recent species can be recognized in the State, even if they cannot always be named correctly. They represent two genera based wholly on differences in genitalia. In *Succinea* the penis lacks a diverticulum (appendix) and is mostly enveloped in a sheath, which covers also all or part of a well-differentiated epiphallus; the vagina varies in length with the species. In *Catinella* W. H. Pease (1871; *Quickella* C. R. Boettger, 1939, is a synonym) the penis is provided with a diverticulum (appendix), peculiar in size, shape, and position for each species, but a sheath and distinct epiphallus are lacking; the vagina is always very short. All North American *Catinella* are of subgenus *Mediappendix* H. A. Pilsbry (1948), with a large, sac-like diverticulum, arising from the penis at about mid-length or nearer the base. Succineids are common Pleistocene fossils in the State, mostly with shells similar to either *Succinea* or *Catinella*, but at present not referable to living species; some differ markedly from those of the Recent Arizona species. *Oxyloma*, a genus with a distinctive shell as well as anatomy, is widespread in the Recent Nearctic fauna, but known in Arizona only from Late Pleistocene fossils of San Pedro Valley at Lehner Mammoth and Murray Springs Sites; living specimens from New Mexico (Tularosa Riv valley, S side of Mescalero, Otero Co., A. L. Metcalf, in litt., 1970) were dissected by D. S. Franzen.

* **Succinea (Calcisuccinea) campestris** T. Say, 1817. Originally described from Sea Islands and Cumberland I, Georgia, and Amelia I, NE Florida; T.L. not selected thus far. Not indigenous in Arizona. Introduced in Pima Co. in the Tucson area: first record for the State at College of Agriculture Farm of Univ. of Arizona on N Campbell Ave, just N of Tucson city limits (G. D. Butler, Aug. 18, 1964); the anatomy of snails from Tucson proper, dissected by junior author, was that of *campestris;* a dead, fresh, adult shell seen from San Xavier Mission area near South Tucson (May 1968, Mr. Greenleaf). Native in SE United States (North Carolina, South Carolina, Georgia, Florida, and Alabama); introduced also in Oregon (G. D. Hanna, 1966:27).

? 99. **Succinea (C.) luteola** A. A. Gould, 1848. T.L.: Texas, without more precise locality; W. G. Binney's (1885:497) listing of 10 shells (at U.S. Nat. Mus. No. 39757) from Corpus Christi as "type, original lot," seems to preempt Pilsbry and Ferriss' (1906:158) selection of Galveston as T.L.; R. I. Johnson (1961:106) mentions syntypes at Mus. Comp. Zool., but does not select a holotype or a T.L. We have not recognized it thus far from live snails in Arizona; it is included here on the strength of Pilsbry's (1948:830, fig. 450g) record and figure of a shell from Laguna Canyon near Betatakin Ruin, Coconino Co. Other published records from the State may have been based on misidentifications: near Willcox, on "dry alkali flats," Cochise Co., R. E. C. Stearns (1891:100); and drift of Santa Cruz Riv at Tucson, Pima Co., W. H. Dall (1897a:365). Specimens seen by us from Stations at or near Willcox Playa and from drift at Tucson were not *luteola* and are tentatively called *S. grosvenorii*. True *S. luteola* is widespread in Mexico and in some southern States (Texas, Louisiana, and New Mexico).

Succinea (Sect. *Succinea*) *lineata* variety β *sonorensis* P. Fischer and H. Crosse, 1878:662, Pl. 27, figs. 8 and 8a-b, with T.L. at Yaqui River, Sonora, may be a recognizable subsp. of *S. luteola*. We refer to it (as *S. luteola sonorensis*) dead shells from Sonora: 33 mi W of Santa Ana; 1 mi N of Santa Ana; and 2 mi N of Granados, on Rio Bavispe. Possibly this was the form recorded as *S. luteola* from Sonora by B. A. Branson, et al (1964:104), in drift of Rio Bavispe, 21 mi S of Agua Prieta; and from Chihuahua by W. H. Dall (1897:365), at Lake Palomas in Mimbres Valley. No Sonora or Chihuahua specimens have ever been dissected.

100. **Succinea (Novisuccinea) avara** T. Say, 1824. T.L.: "Northwest Territory" [Minnesota], without more precise T.L.; none selected thus far. Probably common in Arizona, but there are few trustworthy records based on dissections. Pilsbry (1948:818 and 840, figs. 442 B-b and C-c) figured genitalia of true *avara* from Coconino Co. (Warm Spring Canyon, Kaibab Plateau) and Mohave Co. (Hurricane Fault near Utah border). Junior author dissected lots from Pima Co. (Tucson, at 4710 E 22nd St, and at 1207 E Adelaide Drive), Santa Cruz Co. (Sycamore Canyon, Atascosa Mts, 4,000 ft), and Cochise Co. (Parker Canyon Lake, 8 mi SE of Canelo, 5,500 ft), their genitalia agreeing with H. A. Pilsbry's (1948:818) figures of *avara*. Dead shells, mostly from drift, similar to dissected *avara*, seen or recorded from the following counties: Mohave, Coconino, Yavapai, Apache, Graham, Cochise, Pima, Santa Cruz, Yuma, and Gila; at 3,500 to 7,200 ft. Shells like Recent *avara* common in Pleistocene of San Pedro Valley (Lehner Mammoth and Murray Springs Sites), but their true identity uncertain.

101. **Succinea (N.) grosvenorii** I. Lea, 1864. T.L.: described from Santa Rita Valley, Kansas, and Alexandria, Louisiana; no precise T.L. selected thus far. Probably widespread Recent in Arizona. Pilsbry (1948:820-821, figs. 441) refers to *grosvenorii* shells from Mohave Co. (Mt Trumbull, 3 figs. 441e; Antelope Valley, 3 figs. 441f), Coconino Co. (Snake Gulch at Big Spring, Kaibab Plateau, 6,750 ft; Walnut Creek near Winona), and Navajo Co. (Willow Spring, 80 mi NE of Flagstaff). We have seen similar shells from Coconino Co. (Trail Canyon, between Jacob Lake and Houserock Ranch, ca 7,000 ft, F. G. Werner, June 1965; Kaibab National Forest, ca 11 mi W of Jacob Lake, on U.S. Hwy 89, R. H. Russell, Aug. 1968), Navajo Co. (N side of Winslow, 4,800 ft, R. H. Russell, 1968), Cochise Co. (Cochise at RR overpass of Hwy 666, 1 mi W of Willcox Playa, senior author, 1963-1969; NE side of Willcox Playa, 2 mi SE of Willcox, 4,150 ft, E. H. Erickson, 1968), and Pima Co. (Tucson, at Cherry Ave, between Broadway and 22nd St, C. J. May, 1963). No Arizona specimens dissected thus far. Not known fossil. Synonym: *Succinea lineata* W. G. Binney, 1857 (T.L.: "on Yellowstone Riv near Fort Union, Nebraska" [actually near junction of Yellowstone Riv and Missouri Riv, North Dakota]) (not *Succinea ovalis* variety A, *lineata* J. S. Dekay, 1844).

102. **Catinella (Mediappendix) vermeta** (T. Say, 1829, as *Succinea*). T.L.: New Harmony, Posey Co., Indiana. Possibly widespread in Arizona, but thus far positively in the State at only one Station in Santa Cruz Co.: Monkey Spring, at Rail X Ranch, 5.5 mi SW of Sonoita, 4,400 ft, senior author, 1967; these snails, dissected by junior author, agree in genitalia with Pilsbry's (1948:843-845, figs. 456A-C) description and figures of his *Quickella vagans.* Dead shells similar to the Monkey Spring lot seen from Pima Co., Santa Cruz Co., and Apache Co., but of uncertain identity for want of dissections. True *C. vermeta*, dissected by D. S. Franzen, were found by A. L. Metcalf (in litt. 1970) in New Mexico: San Simon Cienega, Hidalgo Co., ca one mi E of Arizona border. Synonym: *Succinea campestris vagans* H. A. Pilsbry, 1900b:74 (T.L.: Cape May Point, Cape May Co., New Jersey, shell only described); of Pilsbry's three dissections of 1948, that of fig. 456A-a, from a syntype at T.L., was selected by H. B. Baker (1961, *Nautilus* 74:123, January issue) as that of the holotype, thus definitely connecting the trivial name *vagans* with a species of *Catinella* subgenus *Mediappendix,* as Pilsbry intended; it is now generally held that the earlier *S. vermeta* T. Say is a *Catinella* and cospecific with *vagans.*

Family Pupillidae (including Chondrinidae and Vertiginidae)

103. **Gastrocopta (Vertigopsis) pentodon** (T. Say, 1822, as *Vertigo*). T.L.: "Pennsylvania," without precise locality; no T.L. seems to have been selected thus far. Pilsbry (1948:889) claims, for the synonym *tappaniana*, that it is "a rather common species in central Arizona"; we have found it there common only in drift, usually as a washed-up fossil, below 4,000 ft. All known, fully reliable Stations are at ca 4,500 to 6,500 ft elevation in Santa Cruz Co. (Fort Crittenden, Site at E foot of Santa Rita Mts, "under logs," E. H. Ashmun, 1899:15), Cochise Co. (Huachuca Mts: Tanner [now Garden] Canyon, 6,000 ft, Pilsbry and Ferriss, 1910a:515; Sylvania Springs, NW side of Range, 6,150 ft, R. H. Russell and senior author, 1968; Huachuca Canyon, 5,250 ft, senior author, 1967. Cienega of Babocomari Creek, at Babocomari [now Brophy] Ranch, 4,500 ft, senior author, 1969), Navajo Co. (drift of Little Colorado Riv, 3 mi E of Winslow, 4,810 ft, one shell, dead but fresh, R. H. Russell and senior author, 1968), and Yavapai Co. (Jerome, Pilsbry, 1948:889, as *G. tappaniana,* westernmost Station of species). Fossil in Late Pliocene to Pleistocene in Cochise Co. (San Pedro Valley: El Paso Pipeline Site, 7 mi S of Benson, R. S. Gray, 1965, D. W. Taylor, 1966:94, as *G. tappaniana;* Lehner Mammoth and Murray Springs Sites, associated with mammoth remains 10,000 to 11,000

yrs old). Recent in S Canada from Prince Edward I to British Columbia, and in most of E and central United States, W to Montana, E Colorado, and Arizona; also in Mexico (Puebla, Tamaulipas, San Luis Potosi, Nuevo Leon, Chihuahua) and Guatemala; not known from Sonora. Synonyms: *Pupa tappaniana* C. B. Adams, 1842, in Z. Thompson's *History of Vermont:*158 (described without locality; T.L. not in Vermont, as was assumed, but Roscoe, Coshocton Co., Ohio, according to holotype at Mus. Comp. Zool., Cambridge, Mass.); stated by W. J. Clench (1965, *Nautilus* 78:106) not to be separable subspecifically from *G. pentodon*, with which it was synonymized before by A. A. Gould (1844) and W. G. Binney (1885: 323). *Pupa curvidens* A. A. Gould, 1841, *Report Invertebrata Massachusetts:*189, fig. 120 (T.L.: Phillips Point at Lynn, Massachusetts).

104. **Gastrocopta (V.) pilsbryana** (V. Sterki, 1890, as *Pupa*). T.L.: "Colorado River," without more precise locality, but presumably in Arizona, since it is common in Grand Canyon of Colorado Riv. Widespread, living in Arizona highlands, mainly in Transition and Canadian Life-Zones at 3,000 to 10,000 ft; below 4,000 ft usually in riparian drift. Known Recent from Mohave Co. (Mt Trumbull; Hualapai Peak, at 8,000 ft), Coconino Co. (Bill Williams Mt; several Stations in Grand Canyon and on Kaibab Plateau; San Francisco Mtn at Lockett Meadow, W slope of Sugarloaf Mtn, 8,500 ft; Rio de Flag, 2 mi NW of Flagstaff, 7,000 ft; Mahan Mtn; Woody Mtn, 10 mi SW of Flagstaff, 8,000 ft; Walnut Creek near Winona, 13 mi E of Flagstaff, 6,450 ft; Padre Canyon, 25 mi E of Flagstaff; Mt Elden, 7,900 ft; drift of Oak Creek Canyon, 3,500 ft), Navajo Co. (drift of Carrizo Creek, on Hwy 60, 1 mi SW of Carrizo, 5,200 ft; dead but fresh shells in drift of Little Colorado Riv, 3 mi E of Winslow, 4,850 ft), Apache Co. (Alpine Divide, 3 mi NW of Alpine, 8,100 ft; drift of creek 12 mi N of Fort Defiance, 7,000 ft), Maricopa Co. (drift in Sycamore Canyon near Sunflower), Gila Co. (Pinal Peak at 7,300 to 7,500 ft; Sierra Ancha, widespread at 5,000 to 6,500 ft; litter on Pine Creek at Tonto Natural Bridge, 4,500 ft), Pinal Co. (Galiuro Mts at Whitlock Ranch; litter on S bank of Salt Riv, at Hwy 60, 4 mi N of Seneca, 3,000 ft), Pima Co. (Kitt Peak at 6,300 to 6,700 ft; Baboquivari Mts at 3,500 to 4,000 ft; Santa Catalina Mts, widespread at 6,000 to 8,500 ft, and in drift of lower Sabino Canyon at 2,800 to 3,000 ft; Rincon Mts at Spud Rock, 8,600 ft, and Manning Camp, 8,000 ft; drift of San Pedro Riv at Redington), Santa Cruz Co. (Santa Rita Mts, 5,200 to 6,200 ft; foot of Sanford Butte on Sonoita Creek, 3,800 ft; Pajaritos Mts, 1.5 mi S of Pena Blanca Lake, 4,000 ft), Cochise Co. (Dos Cabezas Mts, ca 8,000 ft; Whetstone Mts at 6,000 to 6,800 ft;

Dragoon Mts at 4,700 ft; Huachuca Mts, widespread at 5,200 to 9,200 ft; Chiricahua Mts, widespread at 5,600 to 7,600 ft; drift of San Pedro Riv at The Narrows, Fairbank, Hereford, and Palominas), Graham Co. (Pinaleno Mts: Mt Graham at 5,000 to 9,000 ft), and Greenlee Co. (Cosper's Ranch on Blue Riv, 5,060 ft; Cosper's Pasture in Blue Mts, ca 10,000 ft; Fish Creek; drift of Eagle Creek, 6 mi SW of Morenci, 3,550 ft). Not found fossil thus far in Pleistocene of San Pedro Valley; only fossil record from Arizona at Winona Site, 13 mi E of Flagstaff, R. D. Reger and G. L. Batchelder, 1971:191-193) [perhaps based on a dead Recent shell?]. Recent also in New Mexico, S Utah (1 record), trans-Pecos Texas (Guadalupe Mts in South McKittrick Canyon, 5,300 ft, Culberson Co., senior author, 1966), and Mexico (Sonora, Chihuahua, Tamaulipas). Synonym: *Gastrocopta pilsbryana amissidens* H. A. Pilsbry, 1934 (T. L.: Mahan Mtn, Coconino Co.).

105. **Gastrocopta (Staurotrema) quadridens** H̄. A. Pilsbry, 1916. T.L.: Capitan Mts, Lincoln Co., ca 35° 35' N, 105° 20' W, New Mexico, easternmost Station of species. Sporadic in Canadian and Hudsonian Life-Zones of Arizona at 8,000 to 10,000 ft. Known from Coconino Co. (Bill Williams Mtn, 9,200 ft, 35° 10' N, 105° 20' W, northwesternmost Station of species), Yavapai Co. (Mingus Mtn), Pima Co. (Santa Catalina Mts, 9,000 to 9,500 ft), Cochise Co. (Huachuca Mts, Wickersham Rock near Miller Peak, 8,500 ft, ca 31° 23' N, 110° 17' W, southernmost Station of species. Chiricahua Mts, Barfoot [or Bearfoot] Park, 8,500 ft [not 10,000 ft as given], ca 31° 55' N, 109° 10' W), and Graham Co. (Pinaleno Mts: Hospital Flat, at S foot of Mt Graham, 9,000 ft, senior author, 1960). Restricted to Arizona, New Mexico, and Utah (Fish Lake, Sevier Co., ca 38° 35' N, 111° 45' W, northernmost Station of species). Not known fossil. Synonym: *Bifidaria quadridentata* V. Sterki, 1899 (same T.L. as *quadridens;* not *Pupa quadridentata* von Klein, 1833, now placed also in *Gastrocopta*).

106. **Gastrocopta (Immersidens) ashmuni** (V. Sterki, 1898, as *Bifidaria*). Originally described from Arizona (Santa Rita Mts) and New Mexico (Cooks Peak, Luna Co., and Dripping Springs in Organ Mts, Dona Ana Co.); the Station in Santa Rita Mts, herewith selected as T.L., was defined by the collector, E. H. Ashmun (1899:14), as on E slope of Range, 20 mi W of Crittenden [Site of a former Army fort, 3 mi NE of Patagonia], Santa Cruz Co. The nominate, dextral form widespread in Arizona, mainly in Upper Sonoran and Transition Life-Zones, usually at 3,000 to 8,000 ft. Known alive in well-sheltered organic litter from Mohave Co. (Mt Trumbull, 8,000 ft, 36° 25' N, 113° 20' W, northwesternmost Station

of species), Coconino Co. (Oak Creek Canyon, at Owenby's Ranch and Page's Ranch, ca 4,500 ft; Rio de Flag, 2 mi N of Flagstaff, 7,000 ft; Grand Canyon of Colorado River, Bright Angel Trail, 100 ft below South Rim; Powell-Kaibab Saddle, N of Grand Canyon, ca 20 mi N of El Tovar; Walnut Creek, 13 mi E of Flagstaff, 6,450 ft), Yavapai Co. (Mingus Mtn: Hull Canyon, 6,000 ft; Walnut Canyon; Mescal Canyon, 6,000 ft; Kirwagen Ranch), Gila Co. (S bank of Salt Riv, 4 mi N of Seneca, 3,000 ft; Mogollon Rim at W side of Milk Ranch Point, 2.5 mi NE of Pine, 6,600 ft; Sierra Ancha, upper Parker Creek, 5,080 ft), Pinal Co. (Galiuro Mts: Copper Creek at Site of P.O., 3,800 ft; N slope at Whitlock Ranch), Pima Co. (Ajo Mts, Arch Canyon, 3,500 ft, 32° 2′ N, 112° 42′ W, southwesternmost Station of species. Baboquivari Mts, W slope at mouth of Baboquivari Canyon, 3,500 ft. Sierrita Mts, S slope on road to Harris Ranch. Santa Catalina Mts: Rattlesnake Canyon, off Sabino Canyon, 3,100 ft; SW slope above Molino Canyon, 4,800 ft; Marble Peak, 7,670 ft [not 8,000 as given]. Rincon Mts, SW slope in Madrona Canyon, 3,500 to 4,000 ft. Santa Rita Mts, S end at Rosemont and Helvetia, 5,500 to 5,600 ft), Santa Cruz Co. (Ephraim Canyon in Nogales, 3,800 ft. NE foot of Tumacacori Peak, 3,720 ft. Atascosa Mts, Sycamore Canyon, 4 mi SE of Ruby, 4,000 ft. Pajaritos Mts, Pena Blanca Canyon. Santa Rita Mts: Temporal Gulch, 4,000 ft; Madera Canyon, 4,800 to 6,000 ft; Adobe Canyon, 5,200 ft; Gardner Canyon, near Onyx Cave, 5,400 ft. Sanford Butte at Sonoita Creek, 3,800 ft. Patagonia Mts: mouth of cave on Bald Mtn; Harshaw Creek at Hermosa Hill, 4,850 ft. Canelo Hills: 4 mi S of Canelo, 5,500 ft; limestone hill W of Duquesne Rd; NW slope of Lookout Knob, 5,400 ft), Cochise Co. (Whetstone Mts: French Joe Canyon, 5,400 ft; N of Two-Peak mining camp. Mustang Mts, N slope of Eastern Dome, 5,200 to 5,400 ft. Huachuca Mts: Garden Canyon, 6,000 ft; Ash Canyon; Carr Canyon, 6,600 ft; Miller Canyon, 6,000 to 6,500 ft; Ramsey Canyon 5,800 to 6,000 ft; Manila Mine; Jack Wakefield Mining Camp, 5,600 ft; Copper Canyon, 6,000 to 6,200 ft; W slope at Sylvania Springs in Scotia Canyon, 6,150 ft. Dragoon Mts: Fourr Canyon, 5,500 to 5,700 ft; 9 Stations in Tweed [now Stronghold West] Canyon, 4,700 to 5,700 ft; Stronghold Canyon East, 4,700 to 4,800 ft; Bear Gulch branch of Stronghold Canyon East; mouth of Wood Canyon, 2 to 3 mi E of Dragoon, 4,800 ft. Chiricahua Mts: Whitetail Canyon, below mouth of Indian Creek, 5,500 ft; N slope on Paradise-Portal road, 5,600 to 5,800 ft; Big Emigrant Canyon, 4,800 to 5,400 ft; Limestone Mtn; Pinery Canyon, 5,000 to 7,000 ft; Crystal Cave, 0.5 mi SW of Southwestern Research Station; Cave Creek at Herb Martyr Dam, 5,600 ft; Tex

Canyon, 5,800 ft; Rucker Canyon, 5,900 to 6,000 ft; etc. Guadalupe Canyon, 18 mi E of San Bernardino Ranch, 4,200 ft), and Greenlee Co. (Cosper's Ranch on Blue Riv, 5,060 ft; Eagle Creek at Ole Hagen's Ranch). Common in riparian drift in Yuma Co. (Bill Williams Riv at Browns Crossing, 1,500 ft, 30 mi N of Wenden, 34° 25′ N, 113° 25′ W), Coconino Co. (Oak Creek, 7 mi N of Sedona, 3,500 ft), Navajo Co. (Little Colorado Riv at Winslow, 4,800 ft, and Holbrook, 5,075 ft; Carrizo Creek, 1 mi SW of Carrizo, 5,200 ft), Apache Co. (Puerco Riv at Adamana; Chinle Wash), Yavapai Co. (Big Chino Wash, 8 mi S of Seligman; Beaver Creek at Montezuma Castle National Monument; Verde Riv at Clarkdale, 3,500 ft), Gila Co. (East Verde Riv, 5 mi NW of Payson, 4,500 ft; creek 6 mi SE of Roosevelt), Maricopa Co. (Salt Riv at Tempe), Pima Co. (Santa Cruz Riv at Tucson and Continental; Pantano Wash near Vail; Rillito Creek at Tucson city limits; Agua Verde Creek, 5 mi E of Vail, 3,300 ft; Posta Quemada Canyon near Colossal Cave, 3,300 to 3,500 ft; San Pedro Riv at Redington; Cienega Creek, 6 mi NE of Empire Ranch, 4,200 ft), Santa Cruz Co. (drift of Santa Cruz Riv at Amado, Calabasas, and 5.5 mi NE of Nogales; Pena Blanca Lake, 3,900 ft; Oro Blanco Wash, 2.5 mi W of Ruby, 4,000 ft; wash 2.5 mi W of Pena Blanca Lake, 4,000 ft; Sonoita Creek at Patagonia, 4,044 ft), Cochise Co. (Walnut Gulch, 1 mi E of Tombstone, 4,500 ft. San Pedro Riv at Palominas, Hereford, Benson, and The Narrows 10 mi N of Benson. Dragoon Mts: creeks of W slope in Fourr Canyon, 5,500 to 5,600 ft. Stronghold Canyon East, 4,700 to 4,800 ft), Graham Co. (Turkey Creek, E slope of Galiuro Mts, 10 mi NW of Klondyke), and Greenlee Co. (Eagle Creek, 6 mi E of Morenci, 3,550 ft).

Dextral form also in New Mexico (in leaf litter: Grants and San Rafael, Valencia Co., ca 35° 10′ N, 108° W, northeasternmost Stations of species; Oscura Mts, Socorro Co.; White Oaks, Lincoln Co.; Iron Creek in Black Mts, Grant Co.; Organ Mts, Dona Ana Co.; Little Palomas Creek, Sierra Co.; Florida Mts and Cooks Peak, Luna Co.; Big Hatchet Mts, Hidalgo Co.), trans-Pecos Texas (in soil rich in leaf debris at Hueco Tanks, ca 20 mi E of El Paso, El Paso Co., A. L. Metcalf and W. E. Johnson, 1971:89; South McKittrick Canyon of Guadalupe Mts, 5,500 ft, 31° 50′ N, 104° 26′ W, Culberson Co., A. R. Mead et al, 1969, easternmost Station of species; also in riparian drift of Rio Grande at Vinton, El Paso Co., and 11 mi SE of Redford, Presidio Co., and of Fresno Creek, 25 mi SE of Redford, Presidio Co., all 3 by senior author, 1961), Sonora (in leaf litter: Nogales, Pilsbry and Ferriss, 1906:144. Five Stations by junior author, 1964-1965: one in Sierra del Santo Niño,

N of Mina El Milagro, on road to Sahuaripa, at 4,000 ft, ca 29°
N, 109° 30′ W; two near Magdalena, at 3,650 and 4,250 ft, ca 30°
55′ N, 111° W; and two near Nacozari at 3,550 and 4,000 ft, ca
30° N, 110° W. Rancho Pinos Altos, Sierra Nacori, 5,800 ft, 29° 45′
N, 108° 30′ W, M. D. Robinson, 1969), and Chihuahua (in leaf
litter, Rio Piedras Verdes, ca 9 km above Colonia Juárez, ca 30°
15′ N, 108° W, Pilsbry, 1953:162). Not definitely known fossil thus
far. H. G. Richards (1936:371) recorded a dead specimen from the
Lindenmeier Site of Early Man, N of Fort Collins, Larimer Co.,
Colorado, some 300 mi N of northernmost know Recent Stations
in Arizona and New Mexico; this may require verification, since
there are no Recent or fossil records from Colorado or Utah.
Synonyms: *Bifidaria ashmuni* form *minor* V. Sterki, 1898 (T.L.:
Ephraim Canyon in Nogales, Santa Cruz Co.). *Gastrocopta per-
versa* form *sana* H. A. Pilsbry, 1918 (T.L.: drift of Salt Riv at
Tempe, Maricopa Co.), *Gastrocopta ashmuni imperfecta* H. A.
Pilsbry and J. H. Ferriss, 1923 (T.L.: Mustang Mts, at Site of former
Dan Mathew's Ranch, ca 1.5 mi SE of Mustang Peak, 5,000 ft,
Santa Cruz Co.); not figured, but probably based on juvenile snails
with incomplete columellar lamella.

Many lots of nominate, dextral *ashmuni* contain also sinistral
shells, described as a distinct species, *Gastrocopta perversa* (V.
Sterki, 1898, as *Bifidaria;* T.L.: Ephraim Canyon in Nogales, Santa
Cruz Co., 3,800 ft; found with dextral snails). Otherwise, *perversa*
agrees with dextral *ashmuni,* sharing the diagnostic straight vertical
extension of the columellar lamella alongside the parietal lamella
(shown by Pilsbry and Ferriss, 1910c:137, fig. 29); it can therefore
not rank as a valid subspecies, but is merely a recurrent mutation
in many populations of the dextral form. In Arizona, it seems restricted to the southern half of the State, where it is usually less
abundant than the dextral form, which may explain the more
sporadic records. Known alive in leaf litter from Gila Co. (S bank
of Salt Riv, 4 mi N of Seneca, 3,000 ft), Pinal Co. (Aravaipa Creek
Canyon, 15 mi NE of Mammoth, 2,500 ft), Pima Co. (Baboquivari
Mts: W slope at mouth of Baboquivari Canyon, 3,500 to 4,000 ft;
SE slope, in Sycamore Canyon, 3,700 to 4,000 ft, and Brown Canyon,
4,000 ft. Sierrita Mts, at S side on road to Harris Ranch. Cerro
Colorado, 3 mi NW of Santa Lucia Ranch, 4,100 ft. Santa Rita
Mts: NE slope in Florida Canyon, 4,400 ft; E slope in Temporal
Gulch, 4,000 ft, and Gardner Canyon, near Onyx Cave, 5,400 ft;
W slope in mouth of Agua Caliente Canyon, 3,800 ft. Agua Verde
Creek, 5 mi E of Vail, 3,300 ft. Empire Mts, 1.5 mi N of Total
Wreck Mine, 4,800 ft. Coyote Mts, E slope in upper Mendoza

Canyon, 3,500 ft), Santa Cruz Co. (Nogales, 3,865 ft. Atascosa Mts, Sycamore Canyon, 4 mi E of Ruby, 4,000 ft. Pajaritos Mts, Pena Blanca Canyon, 1.5 mi S of Pena Blanca Lake, 4,000 ft. Foot of Sanford Butte on Sonoita Creek, 3,800 ft. Patagonia Mts: SW slope 10 mi NE of Nogales; mouth of cave on Bald Mtn), and Cochise Co. (Huachuca Mts: Ash Canyon; Carr Canyon, 6,600 ft; Manila Mine; Sylvania Springs in Scotia Canyon, 6,150 ft. Mustang Mts, N slope of Eastern Dome, 5,200 to 5,400 ft. Dragoon Mts: Tweed Canyon [now Stronghold Canyon West]; Stronghold Canyon East, 4,700 ft; N slope in mouth of Wood Canyon, 4,800 ft. Chiricahua Mts: Big Emigrant Canyon, 5,400 ft; Pinery Canyon, 5,000 ft; Rattlesnake Hill, 0.25 mi SW of Portal, 5,800 ft; Crystal Cave, 0.5 mi SW of Southwestern Research Station, 5,500 to 5,600 ft; East Turkey Creek, below Paradise; Whitetail Canyon, 5,500 to 5,600 ft; Tex Canyon, 5,800 ft). Also in riparian drift in Pinal Co. (Gila Riv, 1 mi N of Florence), Pima Co. (Cienega Creek, 6 mi NE of Empire Ranch; Posta Quemada Canyon, near Colossal Cave, 3,300 ft; Rillito Creek at Tucson city limits; Pantano Wash, 4.5 mi E of Vail; Santa Cruz Riv at Tucson and Continental; San Pedro Riv at Redington), Santa Cruz Co. (Santa Cruz Riv at Amado, Calabasas and 4.5 mi E of Nogales; Sonoita Creek at Patagonia; wash 2.5 mi W of Pena Blanca Lake), Cochise Co. (San Pedro Riv at Fairbank, Benson, and The Narrows 10 mi N of Benson), and Graham Co. (Turkey Creek of Galiuro Mts, 10 mi NW of Klondyke, 3,400 ft). Form *perversa* occurs also in New Mexico, Sonora (Rio Nacozari, 7 mi S of Nacozari, in drift, B. A. Bransom, 1964:104; two Stations near Nacozari, at 3,550 and 4,200 ft, in leaf litter, junior author, 1965), and Chihuahua (S side of Rio Piedras Verdes, 9 km above Colonia Juarez, Pilsbry, 1953:162; Presa Chihuahua, 7 mi S of city of Chihuahua, 4,800 ft, 28° 35′ N, 106° 5′ W, southernmost Station of species, R. H. Russell, 1971).

107. **Gastrocopta (I.) cochisensis** (H. A. Pilsbry and J. H. Ferriss, 1910, as *Bifidaria*). T.L.: Tanner [now Garden] Canyon, Huachuca Mts, 6,000 ft, Cochise Co. Sporadic in SE Arizona at 4,000 to 7,200 ft, mostly in canyons of Upper Sonoran Life-Zone. Known from leaf litter in Gila Co. (upper Cherry Creek, 1 mi E of Young, 5,100 ft), Pima Co. (Baboquivari Mts, W slope in Baboquivari Canyon, 3,500 to 4,000 ft. Quinlan Mts, Kitt Peak, 6,300 to 6,700 ft. Santa Catalina Mts: Rose Canyon, 7,000 ft; Alder Springs; SW ridge from Marble Peak, 7,200 ft [not 8,000 as given]. Santa Rita Mts, NE slope, 1.5 mi S of Helvetia, 4,500 ft. Rincon Mts, N slope of Rincon Peak, 6,100 ft), Santa Cruz Co. (Santa Rita Mts: Gardner Canyon on E slope, near Onyx Cave, 6,000 to 6,200

ft; Madera Canyon on N slope, 5,200 to 6,200 ft; Temporal Gulch on E slope, 4,000 ft. Patagonia Mts: ravine on SE slope, 1 mi W of pass to Washington Camp, 5,400 ft; E slope near Mowry, 9 mi SE of Patagonia, 5,200 ft), Cochise Co. (Huachuca Mts: Garden Canyon, 6,000 ft; Ramsey Canyon, 5,800 ft; Brown Canyon; Miller Canyon, 5,800 to 6,500 ft; Scotia Canyon on W slope, at Sylvania Springs, 5,200 ft; Huachuca Canyon, 6,000 ft. Chiricahua Mts: Morse Canyon at head of West Turkey Canyon, 7,200 ft; South Fork of Cave Creek, 4 mi SW of Southwestern Research Station, 5,100 to 5,800 ft; Crystal Cave off Cave Creek, 0.5 mi SW of Southwestern Research Station, 5,500 to 5,600 ft; Rucker Canyon, 5,900 to 6,000 ft; Limestone Mtn near mouth of Rucker Canyon; lower Pinery Canyon, 5,000 ft; Whitetail Canyon, 5,500 ft; Big Emigrant Canyon, at N foot of Rough Mtn, 5,400 ft), and Graham Co. (Pinaleno Mts: Wet Canyon on SE slope of Mt Graham, 6,300 ft). Dead in riparian drift of Pima Co. (Gardner Canyon, 5 mi N of Sonoita, 4,700 ft), Santa Cruz Co. (Santa Cruz Riv at Amado), Cochise Co. (Dragoon Mts in Stronghold Canyon West, 5,000 ft. San Pedro Riv at Palominas and Hereford), and Navajo Co. (Little Colorado Riv at Winslow, 4,810 ft; Carrizo Creek, 1 mi S of Carrizo, 5,200 ft). Not recorded from New Mexico or trans-Pecos Texas. Known from Sonora (arroyo 8 km S of Guaymas, Pilsbry, 1953:162), Chihuahua (Rio Piedras Verdes, below Pacheco, 5,900 ft; Sierra de la Breña, ca 17.5 km from Pearson [Mata Ortiz], on road to Pacheco, 7,000 ft; both Stations Pilsbry, 1953:162), and Sinaloa (Rio Fuerte at San Blas, Pilsbry, 1953:162). Not known fossil.

108. **Gastrocopta (I.) prototypus** (H. A. Pilsbry, 1899, *Proc. Acad. Nat. Sci. Philadelphia* 51:400, as *Bifidaria*; 1904, ibid. 55 [for 1903]:766, figs. 7-7a. [copied by Pilsbry and Ferriss:1910:142, 2 figs. 35]; 1916, in 1916-1918:47, Pl. 7, figs. 1-5). T.L.: Huingo, near Lago de Cuitzco, 40 km NW of Morelia, ca 19° 50′ N, 100° 50′ W, ca 4,500 ft, Michoacan, Mexico. Recent in Arizona at a few Stations in Upper Sonoran Life-Zone of SE mountain canyons at 4,700 to 5,300 ft; in drift at lower elevations. Alive in leaf litter in Santa Cruz Co. (Patagonia Mts, S slope at 0.5 mi E of Mowry, 9 mi SE of Patagonia, 5,200 ft), and Cochise Co. (Huachuca Mts: SE slope in Ash Canyon, 5,000 ft; W slope in Scotia Canyon, at Sylvania Springs, 5,200 ft. Chiricahua Mts, W slope in lower Rucker Canyon, 4,700 to 5,300 ft, Pilsbry, 1916, in 1916-1918:46). Dead shells, mostly washed-up fossils in riparian drift in Gila Co. (Salt Riv, 4 mi N of Seneca, 3,000 ft; tributary of East Verde Riv, 5 mi NW of Payson, 4,500 ft, ca 34° 10′ N, 111° 35′ W, northernmost Station of species), Pima Co. (Santa Cruz Riv at South Tucson, 2,450 ft),

Santa Cruz Co. (Santa Cruz Riv at Amado, 3,050 ft, at 5.5 mi NE of Nogales, 3,600 ft, and at 2.5 mi E of Lochiel, 4,600 ft), and Cochise Co. (San Pedro Riv at Hereford, 4,100 ft, Fairbank, 3,840 ft, and Benson, 3,570 ft. Dragoon Mts, wash in Stronghold Canyon West, 4,700 to 5,300 ft). Fossil in Late Pleistocene of San Pedro Valley at Lehner Mammoth and Murray Springs Sites, Cochise Co., associated with mammoth remains 10,000 to 11,000 yrs old. Known in New Mexico from riparian drift only, in Luna Co. (Mimbres Riv at Deming) and Dona Ana Co. (Rio Grande at Mesilla). Also in Mexico from Jalisco (Guadalajara, 5,220 ft, 20° 20′ N, 103° 20′ W) and Michoacan (T.L.), and in Guatemala (Guatemala City, 4,850 ft, 14° 37′ N, 90° 30′ W); not thus far in Sonora or Chihuahua. Synonyms: *Bifidaria* (Sect. *Immersidens*) *cochisensis oligobasodon* H. A. Pilsbry and J. H. Ferriss, 1910:141, figs. 34A-C [copied poorly by Pilsbry, 1948:901, figs. 487A-C, the basal folds, small but distinct in originals, being unclear] (T.L.: Ash Canyon, SE slope of Huachuca Mts, Cochise Co.; elevation not given, but ca 5,000 ft); *oligobasodon* raised to specific rank by Pilsbry, 1916, in 1916-1918: 45, figs. 16A-C [copies of 1919 figs.] and Pl. 7, figs. 8-11 (Ash Canyon cotypes); Pilsbry, 1948:900, figs. 480 (6, 8, and 9) [copies of figs. 8-11 of 1916] and figs. 487A-C [poor copies of 1910 figs.]. *Gastrocopta* (*Immersidens*) *prototypus basidentata* H. A. Pilsbry, 1916, in 1916-1918:48, Pl. 7, figs. 6-7 (T.L.: Guatemala City, southernmost Station of species). The complex of *B. prototypus* (1899), *B. cochisensis oligobasodon* (1910), and *G. prototypus basidentata* (1916) is here treated as a single species without recognized subspecies. All three forms agree in an essential character of the aperture; the parietal and angular lamellae, although fully separated over much of the length, are reduced in size and more simplified than in other members of subg. *Immersidens*—a feature separating it in particular from *cochisensis* when full-grown shells are compared. Other supposed differences between the three forms are unreliable, such as the presence or absence of the basal apertural lamella (usually but not always present, though small in *oligobasodon;* said to be always absent in *prototypus*, but well-developed in *basidentata*), or the shape of the columellar lamella, are too variable in shells seen from Arizona. Present discontinuous general distribution discussed in section on Zoogeography.

109a.   **Gastrocopta (I.) d. dalliana** (V. Sterki, 1898, as *Bifidaria*). T.L.: Nogales, Santa Cruz Co.; more precisely Ephraim Canyon in Nogales, ca 3,800 ft, by V. Sterki, 1899:14. Nominate subsp. widespread in Lower and Upper Sonoran Life-Zones of S half of Arizona at 2,800 to 6,500 ft, rarely more (7,600 ft in Chiricahua Mts, 9,300 ft on San Francisco Mtn; etc); near sea level in Sonora and Baja California, perhaps as human introductions. Alive in leaf litter in Coconino Co. (San Francisco Mtn, SE slope near work camp on Watershed Rd, 9,300 ft, 10 mi N of Flagstaff, ca 35° 30′ N, 110° 40′ W, northernmost Station of species, R. H. Russell and senior author, 1969), Yavapai Co. (Mingus Mtn at Kirwagen's Ranch, 5 mi S of Jerome, Pilsbry and Ferriss, 1910:143, as *dalliana*; H. A. Pilsbry, 1916, in 1916-1918:50, as *dalliana media*), Pima Co. (Baboquivari Mts, W slope in Baboquivari Canyon, 3,500 to 5,000 ft. Quinlan Mts, Kitt Peak, 6,500 ft, northwesternmost Station of subspecies. Santa Rita Mts: N slope in Florida Canyon, 4,400 ft. Las Guijas Mts, N slope, 3,500 to 4,000 ft. Santa Catalina Mts: widespread, Sabino Canyon, etc, 2,800 to 6,400 ft. Rincon Mts, N slope of Rincon Peak, 6,100 ft. Empire Mts, 2 mi NW of Total Wreck Mine, 4,800 ft. Agua Verde Canyon, 4 mi NE of Vail, 3,300 ft), Santa Cruz Co. (Nogales, 3,800 ft. Pajaritos Mts: Pena Blanca Canyon, 4,000 ft, etc. Atascosa Mts, Sycamore Canyon, 4 mi E of Ruby, 4,000 ft. NE foot of Tumacacori Peak, 3,700 ft. Santa Rita Mts: Madera Canyon, 4,800 to 5,800 ft; etc. Patagonia Mts: Harshaw Creek at Hermosa Hill, 9 mi SE of Patagonia, 4,650 ft; etc. Canelo Hills: Lookout Knob, 5,400 ft; etc), Cochise Co. (Huachuca Mts: SW slope in Copper Canyon, 6,000 to 6,200 ft; etc. Mustang Mts, N slope of Eastern Dome, 5,200 to 5,400 ft. Whetstone Mts: SE slope of French Joe Peak, 6,000 to 6,800 ft; etc. Dragoon Mts: Stronghold Canyon East, 4,700 to 5,300 ft, some with divided basal lamella; etc. Chiricahua Mts: widespread, Cave Creek Canyon, etc, 4,500 to 6,800 ft; Onion Saddle at 9,300 ft; NW end of Range on Quartzite Peak, 1 mi S of Site of old Fort Bowie, 6,000 ft. Guadalupe Mts: canyon 30 mi NE of Douglas, 4,200 ft, 31° 28′ N, 109° 5′ W, northeasternmost Station of species), Graham Co. (Pinaleno Mts: SE slope of Mt Graham at Noon Creek, 5,100 ft; Wet Creek, 6,300 ft), and Greenlee Co. (foot of Copper King Mtn near Harper's Place, Blue Mts. Eagle Creek at Ole Hagen's Ranch). Common in riparian drift in Yavapai Co. (Verde Riv at Clarkdale, 3,300 ft; etc), Maricopa Co. (Salt Riv at Tempe, 1,150 ft), Pinal Co. (Queen Creek, 10 mi E of Superior, 3,700 ft; etc), Gila Co. (tributary of East Verde Riv, 5 mi NW of Payson, 4,500 ft), Pima Co. (Alambre Wash, N foot of Saucito Mtn, 3.5 mi SE of Kitt Peak, 3,950 ft; Madrona Canyon, SW slope of Rincon Mts; Gardner Canyon at Hwy 83, 6 mi N of Sonoita; San Pedro Riv at Redington, 2,890 ft; Santa Cruz Riv at South Tucson and Continental; Posta Quemada Canyon, near Colossal Cave, 3,500 ft; Rillito Creek at Tucson city limits; Pantano Wash; etc), Santa Cruz Co. (Sonoita Creek at Patagonia, 4,000 ft, and at foot of Mt Shibell, 4,000 ft; Santa Cruz Riv at Amado, Calabasas, and 5.5 mi NE of Nogales, 3,600 ft; etc), Cochise Co. (Walnut Gulch, 1 mi E of Tombstone, 4,500 ft. San Pedro Riv at Hereford, 4,100 ft, Benson, 3,570 ft, and The Narrows 10 mi N of Benson, 3,340 ft. French Joe Canyon, Whetstone Mts, 4,700 ft. Chiricahua Mts, in lower Pinery Canyon, 5,000 ft, etc), Graham Co. (Turkey Creek on NE slope of Galiuro Mts, 10 mi NW of Klondyke, 3,400 ft), and Greenlee Co. (Eagle Creek, 6 mi SW of Morenci, 3,550 ft). Not recorded from New Mexico or trans-Pecos Texas. Not known fossil.

Alive in litter in Sonora (spring-fed marsh on Mex. Hwy 15, 60 mi S of Nogales, B. A. Branson, 1964:104, as *G. bibasidens*. Near sea level on San Carlos Bay, 10 mi N of Guaymas, under mesquite, 28° N, 110° 50' W, southwesternmost mainland Station of species, lot of 510 *dalliana* and 114 *G. pellucida,* R. H. Russell, 1970. Three Stations by junior author, 1964: S end of Sierra Purica, 6,300 ft, 30° 31' N, 109° 45' W; mountain 5 mi S of Magdalena, 4,250 ft, ca 30° 35' N, 111° W; and NW side of Rio Nacozari, 1 mi E of Nacozari, 4,200 ft, 30° 20' N, 109° 40' W), NW Chihuahua (cliff talus on N side of Rio Piedras Verdes, 8 to 9 km above Colonia Juárez, 5,900 ft, ca 30° 10' N, 108° 10' W, easternmost Station of species, Pilsbry, 1953:162, as *G. dalliana* and *G. bibasidens*), and Baja California Territorio Sur (San Ignacio, 500 ft, 27° 20' N, 112° 50' W; San José de Comondú, 1,500 ft, ca 26° 5' N, 111° 50' W; San Bartolo, 500 ft, 23° 50' N, 110° W, southwesternmost Station of species; all junior author, 1970; possibly human introductions from Sonora).

Synonyms: *Gastrocopta (Immersidens) dalliana media* H. A. Pilsbry, 1916, in 1916-1918:50, Pl. 8, figs. 10-11 [copied by Pilsbry, 1948:902, figs. 490 (10-11)]; (T.L.: Montezuma Well near Rimrock, Yavapai Co.); based on transitions of *G. d. dalliana* to *G. d. bilamellata. Gastrocopta bibasidens* H. A. Pilsbry, 1953:162, figs, 2-3 (T.L.: cliff talus on N side of Rio Piedras Verdes, 8 to 9 km above Colonia Juárez, 4,900 ft, ca 30° 10' N, 108° 10' W, Chihuahua); figs. and description show the essential features of subgenus *Immersidens:* angular and parietal lamellae separated in front (outside), partly united behind (inside), a combination like that of many *G. d. dalliana,* in which it varies in SE Arizona; lower palatal fold deeply immersed; basal fold not entering but transverse, sometimes divided into two tubercles (this twinning also in some Arizona *dalliana*); peristome not thickened within; *bibasidens* therefore not related to *G. procera* of subgenus *Gastrocopta,* sensu stricto, as Pilsbry states; moreover, he records both *G. dalliana* and his *G. bibasidens* from the same Station (perhaps even the same population).

109b. Gastrocopta (I.) dalliana bilamellata (V. Sterki and G. H. Clapp, 1909, as *Bifidaria bilamellata*). T.L.: Foothills of Plomosa [misspelled Plumosa] Mts, 8 mi E of Quartzsite, in drift, Yuma Co., 33° 40′ N, 114° W, westernmost Station of species; elevation not given, but found by senior author (1969) at 1,500 ft, in a nearby wash of Plomosa Pass. Replaces nominate *G. dalliana* in Lower Sonoran Life-Zone of SW Arizona (Yuma Co., and adjoining areas of Maricopa Co. and Pima Co.), at 1,500 to 3,500 ft; farther east, in Maricopa, Yavapai, Pinal, and Pima counties, *G. d. dalliana* gradually supplants *G. d. bilamellata* at 2,500 to 6,000 ft; the two subspecies and intergrades sometimes live together, some Stations being listed for both. Known alive from leaf litter in Yuma Co. (Sierra Pinta at Heart Tank, 32° 20′ N, 113° 15′ W, at 1,300 ft, westernmost Station of species, senior author, 1963), Pima Co. (Ajo Mts: Arch Canyon, 3,500 ft, 32° 42′ N, 112° 42′ W, and Dripping Springs, 2,000 ft, both in Organ Pipe Cactus National Monument, senior author, 1963-1968; Walls Well, 32° 10′ N, 112° 35′ W, Pilsbry and Ferriss, 1923:55. Quijotoa Mts, between Poso Blanco and Covered Wells, 2,500 ft, ca 32° 10′ N, 112° 10′ W, Pilsbry and Ferriss, 1923:55. Quinlan Mts, Kitt Peak, 5,500 to 6,000 ft. Coyote Mts: upper Mendoza Canyon, 3,500 ft; etc. Roskruge Mts, NE end at 32° 15′ N, 111° 27′ W, 2,400 ft. Picacho Mts, canyon at SE end, 2,500 ft. Tortolita Mts [main Range in Pinal Co.], S end in Ruelas Canyon, 3,050 ft. Tucson Mts: widespread at 2,400 to 3,000 ft; Contzen Pass, 4 mi SW of Cortaro, 2,500 ft; etc. Santa Catalina Mts: widespread at 2,800 to 5,000 ft; Sabino Canyon, 2,800 to 3,000 ft; S foothills 1 to 4 mi N of Tucson city limits, 3,200 to 3,400 ft; etc. Rincon Mts, SW slope in Chimenea Canyon, 3,800 ft. Tanque Verde Hills, Saguaro National Monument, 6 mi E of Tucson, 3,500 ft), and Pinal Co. (Aravaipa Creek Canyon, 15 mi NE of Mammoth, 2,500 ft). Dead shells from riparian drift in Yuma Co. (at T.L.; and Bill Williams Riv, 30 mi N of Wenden, 1,500 ft), Pima Co. (Cuerda de Lena and Alamo washes, just N of Organ Pipe Cactus National Monument, 1,800 ft. Pantano Wash. Rillito Creek. Tanque Verde Creek, Santa Cruz Riv at South Tucson and Marana. San Pedro Riv at Redington), Maricopa Co. (Canyon Lake of Salt Riv. New River, 29 mi N of Phoenix. Hassayampa Riv: 3 mi W of Palo Verde; and 6 mi SE of Wickenburg, 2,100 ft), Pinal Co. (Aravaipa Creek, 10 mi N of Mammoth. Gila Riv: 1 mi N of Florence, B. A. Branson et al, 1966:149; Sacaton; 5 mi NE of Winkelman. Galiuro Mts, NW slope in Scanton Canyon, near Sombrero Butte, 4,000 ft, ca 32° 40′ N, 110° 28′ W, northeasternmost record of subspecies), and Gila Co. (Pinto Creek, 6 mi SE of Roosevelt). Also in Sonora (S side of Rio Sonora near Hermosillo, 900 ft, ca 29°

N, 111° W; Arroyo San Rafael at San Bernardo, 900 ft, ca 27° 25′ N, 108° 54′ W; drift in N foothills of Cerro Zaporxa, E of Cajeme [Ciudad Obregón], ca 28° N, 109° W, southeasternmost record of subspecies; all three by Pilsbry, 1953:162. Drift in wash W of Pinacate Peak, ca 45 mi NW of Sonoyta, 1,200 ft, ca 31° 50′ N, 112° 50′ W, F. G. Werner, 1968. Drift of Rio Sonoyta at Sonoyta, ca 1,500 ft, ca 31° 50′ N, 112° 50′ W, R. H. Russell, 1968). Not known fossil.

G. bilamellata, described as a species and treated as such by H. A. Pilsbry (1948:903), is here made a subspecies. It is most typical in the very arid section of SW Arizona and NW Sonora (W of long. 113°), where it is best defined in full-grown shells by the continuous parietal edge of the peristome, slightly detached from the body-whorl, an adaptation for survival in an extreme arid environment. As a rule, the parietal and angular lamellae are more completely separated and farther apart than in *G. d. dalliana*; but this and differences in size, shape, and details of columellar lamella, though sometimes helpful, vary too much in populations of either form to be fully diagnostic. In south-central Arizona, where the two subspecies overlap and are sometimes even sympatric, transitional specimens are common, suggesting frequent interbreeding. *G. dalliana media,* which we regard as based on hybrids of the two subspecies, cannot be ranked as a distinct taxon nor strictly synonymized with either of its parents; merely for convenience we have disposed of the name in the synonymy of *G. d. dalliana.* Further discussion in section on Zoogeography.

110. **Gastrocopta (Gastrocopta) cristata** (H. A. Pilsbry and E. G. Vanatta, 1900, as *Bifidaria procera cristata*). T.L.: Camp Verde, Yavapai Co.; elevation not given, but ca 3,200 ft. Common in S half of Arizona, mostly in Lower and Upper Sonoran Life-Zones of moderately arid counties, at 2,500 to 4,500 ft; evidently a pronounced thermophile. Reliable Recent records of living snails from Yavapai Co. (Camp Verde, the T.L., northwesternmost Station, Pilsbry, 1948:911), Pinal Co. (Aravaipa Creek Canyon, 15 mi NE of Mammoth, 2,500 ft), Maricopa Co. (cultivated area near Chandler; Scottsdale at Jokoke Drive, 1,100 ft), Pima Co. (gardens in Tucson, 2,300 ft; Agua Verde Canyon, 3 mi NE of Vail, 3,300 ft; Cienega Creek, 6 mi NE of Empire Ranch, 4,200 ft; La Canoa Ranch, 5 mi S of Continental, 3,000 ft), Santa Cruz Co. (Sonoita Creek at Sanford Butte, 15 mi NE of Nogales, 3,800 ft), and Cochise Co. (floodplain of San Pedro Riv at Hereford, 4,100 ft, A. L. Metcalf, 1967:18; cienega of Babocomari Creek at Babocomari [now Brophy] Ranch, 4,500 ft). Some of these records, particularly those of the cities, may be artificial, due to human introductions, or maintained in man-made

environments. Widespread in drift, sometimes as a mixture of Recent and fossil shells, in Yavapai Co. (Agua Fria Riv, 6 mi E of Bumblebee, 3,500 ft; Verde Riv at Clarkdale), Navajo Co. (Holbrook; Little Colorado Riv, 3 mi E of Winslow, 4,860 ft), Apache Co. (Navajo Springs, 50 mi N of St Johns, northeasternmost Station in State; Zuni Riv, N of St Johns, B. A. Branson et al, 1966:149), Gila Co. (Gila Riv, 5 mi NE of Winkelman, 2,000 ft; Salt Riv, 4 mi N of Seneca, 3,000 ft), Pinal Co. (Gila Riv, 1 mi N of Florence, B. A. Branson et al, 1966: 149; Gila Riv at Sacaton; San Pedro Riv at Mammoth, 2,300 ft), Maricopa Co. (Salt Riv at Tempe; Hassayampa Riv, 3 mi W of Palo Verde, southwesternmost Station), Pima Co. (Santa Cruz Riv at Tucson, 5 mi S of Marana, and 1 mi W of Continental; Rillito Creek at Tucson; Pantano Wash near Vail, 2,400 ft; Brawley Wash at State Hwy 286, 15 mi S of Robles Junction; San Pedro Riv at Redington; Upper Rincon Creek, 3,100 ft), Santa Cruz Co. (Sonoita Creek at Patagonia, 4,000 ft; Santa Cruz Riv at Amado, Tumacacori, Calabasas, and 5.5 mi E of Nogales), and Cochise Co. (Babocomari [misspelled "Barbacoma"] Creek, N foothills of Huachuca Mts, H. A. Pilsbry, 1916-1918:68; San Pedro Riv at Benson [one sinistral and 222 dextral shells], Fairbank, Hereford, and Palominas). Fossil in Pliocene or Pleistocene of Cochise Co. (San Pedro Valley Sites: Post Ranch, 6 mi, California Wash, 9.5 mi, and El Paso Pipeline, 7 mi S of Benson, R. S. Gray, 1965, D. W. Taylor, 1966b:94; Lehner Mammoth and Murray Springs, associated with mammoth remains 10,000 to 11,000 yrs old). Recent also in Nebraska, Kansas, Oklahoma, western Texas (E to Robertson, Brazos, and Austin counties), New Mexico (widespread), and Sonora (drift of Rio Bavispe, 21 mi S of Agua Prieta; drift of Rio Yaqui, 4 mi N of Ciudad Obregón, southernmost Station of species; both B. A. Branson et al, 1964:104; drift of Rio Sonoyta at Sonoyta, R. H. Russell, 1968). Synonym: *Pupa hordeacea* W. G. Binney and T. Bland, 1869:241, fig. 417; W. G. Binney, 1878, *Bull. Mus. Comp. Zool.* 4:205, fig. 109 (not *Pupa hordacea* W. M. Gabb, 1866, which is a *Pupoides*).

? 111. **Gastrocopta (G.) procera** (A. A. Gould, 1840, as *Pupa*). T.L.: Baltimore, Maryland. Doubtful as a truly Recent (pre-Columbian) native in Arizona; included as such in Check List with misgivings. We have seen only a few, mostly small lots of living specimens, from Pima Co. (Tucson, 8 live *procera* with 2 live *cristata,* at cactus roots in Univ of Arizona greenhouse, Apr. 1964) and Maricopa Co. (Mesa, one live snail sucked mechanically from Bermuda grass lawn by G. D. Butler, Oct. 1960; Scottsdale, in irrigated lawn at Jokoke Drive, 1,100 ft, with *G. cristata,* Jan. 1969; picnic area of Hassayampa Riv, 5 mi SE of Wickenburg, many alive, Jan.

1969). These occurrences probably all to be traced to introductions by man from outside the State, at roots of plants or mixed with seeds. Similar sources or transport from fossil Sites might also account for the few records of dead drift shells from Apache Co. (drift near Adamana, J. H. Ferriss, 1920:14, as *G. procera mcclungi;* drift at Fort Defiance, B. Walker, 1915:2, as *Bifidaria procera*), Navajo Co. (Holbrook, Pilsbry, 1948:910, as *G. procera mcclungi;* drift of Little Colorado Riv, 3 mi E of Winslow, 51 *procera* associated with 160 *cristata,* senior author and R. H. Russell, Oct. 1968), Gila Co. (drift of Salt Riv at U.S. Hwy 60, 4 mi N of Seneca, 1 *procera* with 1 *cristata,* B. A. Branson et al, 1966:149), and Cochise Co. (drift of San Pedro Riv: at Hereford, 58 *procera* with 123 *cristata,* B. A. Branson et al, 1966:149; 1 *procera* with 250 *cristata* at Palominas, senior author, 1962). *G. procera* is known from Pleistocene of Texas and from modern floodplain alluvium in New Mexico (A. L. Metcalf, 1967:41); not found fossil thus far in situ in Arizona, although *G. cristata* is common in Pleistocene of San Pedro Valley. Recent *procera* ranges over SE United States, from Maryland, North Carolina, Arkansas, and Missouri S to Alabama and Texas, W to South Dakota, Kansas, Oklahoma, New Mexico (living in Sacramento Mts, Otero Co., at 7,000 to 8,000 ft, A. L. Metcalf, 1970b:42), and Colorado. Not in Sonora or Chihuahua; one shell found in Baja California Territorio Sur at La Purisima by junior author, 1970, presumably introduced accidentally by man with nursery plants. Synonyms: *Bifidaria duplicata* V. Sterki, 1912 (T.L.: Paluxy Creek at Glenrose, Somervell Co., Texas; renamed *Gastrocopta procera sterkiana* by H. A. Pilsbry, 1917). *Bifidaria mcclungi* G. D. Hanna and E. C. Johnston, 1913 (T.L.: Dog Creek, Phillips Co., Kansas, as a Pleistocene fossil).

112. **Gastrocopta (G.) pellucida** (L. Pfeiffer, 1841, as *Pupa*). T.L.: Cuba, without more precise locality. Common in Arizona in Lower and Upper Sonoran Life-Zones, at 1,100 to 6,500 ft; known from all counties; lives in Tucson. A Late Pleistocene fossil in San Pedro Valley at Lehner Mammoth and Murray Springs Sites, associated with mammoth remains 10,000 to 11,000 yrs old. Recent distribution covers a narrow Atlantic Coast strip from S of Cape May, New Jersey, to Florida; also in SW United States from Texas to SE California, Mexico, and Central America, S to Panama and Antilles. Synonyms: *Pupa hordeacella* H. A. Pilsbry, 1890 (T.L. not with original description; New Braunfels, Comal Co., Texas, selected by Pilsbry, 1916, in 1916-1918:79). *Bifidaria hordeacella* var. *parvidens* V. Sterki, 1899 (T.L.: Jerome, Yavapai Co.); based on depauperate shells with underdeveloped apertural teeth, found

in pure colonies or mixed with normal shells over much of the State, mostly in N counties; also in SE California (S. S. Berry, 1922:97, 3rd fig. from left), Sonora, and Baja California. Discussed further in section on Zoogeography.

Gastrocopta (*Albinula*) *armifera* (T. Say, 1821, as *Pupa*) does not now live in Arizona, but is fossil in Late Pleistocene of San Pedro Valley at Lehner Mammoth and Murray Springs Sites, associated with mammoth remains 10,000 to 11,000 yrs old. It was found living in New Mexico on the Tularosa Riv in Sacramento Mts, Otero Co., at 7,200 ft, by A. L. Metcalf (1970b:42), the southernmost Recent Station of the species. *G. armifera ruidosensis* (T. D. A. Cockerell, 1899:36) was described from Lincoln Co., New Mexico.

Chaenaxis H. A. Pilsbry and J. H. Ferriss, 1906.

The anatomy of this genus was unknown thus far. Junior author dissected *C. tuba* from live snails found by R. H. Russell under decayed sotol at N end of Dragoon Mts, 3 mi E of Dragoon, and others found by J. R. Hershey in Palm Canyon, Kofa Mts. Basic anatomy of Pupillidae, the snail orthurethrous; penis extremely small, provided with an appendix; a pair of tentacles; these features place *Chaenaxis* in subfamily Pupillinae, not in Gastrocoptinae as H. A. Pilsbry (1918, in 1916-1918:xi; 1948:870) surmised.

The hollow axis of the shell, broadly open at umbilicus, is diagnostic for the genus and nearly unique in Pupillidae. In addition, some shells have a cord-like supracolumellar lamella on the axis, entering over the last one to two whorls. Size and shape of shell vary greatly, even in one population, with occasional bizarre abnormalities, as described for *C. tuba intuscostata* by G. H. Clapp (1909:96, Pl. 7, figs. 5-8) and Pilsbry (1948:919, figs. 498[5-8]). We treat the genus as monotypic, the names based on variants being listed as synonyms. We also agree with B. A. Branson (1964:104) that *C. sonorensis* of Sonora is not separable from Arizona *C. tuba*. Although small (3 to 4.5 mm high, 1.5 to 2 mm wide), *Chaenaxis* is one of the most distinctive snails of the Southwestern arid fauna, restricted to the southern counties of Arizona and much of Sonora, SW to the Gulf of California (map, Fig. 7).

113. **Chaenaxis tuba** (H. A. Pilsbry and J. H. Ferriss, 1906:145, 4 figs. 6 and fig. 7, as *Bifidaria* subgenus *Chaenaxis*) [*Infundibularia tuba* J. H. Ferriss, 1904:52, from drift to San Pedro Riv at Benson, is nomen nudum]. T.L.: based on dead shells from riparian drift of San Pedro Riv, 2 mi E of Benson, Cochise Co. [not living near there, the closest known natural Station of live *tuba* in San Pedro drain-

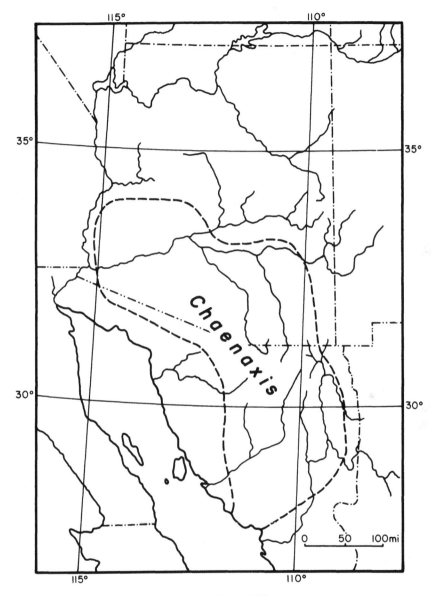

Fig. 7. Distribution of *Chaenaxis*.

age being ca 15 mi to the E, at N end of Dragoon Mts]. Alive in leaf litter of S Arizona canyons at 1,500 to 5,000 ft, mainly in Lower Sonoran Life-Zone, preferably associated with limestone; but dead shells more common in riparian drift. Not known fossil.

Arizona records to date (1971), published or new, of living or dead snails, are from Yuma Co. (drift in Plomosa Mts, 8 mi E of Quartzsite, 1,500 ft, 33° 40′ N, 114° 5′ W, T.L. of *C. tuba intuscostata,* westernmost published Station of species, collected by G. S. Hutson; drift in wash W of Plomosa Pass, 10 mi E of Quartzsite, 1,500 ft, senior author, 1969; litter in Kofa Mts, 25 mi S of Quartzsite, ca 2,000 ft, "in moist places among piles of loose rock covered by decaying cactus," collected by G. S. Hutson [as from "Short Horn Mts," a local name of Kofa Mts, as explained by us for *Eremarionta r. hutsoni*], Station mentioned by G. H. Clapp, 1908:77, for "specimens intermediate between the type [of *C. tuba*] and *intuscostata*" and later referred by Pilsbry, 1916, in 1916-1918:5, to his *C. intuscostata* form *brevicostata;* litter in Kofa Mts, in Palm Canyon, 2,500 ft, 33° 20′ N, 114° 5′ W, J. R. Hershey, 1969; litter in Sierra Pinta, near Heart Tank, at 1,300 ft, 32° 20′ N, 113° 15′ W, senior author, 1963; drift on S bank of Bill Williams Riv, 30 mi W of Wenden, 1,500 ft, 34° 25′ N, 113° 25′ W, northernmost Station of species, senior author, 1969), Yavapai Co. (drift of Agua Fria Riv at mouth of Badger Springs Creek, 6 mi NE of Bumblebee, ca 34° 15′ N, 112° 10′ W, senior author, 1969), Maricopa Co. (drift of Salt Riv at Tempe, 1,150 ft [T.L. of *C. intuscostata* form *brevicostata*], 33° 26′ N, 111° 56′ W, collected by E. H. Ashmun, Pilsbry, 1916, in 1916-1918:4), Pinal Co. (drift of Gila Riv, 1 mi N of Florence, ca 33° N, 111° 23′ W, B. A. Branson et al, 1966:149, and at Sacaton, senior author; drift of San Pedro Riv at Mammoth and at mouth of Aravaipa Creek, 2,300 ft, senior author; drift of Mulberry Wash in San Pedro Valley, at Mercer Ranch, 3,000 ft, senior author; litter on N slope of Picacho Peak near Eloy, 2,000 ft, M. L. Walton), Pima Co. (litter in Ajo Mts, on W slope in Arch Canyon, 2,800 to 3,500 ft, senior author; drift of Cuerda de Lena Wash, 13 mi SE of Ajo, senior author; litter in Quijota Mts, Pilsbry and Ferriss, 1923:55, as *C. intuscostata;* litter in dolomite quarry of Robles Hills, E outlier of Roskruge Mts, Pilsbry and Ferriss, 123:55; drift in Roskruge Mts, in wash of NW end of Range, 1.5 mi S of Poso Blanco Ranch, 2,400 ft, 32° 15′ N, 111° 27′ W, senior author; litter in Tortolita Mts, in Ruelas Canyon, 3,000 ft, senior author; Tucson Mts: 0.5 mi NW of Picture Rocks, ca 3 mi S of Picture Rocks, dry wash at Twin Cacti camp, N side of Mtn near Limekiln camp, and 2 mi SE of Mile-Wide Copper Company camp, all 5

Stations by Pilsbry and Ferriss, 1923:55 and 61, as *C. tuba* and *C. intuscostata*, and drift of wash at foot of Golden Gate Mtn, 3,000 ft, senior author; litter under stones on Tumamoc Hill, at Tucson W city limits, 2,700 ft, Pilsbry and Ferriss, 1915b:399; Santa Catalina Mts, drift at dam of lower Sabino Canyon, 2,800 ft, R. H. Russell, 1971; drift of Santa Cruz Riv in Tucson at W end of Congress St, Pilsbry and Ferriss, 1915b:399, and 5 mi SW of Marana, senior author; drift of Rillito Creek at Campbell Ave near Tucson city limits, senior author; drift of Pantano Wash, 1 mi E of Tucson city limits, and near Colossal Cave, senior author; litter in Posta Quemada Canyon, NE of Colossal Cave, 3,300 to 3,800 ft, senior author; litter in Agua Verde Canyon, 4.5 mi NE of Vail, 3,300 ft, senior author; litter in Empire Mts, N side of large Peak at 1.5 mi NW of Total Wreck Mine, 4,800 ft, and on road to Forty-Nine Mining Camp, Pilsbry and Ferriss, 1923:61; drift of Cienega Creek, 1.5 mi N of Cienega Ranch, and at mouth of Mattie Canyon, 4,200 ft, R. H. Russell; Shaw's [now Posta Quemada] Ranch near Colossal Cave, 3,500 ft, Pilsbry and Ferriss, 1919a:303; drift of San Pedro Riv at Redington, senior author), Santa Cruz Co. (litter, Western Dome of Mustang Mts, Station 306, Pilsbry and Ferriss, 1923:62), Cochise Co. (Huachuca Mts, litter on limestone ridge at N edge of Carr Canyon, Pilsbry and Ferriss, 1923:62; litter at Station 288, on "N side of hill N of Mustang Range," Pilsbry and Ferriss, 1923: 62; drift of San Pedro Riv at Benson [T.L. of *C. tuba*] and Fairbank, senior author, and The Narrows 10 mi N of Benson, R. H. Russell; litter at N end of Dragoon Mts, in limestone quarries 3 mi E of Dragoon, and in arroyo 1.5 mi NE of mouth of Wood Canyon, both ca 4,800 ft, R. H. Russell; litter of limestone hill at cave 1.5 mi W of Dos Cabezas, 4,800 ft, ca 32° 11′ N, 109° 39′ W, northeasternmost Station of species, collected by Mort Wien, Pilsbry and Ferriss, 1910c:144), and Graham Co. (drift in wash on U.S. Hwy 70, 14 mi W of Gila Riv bridge, 33° 10′ N, 110° W, senior author).

Known in Mexico only from Sonora, sometimes at lower elevations than in Arizona. Three records by Pilsbry (1953:163) as *C. sonorensis*: litter of hills on S side of Rio Sonora, 0.75 km S of Hermosillo, 900 ft, ca 29° N, 111° W [T.L. of *C. sonorensis*]; drift of Rio Magdalena, at Magdalena, 2,280 ft, ca 30° 38′ N, 110° 59′ W; litter in N foothills of Cerro Zaporxa, E of Cajeme [Ciudad Obregón], ca 28° N, 109° W, southernmost Station of species. Drift of Rio de Bavispe, 21 mi E of Agua Prieta, on Mex. Hwy 10, ca 31° N, 109′ 20′ W, B. A. Branson, 1964:104. Litter, Cochore near Guaymas, ca 28° N, 110° 50′ W, J. L. Baily, Jr, 1939. Litter, Sierra de Magdalena, N of Magdalena, 3,650 ft, ca 30° 45′ N, 111° W, and litter, Sierra del Santo Niño, N of Mina El Milagro, on road to Sahuaripa, 5,000 ft, ca 29° N, 109° 30′ W, southeasternmost Station of species, both junior author, 1965. Drift of Rio Sonoyta at Sonoyta, 1,500 ft, 31° 30′ N, 112° 50′ W, R. H. Russell, 1968.

Synonyms: *Bifidaria* (*Chaenaxis*) *tuba intuscostata* G. H. Clapp, 1908:76, Pl. 7, figs. 1-10 (T.L.: foothills of Plomosa [misspelled "Plumosa"] Mts, 8 mi E of Quartzsite, Yuma Co.). *Chaenaxis intuscostata* form *brevicostata* H. A. Pilsbry, 1916, in 1916-1918:4, Pl. 9, figs. 4-5 (T.L.: drift of Salt Riv at Tempe, Maricopa Co.). *Chaenaxis sonorensis* H. A. Pilsbry, 1953:163, Pl. 9, figs. 8-8a (T.L.: hills on S side of Rio Sonora, 0.75 km S of Hermosillo, 900 ft, Sonora).

114. Pupoides (Pupoides) albilabris (C.B. Adams, 1841, as *Pupa*). Substitute trivial name for *Cyclostoma marginata* T. Say, 1821, with same T.L.: "Upper Missouri," no more precise T.L. selected thus far. Widespread in Arizona at 1,500 to 5,800 ft, mostly in Lower Sonoran Life-Zone; some local records, particularly in towns, may be due to transport by man, but fossils show that it is a pre-Columbian native. Recent, living in Yuma Co. (Plomosa Mts; Kofa Mts, in Palm Canyon, 2,500 ft; etc), Mohave Co. (Mt Trumbull), Coconino Co. (8 mi S of Seligman; Padre Canyon, 25 mi E of Flagstaff, 5,800 ft), Navajo Co. (canyon 5 mi NE of Winslow), Maricopa Co. (Phoenix), Pinal Co. (Picacho Mts, 2,500 ft), Pima Co. (Tucson Mts; Cerro Colorado Mts, 4,100 ft; Santa Catalina Mts, in Bear Canyon, 2,800 ft; Ajo Mts; Quijotoa Mts; Whetstone Mts, in Mattie Canyon, 4,200 ft; Tucson; etc), Santa Cruz Co. (Rail X Ranch, 5 mi SW of Sonoita, 4,400 ft; Atascosa Mts, in Sycamore Canyon, 4,000 ft; etc), and Cochise Co. (Cochise; Chiricahua Mts, in Cave Creek Canyon, 5,800 ft; etc). Drift shells in Yuma Co. (Bill Williams Riv, 30 mi N of Wenden), Yavapai Co. (Agua Fria at Badger Springs Creek), Maricopa Co. (Gila Riv; Hassayampa Riv), Gila Co. (affluent of East Verde Riv, 5 mi NW of Payson; etc), Pima Co. (Santa Cruz Riv; Pantano Wash; Rillito Creek; San Pedro Riv; etc), Cochise Co. (San Pedro Riv; Stronghold Canyon West, Dragoon Mts, 5,000 ft), and Apache Co. (Fort Defiance). Fossil in Late Pliocene and Pleistocene in Coconino Co. (Winona Site, 13 mi E of Flagstaff, R. D. Reger and G. L. Batchelder, 1971:191 and 193) and Cochise Co. (San Pedro Valley: Sites at Post Ranch, 6 mi, El Paso Pipeline, 7 mi, and California Wash, 9.5 mi S of Benson, R. S. Gray, 1965, D. W. Taylor, 1966b:94; Lehner Mammoth and Murray Springs Sites, associated with mammoth remains 10,000 to 11,000 yrs old). Recent in E North America from Ontario and Maine to Gulf of Mexico in the east, and westward to the Dakotas, Utah and Arizona; a California record in a lawn at Brawley, Imperial Co. (G. D. Hanna, 1966:26), perhaps an introduction by man. Recent, but sometimes only adventive, in Mexico (Baja California, Sonora, Chihuahua, Sinaloa, Nuevo Leon, Tamaulipas, etc), Lesser and Greater Antilles, Bahamas, Bermuda, and South America (details in section on Zoogeography). Synonyms: *Cyclostoma marginata* T. Say, 1821 (not *Cyclostoma marginata* G. Fischer, 1807). *Bulimus nitidulus* L. Pfeiffer, 1839 (T.L.: Matanzas, Cuba; not *Bulimus nitidulus* H. Beck, 1837). *Pupa fallax* A. A. Gould, 1843 (not *Pupa fallax* T. Say, 1824, based on a European snail). *Pupa modica* A. A. Gould, 1848 (T.L.: "Florida," without precise locality). *Pupa (Modicella) arizonensis* W. M. Gabb, 1866:331, Pl. 21, fig. 6 (T.L.: old Fort Grant, at junction of San Pedro Riv and Aravaipa [misspelled "Arivapa"] Creek, 2,100 ft, marked Site, 8 mi N by 5 mi W of Mammoth, Pinal Co.).

115. **Pupoides (Ischnopupoides) hordaceus** (W. M. Gabb, 1866: 331, Pl. 21, fig. 7, as *Pupa*). T.L.: old Fort Grant, at junction of San Pedro Riv and Aravaipa [misspelled "Arivapa"] Creek, 2,100 ft, marked Site, as above, Pinal Co. Widespread in Arizona, Recent at 4,000 to 6,000 ft, mostly in Upper Sonoran Life-Zone. Known from Mohave Co. (Finley's Reservoir near Mt Trumbull; Antelope Valley), Coconino Co. (first collection of live snails, Aug. 19, 1969, by R. H. Russell, Padre Canyon, N of Interstate Hwy 40, 25 mi E of Flagstaff, 5,800 ft, in litter of an arid biotope, under shrubs and low trees, near the top of a steep rocky westside bluff of the canyon with eastern exposure, associated with *Hawaiia minuscula, Retinella indentata paucilirata, Helicodiscus singleyanus, Succinea* sp., *Gastrocopta pilsbryana, G. pellucida, Pupoides albilabris, Pupilla hebes, Vallonia cyclophorella, V. perspectiva,* and *Cochlicopa lubrica;* Seligman, in ant nest; Canyon Diablo at Two Guns, dead but very fresh, in same biotope as at Padre Canyon), Navajo Co. (drift of Little Colorado Riv, 3 mi E of Winslow, 4,860 ft; drift in Jacks Canyon, 5 mi SE of Winslow, 4,850 ft), Apache Co. (Adamana; Zuni Riv near St Johns; Navajo Springs, ca 50 mi N of St Johns; Chinle Creek near Utah border; Fort Defiance), Yavapai Co. (drift of Verde Riv near Jerome), Pinal Co. (at T.L.), Pima Co. (drift of San Pedro Riv at Redington), and Cochise Co. (drift of San Pedro Riv at Fairbanks and Benson; drift of Walnut Gulch, 1 mi E of Tombstone). Fossil in Late Pleistocene of Coconino Co. (Meteor Crater Site, 20 mi W of Winslow [only 12 mi SE of the living population in Padre Canyon], R. D. Reger and G. L. Batchelder, 1971: 192-193) and Cochise Co. (San Pedro Valley: Murray Springs Site,

two washed-up fossils in drift of arroyo, senior author, 1965; one recognizable fragment [in Unit Z of C. V. Haynes, Jr., 1968, below Boquillas Formation], ca 20,000 yrs old or more, E. H. Lindsay, 1968). General distribution in section on Zoogeography; known Recent, but mostly from dead shells, also in Wyoming, Utah, Colorado, Kansas, and New Mexico (living in Echo Amphitheater, Rio Arriba Co., A. L. Metcalf, 1970b:43); fossil also in Late Pleistocene of Kansas, west Texas, and New Mexico. Synonyms: *Pupa arizonensis* W. G. Binney and T. Bland, 1869:240, not of W. M. Gabb, 1866. *Pupa hordeacea* H. A. Pilsbry, 1889:207, error for *hordacea*. *Pupa gabbii* W. H. Dall, 1897a:367, new name for *Pupa arizonensis* W. G. Binney and T. Bland, 1869, not of W. M. Gabb, 1866. *Pupa gabbii* var. *mexicanorum* T. D. A. Cockerell, 1897b:143 (T.L.: drift of Rio Grande at Mesilla, Dona Ana Co., New Mexico).

Pupoides (*Ischnopupoides*) *inornatus* E. G. Vanatta, 1915:95 (T.L.: drift of White Riv, Shannon Co. [former Washington Co.], South Dakota), is not known at present from Arizona, either living or fossil. Recorded presumably Recent, but mostly for dead riparian drift shells, from South Dakota, Nebraska, Colorado, and New Mexico (Four Mile Hill and Arroyo Pecos near Las Vegas, San Miguel Co.); definitely known alive only from Larimer Co., Colorado, on U.S. Hwy 287, "opposite the turnoff to Red Feather Lakes" (D. E. Beetle, 1960, *Nautilus,* 73:160). Fossil in Upper Pliocene or Lower Pleistocene of Nebraska, Kansas, Oklahoma, and Texas.

116. **Pupilla blandii** E. S. Morse, 1865. T.L.: riparian drift of Missouri Riv, near Site of Fort Berthold, McLean Co., North Dakota. Recent in Arizona at 5,000 to over 9,000 ft in Mohave Co. (Mt Logan), Coconino Co. (Grand Canyon of Colorado Riv), Apache Co. (Reservation Creek, 6,800 ft; Phelps Botanical Area, 8 mi SW of Greer, 9,200 ft; head of Black Riv; 15 mi SW of Eagar, on State Rd 273, in aspen grove, 8,500 ft, R. M. Russell, 1970; Lukachukai Canyon, Chuska Mts, 4 mi NE of Lukachukai, 7,300 ft, R. H. Russell, 1970), Greenlee Co. (Cosper's pasture, Blue Mts, ca 10,000 ft), and Graham Co. (Pinaleno Mts, Mt Graham at 6,300 to 9,000 ft). Dead, partly washed-up fossils, in riparian drift in Pima Co. (Pantano Wash at Tucson city limits), Cochise Co. (San Pedro Riv at Benson; W side canyons of Dragoon Mts), Navajo Co. (Holbrook), and Apache Co. (Adamana; Chinle Wash near Utah border). In Late Pleistocene of Cochise Co. (San Pedro Valley: Murray Springs Site, associated with mammoth remains 10,000 to 11,000 yrs old). Recent mainly in Rocky Mts, from Alberta, Idaho, and Montana S to New Mexico and Arizona, W to Nevada, E to Colorado and per-

haps South Dakota; not known in Mexico. In Late Pliocene, Pleistocene, or sub-Recent (alluvium) of Missouri, Nebraska, Kansas, Oklahoma, Texas, New Mexico, and Arizona. Synonyms: *Pupilla blandi* form *obtusa* T. D. A. Cockerell, 1892 (T.L.: Micawber Mine, Custer Co., Colorado). *Pupilla blandi pithodes* H. A. Pilsbry and J. H. Ferriss, 1917 (T.L.: shoulder of Mimbres Peak of Black Range, at Sierra Co.-Grant Co. line, New Mexico).

117. **Pupilla muscorum** (C. von Linné, 1758, as *Turbo*). T.L.: Europe, without precise locality, but holotype, now in London, from Sweden. According to Pilsbry (1948:934), and C. W. Hibbard and D. W. Taylor (1960:132), the Recent American range includes northern Arizona, where it probably lives sporadically at suitable high elevations N of Mogollon Rim, although the related *P. hebes,* readily confused with *muscorum,* is more common there. However, there are hardly any published State records based positively on living or fresh specimens, most of them being for dead shells from drift, possibly washed-up Pleistocene fossils. The first mention for Arizona by W. H. Dall (1897a:367) was without precise locality, for shells collected by E. J. Palmer. Pilsbry and Ferriss' (1906:143) Benson locality was presumably of a drift shell, since a *Pupilla* cannot be expected to live now near there, and dead shells were found there in drift of San Pedro Riv by senior author (1964). A. G. Smith's (1958:8) shells were from drift of Beaver Creek at Montezuma Castle National Monument, Yavapai Co. Pilsbry's (1920-1921:158; 1948:935) two shells from Holbrook, Navajo Co., probably came from drift; but they could have been Recent, washed down by floodwater from the highland drainage of Little Colorado Riv. At present our only fully trustworthy Recent Arizona *P. muscorum* are from Coconino Co., on Agassiz Peak of San Francisco Mtn, where R. H. Russell found many live snails above timberline in Alpine Life-Zone at 12,100 ft, Aug. 11, 1970, associated with live *Discus shimekii* and *Vertigo modesta ingersolli.* Dead shells, mostly washed up from Pleistocene, were seen by us from Pima Co. (drift of Rillito Creek at Tucson city limits; drift of Pantano Wash at Vail; drift of San Pedro Riv at Redington; all by senior author), Cochise Co. (drift of San Pedro Riv at Benson and Fairbank, both by senior author), and Navajo Co. (drift of Little Colorado Riv, 3 mi E of Winslow, 4,810 ft, R. H. Russell and senior author, 1969, some very fresh, and perhaps from live colonies in the highland drainage of White Mts). Fossil in Late Pleistocene of Cochise Co. (San Pedro Valley at Lehner Mammoth and Murray Springs Sites, associated with mammoth remains 10,000 to 11,000 yrs old) and Graham Co. (San Simon Valley at 111 Ranch Site, 14 mi SE of Safford, P. Seff,

1960:139). A Recent Holarctic, circumpolar snail, the distribution discussed in section on Zoogeography; also known as a Late Cenozoic fossil in Eurasia and North America. Recent in America from Alaska, British Columbia, Alberta, Anticosti I, and Newfoundland S to New Jersey in the E, and to Oregon, Utah, Colorado, central New Mexico (Torrance, Socorro, and Lincoln counties), and N Arizona in the W; not known from California, Sonora, or elsewhere in Mexico. Fossil in Pleistocene of Ohio, Iowa, Nebraska, Kansas, Oklahoma, Texas, New Mexico, and Arizona. Synonym: *Pupilla muscorum sinistra* D. S. Franzen, 1946, *Nautilus* 60 (1):24; D. S. Franzen and A. B. Leonard, 1947, *Univ. Kansas Sci. Bull.* 31 Art. II (15):377, Pl. 22, fig. 4, of paratype (T.L.: fossil in Lower Pleistocene of Meade Formation, on Pyle Ranch, Clark Co., Kansas).

118.   **Pupilla hebes** (C. F. Ancey, 1881, as *Pupa*). T.L.: White Pine, White Pine Co., Nevada. Recent in Arizona highlands at 5,000 to 11,000 ft; recorded from Mohave Co. (Hualapai Peak, 8,000 ft, M. D. Robinson and R. L. Bezy, 1967), Coconino Co. (Woody Mtn, 10 mi SW of Flagstaff, 7,500 to 8,000 ft, Martha Noller, July 1964, R. H. Russell and senior author, Aug. 1969. San Francisco Mtn: Walker Lake, 15 mi NW of Flagstaff, 8,250 ft; Snow Bowl area, 8,500 to 9,500 ft; 8 mi E of Kendrick Mtn, 15 mi NW of Flagstaff, 8,000 ft; Mt Elden, 7,900 to 9,300 ft; Lockett Meadow, W side of Sugarloaf Mtn, 10 mi N of Flagstaff, 8,500 ft 9,500 ft. Padre Canyon, 25 mi E of Flagstaff, 5,800 ft. Mormon Crossing, at junction of Chevelon and West Chevelon creeks, 23 mi SE of Heber, 5,500 ft. Bill Williams Mtn. Kaibab Saddle, N of Grand Canyon), Navajo Co. (in aspens at Betatakin Ruin, J. H. Ferriss, 1920:8), Apache Co. (Reservation Creek, 6,800 ft), Greenlee Co. (Cosper's pasture and rim of Blue Mts, ca 10,000 ft, Pilsbry and Ferriss, 1919a:328), Pima Co. (Santa Catalina Mts: Mt Lemmon at ski area, 8,500 to 9,000 ft; Marshall Gulch, 7,600 ft; Bear Wallow, 7,800 ft; Rose Canyon, 7,000 ft; etc. Rincon Mts: Spud Rock, Pilsbry and Ferriss, 1919a:303, form *nefas;* Manning Camp, 1.5 mi S of Spud Rock, 8,000 ft, senior author; NE slope of Rincon Peak, 8,200 ft, J. T. Bagnara and junior author, 1967), Santa Cruz Co. (Santa Rita Mts: E slope in Adobe Canyon, 5,200 ft, R. H. Russell, 1971), and Cochise Co. (Huachuca Mts: Ramsey Canyon, 5,600 ft, senior author; Wickersham Rock near Miller Peak, ca 9,200 ft, Pilsbry and Ferriss, 1910a: 514. Chiricahua Mts: head of Cave Creek, 8,000 ft; Pine Creek, 7,500 ft, form *nefas;* Rucker Canyon; Spring Branch of Rucker Canyon; all Pilsbry and Ferriss, 1910c:135). Dead shells, mostly washed-up fossils, in riparian drift in Coconino Co. (Oak Creek Canyon, 7 mi N of Sedona, 3,500 ft), Navajo Co. (Little Colorado

Riv, 3 mi E of Winslow, 4,810 ft), Pima Co. (Pantano Wash, 4.5 mi E of Vail, 3,200 ft; San Pedro Riv at Redington), and Cochise Co. (San Pedro Riv at Palominas, Hereford, Fairbank, and Benson). Pleistocene fossils in situ in Navajo Co. (Laguna Canyon, J. H. Ferriss, 1920:7) and Cochise Co. (San Pedro Valley at Lehner Mammoth and Murray Springs Sites, associated with mammoth remains 10,000 to 11,000 yrs old). Recent only in W United States in Washington, S Idaho, Nevada, Wyoming, Colorado, New Mexico, Arizona, and Montana, and in north Mexico in Chihuahua (2 Stations mentioned in section on Zoogeography). Synonyms: *Pupilla hebes nefas* H. A. Pilsbry, 1910 (T.L.: Chiricahua Mts, Pine Canyon, 7,500 ft, Cochise Co.); sinistral form, not ranked here as subspecies, often found with dextral snails, Recent as well as fossil, but sometimes in pure colonies. *Pupilla hebes kaibabensis* H. A. Pilsbry and J. H. Ferriss, 1911, not figured (T.L.: Kaibab Saddle, N of Grand Canyon, Coconino Co.); "a stunted or hunger form," according to Pilsbry, 1948:938. *Pupilla hebes* mutant *albescens* J. H. Ferriss, 1920 (T.L.: in aspens at Betatakin Ruin, Navajo Co.); an albino variant.

119. **Pupilla syngenes** (H. A. Pilsbry, 1890, as *Pupa*). T.L.: Arizona, without more precise locality; none selected thus far. In Arizona only in mountains and highlands at 4,500 to 9,000 ft; rarely in drift at lower elevations. Known from Mohave Co. (Mt Logan), Coconino Co. (several Stations in Grand Canyon, at 5,000 to 6,000 ft; Powell Plateau; Kaibab Plateau; San Francisco Mtn, without precise Station; Oak Creek Canyon, at Purtyman's Ranch, 5,000 ft, and in drift at 3,500 ft; Mahan Mtn; Walnut Creek, 8 mi E of Flagstaff, 6,500 ft, in litter; Padre Canyon, 25 mi E of Flagstaff, 5,800 ft, in litter; Red Rock Spring on S slope of Navajo Mtn), Yavapai Co. (Jerome), Navajo Co. (Holbrook; Black Mesa at Marsh Pass, 6,700 ft; Kayenta, 5,600 ft; drift of Little Colorado Riv, 3 mi E of Winslow, 4,880 ft), Apache Co. (branch of Chinle Wash), Maricopa Co. (drift of Sycamore Creek, 1.5 mi N of Sunflower), Gila Co. (drift of Salt Riv, at U.S. Hwy 60; gorge of Pine Creek at Tonto Natural Bridge, 4,500 ft, in litter; drift of tributary of East Verde Riv, 5 mi NW of Payson, 4,500 ft), Cochise Co. (Mustang Mts, N side tower of East Peak. Dragoon Mts, drift in Fourr Canyon, 5,500 ft. Huachuca Mts, Brown Canyon), and Graham Co. (Pinaleno Mts, Hospital Flat, SW slope of Mt Graham, 9,000 ft). Not known fossil before; dextral and sinistral shells with shape and many whorls of *syngenes,* but without apertural teeth, in Late Pleistocene of Murray Springs Site of San Pedro Valley, Cochise Co. (in Unit E of Boquillas Formation and Unit F4 of Lehner Formation of C.

V. Haynes, Jr., 1968), ca 10,000 to 20,000 yrs old. Distribution discussed in section on Zoogeography. Recent also in New Mexico, Colorado, Utah, Wyoming, and Montana; not known from Sonora or elsewhere in Mexico. Synonyms: *Pupa syngenes* form *dextroversa* H. A. Pilsbry and E. G. Vanatta, 1900 (T.L.:San Rafael, Valencia Co., New Mexico); dextral form, nominate form being sinistral; both may occur together in one population. *Pupilla syngenes avus* H. A. Pilsbry and J. H. Ferriss, 1911 (T.L.: upper slope of Grand Canyon, on Bass Trail, ca 200 ft below E Rim, Coconino Co.) *Pupilla syngenes* mut. *nivea* H. A. Pilsbry, 1921 (T.L.: Black Mesa at Marsh Pass, Navajo Co.); albino form.

120. **Vertigo (Alloptyx) hinkleyi** H. A. Pilsbry, 1921, in 1920-1921:234, Pl. 6, figs. 12-16. T.L.: "Cave Canyon, Huachuca Mts, at Station 295, near reservoir"; also two specimens at Station 296; both collected by A. A. Hinkley and J. H. Ferriss, 1919. Although Pilsbry (1948:947) repeats Cave Canyon as T.L., it appears to be an error for Carr Canyon, given as the locality for Stations 295 and 296 of 1919 by Pilsbry and Ferriss in 1923:59 and 63. Alive in Arizona in leaf litter of a few canyons, at 5,600 to 8,000 ft, in Cochise Co. (Huachuca Mts: Carr Canyon, found in 1971 at ca 6,000 ft, 31° 26′ N, 110° 17′ W; Ramsey Canyon, at Bledsoe, 5,600 ft, 31° 27′ N, 110° 18′ W, G. K. MacMillan, 1946:121, and senior author, 1964; Miller Canyon, 6,000 to 6,500 ft, 31° 25′ N, 110° 17′ W, senior author, 1964; Huachuca Canyon, 6,000 ft, 31° 30′ N, 110° 25′ W, junior author, 1967. Dos Cabezas Mts, foot of caprock on N slope of Dos Cabezas Peaks, 8,000 ft, 32° 14′ N, 109° 37′ W, L. K. Sowls, Aug. 1966). One locality in NW Chihuahua (Sierra de la Breña, 17.5 km from Pearson [Mata Ortiz] on road to Pacheco, 7,000 ft, 30° 10′ N, 108° 10′ W, Pilsbry, 1953:164). Not known from New Mexico or Sonora.

121. **Vertigo (Vertigo) milium** (A. A. Gould, 1840, as *Pupa*). T.L.: OakIsland, Chelsea near Boston, Massachusetts. Known alive in Arizona from only two Stations in SE Cochise Co. (Huachuca Mts, Tanner [now Garden] Canyon, 6,000 ft, Pilsbry and Ferriss, 1910a: 515. Chiricahua Mts, head of Pine Canyon, one shell at 7,500 ft, Pilsbry and Ferriss, 1910c:144). Washed-up fossils in riparian drift in Pima Co. (Pantano Wash E of Vail; San Pedro Riv at Redington), Santa Cruz Co. (Santa Cruz Riv at Amado and at 5 mi E of Nogales), Cochise Co. (San Pedro Riv at Palominas, Hereford, Fairbank, Benson, and The Narrows 10 mi N of Benson), Yavapai Co. (Agua Fria Riv at Badger Springs Creek), and Apache Co. (Zuni Riv near St Johns, B. A. Branson et al, 1966:150). Fossil in Late Pliocene and Pleistocene of Cochise Co. (San Pedro Valley: El

Paso Pipeline Site, 7 mi S of Benson, R. S. Gray, 1965, D. W. Taylor, 1966b:94; Lehner Mammoth and Murray Springs Sites, associated with mammoth remains 10,000 to 11,000 yrs old). Recent in E North America from S Ontario, S Quebec, and Maine to Florida and E Texas, and W to North Dakota, N Colorado, and SE Arizona; recorded also from NE Mexico (Tampico) and Greater Antilles.

122. **Vertigo (Vertigo) ovata** (T. Say, 1822, as *Pupa*). T.L.: Philadelphia, Pennsylvania, selected by Pilsbry, 1919, in 1918-1920:83. Rare in Arizona, at 5,000 to 6,000 ft; only six records here accepted as Recent: Cochise Co. (Huachuca Mts: Ash Canyon and Tanner [now Garden] Canyon, 6,000 ft, Pilsbry and Ferriss, 1910a:515; alive at Sylvania Springs, Scotia Canyon, NW side of range, 6,150 ft, R. H. Russell and senior author, Jan. 1968. Cienega of Babocomari Creek at Babocomari [now Brophy] Ranch, 4,500 ft, R. H. Russell and senior author, Feb. 1969), Yavapai Co. (Jerome, westernmost Station in the State, Pilsbry, 1919, in 1918-1920:82), and Greenlee Co. (Blue Mts, foot of Copper King Mtn, 1 mi below Harper's Place, Pilsbry and Ferriss, 1919a:329). Common in drift, mostly washed-up fossils, in Yuma Co. (Bill Williams Riv at Browns Crossing, 30 mi N of Wenden, 1,500 ft), Pima Co. (Santa Cruz Riv at Tucson and 5 mi W of Marana; Pantano Wash near Vail; Rillito Creek at Tucson city limits; Mattie Canyon, W slope of Whetstone Mts; San Pedro Riv at Redington), Santa Cruz Co. (Santa Cruz Riv at Amado; Sonoita Creek at Patagonia), Cochise Co. (San Pedro Riv at Palominas, Hereford, Fairbank, and Benson), Maricopa Co. (Salt Riv at Tempe), Pinal Co. (Gila Riv at Sacaton), Yavapai Co. (Agua Fria Riv at mouth of Badger Spring Creek; Beaver Creek at Montezuma Castle National Monument, A. G. Smith, 1958), Navajo Co. (Navajo Creek near Tso Ranch; Little Colorado Riv, 3 mi E of Winslow), and Apache Co. (Adamana). Fossil in Late Pliocene and Pleistocene of Cochise Co. (San Pedro Valley: Sites of California Wash, 9.5 mi, and El Paso Pipeline, 7 mi S of Benson, R. S. Gray, 1965, D. W. Taylor, 1966b:94; Lehner Mammoth and Murray Springs Sites, associated with mammoth remains 10,000 to 11,000 yrs old) and Navajo Co. (Laguna Canyon near Betatakin Ruin). Recent in North America from Aleutian Is, Alaska, British Columbia, and Labrador S to Florida Keys and Texas in the east, from Oregon, Idaho, Montana, and Nebraska to Colorado and Arizona in the west; barely enters Mexico in NW Sonora (drift of Rio Sonoyta at Sonoyta, 5 mi S of Mexican border, southwesternmost Station of species, R. H. Russell, Sept. 1968); an early record fron Vera Cruz doubtful for Pilsbry, 1919 in 1918-1920:86; doubtfully in Antil-

les. General Recent and fossil distribution discussed in section on Zoogeography.

123. **Vertigo (V.) gouldii** (A. Binney, 1843, ? March, *Proc. Boston Soc. Nat. Hist.* 1:105, as *Pupa;* 1843, ? April, *Boston Journ. Nat. Hist.* 4 [3]:352, cites page of earlier description, Pl. 16, 4 figs. 9). No locality with first description, with second description from New **York State (Troy), Maryland (Baltimore; Worcester Co.),** and **Massachusetts (Cambridge; Roxbury); no precise T.L. selected** thus far. Recent distribution in E North America (three variants) given by Pilsbry (1948:971-976) as from Magdalen I, Prince Edward I, Newfoundland, and Quebec W to Michigan and S to Missouri, Kentucky, Tennessee, and N Alabama. Four precinctive "subspecies" recognized by Pilsbry (1948) from W of 100th Meridian in British Columbia, Montana, Utah, Colorado, Arizona, New Mexico, and Chihuahua; in the Southwest, only at 4,000 to 10,000 ft. Known in Illinois, Kansas, and Texas from Pleistocene only; also fossil in Arizona in Late Pleistocene of Cochise Co. (San Pedro Valley, Lehner Mammoth and Murray Springs Sites, associated with mammoth remains 10,000 to 11,000 yrs old). The western "subspecies," defined by variations of apertural teeth, are in senior author's opinion highly artificial, being based on the assumption that, due strictly to geographical isolation, a distinctive tooth formula prevails in every population, or at least in every mountain Range. In recording *V. gouldii coloradensis* from NW Chihuahua, Pilsbry (1953:164) noted that, although most of the colony was of *coloradensis* (with tooth formula 1-1-2), a few snails agreed in teeth with *V. g. basidens* or *V. g. arizonensis;* he added that "the occurrence of these may perhaps indicate the existence of a highly variable stock in this locality. In Arizona these three subspecies have only been found segregated in separate populations." Even in our limited experience, a population of *V. gouldii* is seldom in Arizona of one type of teeth. We therefore intended originally to treat the four named western forms as variants (morphs) of one geographic subspecies, *V. gouldii coloradensis* (T. D. A. Cockerell). However, such a drastic innovation may be premature. A valid appraisal of the western variants of *V. gouldii* should await a complete, critical revision of all known variants from the entire range of the species, an undertaking obviously beyond the scope of a regional Check List. It is not even clear at present how the eastern forms are to be separated from the western, except by geography.

123a. **Vertigo (V.) gouldii coloradensis** (T. D. A. Cockerell, 1891, *British Naturalist* for 1891 (5):100, as *Pupa coloradensis;* first use of trivial name with description); sometimes misspelled *colo-*

radoensis. T.L.: Swift Creek, Custer Co., Colorado; elevation not given. W. G. Binney (Jan. 1892:191) quoted a description sent to him by T. D. A. Cockerell. H. A. Pilsbry and E. G. Vanatta (1900: 603) published a more extended description, with a drawing by T. D. A. Cockerell of the holotype (at Brit. Mus.) and his remark that he "never observed a second parietal [i.e. angular] tooth in *coloradensis*"; but Pilsbry and Vanatta noted that only two or three specimens were originally taken. However, the drawing, also copied by Pilsbry (1918-1920:115 and 1948:974, fig. 522), shows a weak trace of angular tooth, and V. Sterki (in Pilsbry, 1918-1920:116) mentions that a *coloradensis* he received from T. D. A. Cockerell, through W. H. Binney, had a very small angular. It is possible, therefore, that *coloradensis* was actually based on the form of *gouldii* with a small angular (tooth formula 2-1-2), which Pilsbry and Ferriss later (1900) described as *V. coloradensis arizonensis*; in which case the form normally with the tooth formula 1-1-2 (without angular), if a valid subspecies, may have to be called *V. gouldii utahensis*. The problem can only be solved by studying a new, more adequate sample of topotypes.

Adhering for the time being to Pilsbry's (1948) use of subspecies *coloradensis* for snails with tooth formula 1-1-2, this has been definitely reported from the State at only three Stations in Navajo Co. (TaBiko Canyon near Betatakin Ruin, D. T. Jones, 1940:40) and Cochise Co. (Chiricahua Mts: Pine Canyon, 7,500 ft, Pilsbry, 1918-1920:115; head of Cave Creek Canyon, one shell as *V. columbiana utahensis*, Pilsbry and Ferriss, 1910c:144). Pilsbry's (1920-1921:235) record from Station 295, at Tombstone Reservoir in Carr Canyon, Huachuca Mts, was later transferred to *V. g. inserta* by Pilsbry and Ferriss (1923:63). We have seen thus far only three *V. g. coloradensis* from Arizona, in Gila Co. (Pinal Peak, 7,300 to 7,500 ft, one *coloradensis* with 118 *V. g. arizonensis*. East Verde Riv, 5 mi NW of Payson, 4,500 ft, two *coloradensis* with two *arizonensis*). *V. g. coloradensis* is known also from several Stations in Colorado, one in Utah (the T.L. of *utahensis*), and one in north Mexico (Chihuahua: Rio Piedras Verdes, 4.6 km below Pacheco, 5,900 ft, Pilsbry, 1953:164). Synonym: *Vertigo columbiana utahensis* (nomen nudum of V. Sterki, 1892) H. A. Pilsbry and E. G. Vanatta, 1900:603 and 609, Pl. 23, fig. 10 (T.L.: Box Elder Co., 4,500 ft, Utah); synonymized with *V. g. coloradensis* by Pilsbry (1918-1920: 115).

123b. **Vertigo (V.) gouldii basidens** H. A. Pilsbry and E. G. Vanatta, 1900:604 and 609, as *Vertigo coloradensis basidens*; Pilsbry, 1918-1920:117, Pl. 12, fig. 15; 1948:975, fig. 518, 15 on p. 967.

T.L.: Bland, Bernalillo Co., New Mexico; elevation not given; found with *V. g. arizonensis.* Based on shells with tooth formula 1-1-3 (1 parietal + 1 columellar + 1 basal + 2 palatals; no angular). The only Arizona record thus far, from Cochise Co. (Chiricahua Mts, Rustler Park, elevation not given, Pilsbry and Ferriss, 1910c:144), was not repeated by Pilsbry, who seems to have transferred it later (1918-1920:118) doubtfully to *V. g. inserta.* We have seen only two Arizona lots containing *V. g. basidens,* both from Cochise Co.: Chiricahua Mts, South Fork of Cave Creek, 1.5 mi SE of South-western Research Station, 5,800 ft (one *basidens* with one *V. g. inserta*); Huachuca Mts, NW slope at Sylvania Springs, Scotia Canyon, 6,150 ft (two *basidens*). Reported also from New Mexico, Colorado, Montana, British Columbia, and north Mexico (Chihua-hua: Rio Piedras Verdes, 4.6 km below Pacheco, 5,900 ft, Pilsbry, 1953:164; one *basidens* with many *V. g. coloradensis*). *V. g. basidens* has the same tooth formula as nominate *V. g. gouldii* (figured by Pilsbry, 1948:958, 4, 5, and 8), but the palatals, especially the lower, are said to be decidedly longer, which does not appear from published figures.

123c. **Vertigo (V.) gouldii arizonensis** H. A. Pilsbry and E. G. Vanatta, 1900:604 and 609, Pl. 23, fig. 3, as *Vertigo coloradensis arizonensis;* sometimes misspelled *arizoniensis.* T.L.: top of Mingus Mtn, 7,600 ft [not 8,500 ft as given], Yavapai Co. For shells with tooth formula 2-1-2 (1 parietal + 1 angular + 1 columellar + 2 pala-tals; no basal). Most common variant in Arizona; known from Co-conino Co. (Bill Williams Mtn, Pilsbry and Ferriss, 1911:197; Oak Creek Canyon, Pilsbry, 1918-1920:117, and in drift, 7 mi N of Sedona, 3,500 ft, J. F. Burger, 1965), Yavapai Co. (Mingus Mtn, T.L.), Gila Co. (Sierra Ancha: Workman Creek, 6,500 ft, and Reynolds Creek, 5,600 ft, both senior author. East Verde Riv, 5 mi NW of Payson, 4,500 ft, with one *V. g. coloradensis,* senior author. Pinal Peak, 10 mi SW of Globe, 7,300 to 7,500 ft, junior author, 116 *arizonensis* + one *coloradensis*), Cochise Co. (Dragoon Mts, NW slope at Fourr Ranch, Pilsbry and Ferriss, 1915b:384; [Chiricahua Mts, cited for *arizonensis* by Pilsbry, 1948:975, without precise Station, perhaps by oversight]), Graham Co. (Pinaleno Mts: without precise Station, Pilsbry, 1918-1920:117; SW slope of Heliograph Peak, 8,700 ft, and S slope of Mt Graham at Wet Canyon, 6,300 ft, both senior author), and Greenlee Co. (Horseshoe Bend of Black Riv, White Mts, Pilsbry and Ferriss, 1919a:329. [Red Sack, Graham Mts, of Pilsbry and Ferriss, 1919a:329, could not be located]). J. H. Ferriss (1920:14) mentions *V. g. arizonensis* fossil in Laguna Canyon near Betatakin Ruin, Navajo Co., and from drift of Chinle Wash near

Utah border, Apache Co.; but these Stations are not mentioned elsewhere and should be confirmed. *V. g. arizonensis* is known also from New Mexico and north Mexico (Chihuahua: Rio Piedras Verdes, 4.6 km below Pacheco, 5,900 ft, Pilsbry, 1953:164, one *arizonensis* in a lot of *V. g. coloradensis*. Sierra de la Breña, ca 16 mi SW of Colonia Juárez, 7,000 ft, five adult *arizonensis* and seven immatures, junior author, July 1966; Pilsbry, 1953:164, reports only *V. g. inserta* at this locality).

123d. **Vertigo (V.) gouldii inserta** H. A. Pilsbry, Feb. 1919, in 1918-1920:118, Pl. 12, figs. 10-11, as *Vertigo coloradensis inserta*. T.L.: Santa Catalina Mts, at Bear Wallow; elevation not given, but common there at 7,600 to 7,800 ft. Based on shells with tooth formula 2-1-3 (1 parietal + 1 angular + 1 columellar + 1 basal + 2 palatals). Widespread in Arizona, particularly in the southern highlands, at 5,600 to 9,000 ft; known from Coconino Co. (Mahan Mtn near Mormon Lake, H. A. Pilsbry, 1948:976), Pima Co. (Santa Catalina Mts: Bear Wallow, T.L., Congden Camp, Soldier Camp, Aspen Gulch, 9,000 ft, Alder Spring, and Desert Laboratory Plantation, 8,500 ft, all 6 Stations, Pilsbry and Ferriss, 1919a:305; collected by us at Bear Wallow, 7,800 ft, Marshall Gulch, 7,600 ft, S slope of Marble Peak, 7,400 ft, and Rose Canyon, 7,000 ft, in all 36 *inserta* + 1 *V. g. arizonensis*. Rincon Mts: N side of high Rincon Peaks, Pilsbry and Ferriss, 1919a:305; Manning Camp, 1.5 mi S of Spud Rock, 8,000 ft, senior author), Santa Cruz Co. (Santa Rita Mts, trail from Upper Madera Canyon to Mt Hopkins, 5,800 ft, senior author), and Cochise Co. (Huachuca Mts: 2 Stations, Carr Canyon, near Tombstone Reservoir, Pilsbry, 1920-1921:235, and Pilsbry and Ferriss, 1923:63, and Ramsey Canyon, 5,600 ft, senior author. Chiricahua Mts: Rucker Canyon, Pilsbry, 1948:976; Rustler Park, one shell that "appears to be" *inserta,* Pilsbry, 1918-1920:118; Cave Creek at Herb Martyr Dam, 5,600 ft, and East Fork of Cave Creek, 2.5 mi SW of Portal, 5,800 ft, 1 *inserta* + 1 *basidens,* both Stations by senior author). Dead shells from drift in Pima Co. (Rillito Creek at N Campbell Ave near Tucson city limits; San Pedro Riv at Redington). *V. g. inserta* was recorded thus far only from Arizona and north Mexico (Chihuahua: Sierra de la Breña, ca 16 mi SW of Colonia Juárez, Pilsbry, 1953:164); it occurs also in New Mexico (Sacramento Mts, 2 mi NE of Cloudcroft, 8,600 ft, Otero Co., 4 *inserta* + 48 *arizonensis,* senior author, 1962; junction of N and S Forks of Eagle Creek, 8,000 ft, Lincoln Co., 1 *inserta* + 1 *arizonensis* + 1 *basidens,* A. L. Metcalf, in litt., 1969).

? 124a. **Vertigo (V.) modesta corpulenta** (E. S. Morse, 1865, as *Isthmia corpulenta*). T.L.: Little Valley on E slope of Sierra Nevada,

6,500 ft, Washoe Co., Nevada. There is only one possibly valid record of this subspecies from Arizona, in Cochise Co. (Huachuca Mts, Wickersham Rock near Miller Peak, ca 9,200 ft, for two adult shells called *V. modesta parietalis* by Pilsbry and Ferriss, 1910a:515, and by Pilsbry, 1919-1920:124, text figs. 5-5a; 1948:984, figs. 528, 5-5a; an immature shell from that Station, called *Vertigo concinnula* by Pilsbry and Ferriss, 1910a:515, was perhaps also a young *corpulenta,* since Pilsbry does not cite Wickersham Rock for his *concinnula* in 1948). Moreover, the shells from Wickersham Rock were probably *V. modesta ingersolli* (discussed next). It is also doubtful that *ingersolli* and *corpulenta* can be separated consistently as subspecies. Pilsbry had to admit (1948:983) that the subspecific taxonomy of *V. modesta* (T. Say, 1824) was "more or less arbitrary." *V. modesta corpulenta* has been recorded also from Nevada, Idaho, Utah, and California, Synonym: *Pupa corpulenta* variety *parietalis* C. F. Ancey, 1887 (T.L.: Ogden Canyon, Morgan Co., Utah.

124b.   **Vertigo (V.) modesta ingersolli** (T. D. A. Cockerell, 1891, as *Pupa ingersolli,* with description). T.L.: "San Juan District" [now San Juan Co.], Colorado, the locality of T. D. A. Cockerell's earlier nomen nudum *"Pupa ingersolli"* of 1889. The usual form of *V. modesta* in Arizona, in the highlands at 6,800 to over 12,100 ft; in drift only at lower elevations. Known from Coconino Co. (San Francisco Mtn: SE slope on Watershed Rd, 8,000 to 9,000 ft; NW slope at Snow Bowl, 10 mi NW of Flagstaff, 9,000 to 9,500 ft; Agassiz Peak, above timberline in Alpine Life-Zone, 12,100 ft, R. H. Russell, Aug. 1970; Lockett Meadow, W side of Sugarloaf Mtn, 10 mi N of Flagstaff, 8,500 to 9,500 ft. Bill Williams Mtn. Drift of Oak Creek Canyon, 7 mi N of Sedona, 3,500 ft, J. F. Burger, 1965), Yavapai Co. (Mingus Mtn), Apache Co. (Reservation Creek, 6,800 ft; Alpine Divide, 3 mi NW of Alpine, 8,100 ft, R. H. Russell, Sept. 1969; 19 mi S of Eagar, on State Hwy 273, 8,700 ft, R. H. Russell, 1970; N end of Roger's Reservoir, 6 mi SE of Nutrioso, 8,560 ft, R. H. Russell, 1969; Escudilla Mtn, at "Lookout," 5 mi E of Nutrioso, 10,700 ft, R. L. Bezy, 1969), Greenlee Co. (Blue Mts: Cosper's pasture, ca 10,000 ft, and rim of the Range, Pilsbry and Ferriss, 1919a:304, as *V. m. insculpta*), Graham Co. (Pinaleno Mts: S slope of Mt Graham at Hospital Flat, 9,000 ft; SW slope of Heliograph Peak, 8,700 ft; both senior author), Pima Co. (Santa Catalina Mts: Mt Lemmon at Aspen Gulch, Westfall Mine, Soldier Camp, Congden Camp, Cold Spring, and E side ridge from 9,000 ft to summit at 9,150 ft, Pilsbry and Ferriss, 1919a:304, as *V. m. insculpta*), and Cochise Co. (Chiricahua Mts: head of Cave Creek and Long Park, 8,000 ft, Pilsbry and Ferriss, 1910c:144, as *V. m. parietalis,* cited later as *V. m. in-*

sculpta by Pilsbry, 1919, in 1918-1920:131; 0.75 mi S of Tub's Spring, N slope of Flys Peak, 9,400 ft, J. R. Hershey, 1969. Huachuca Mts: Miller Peak, elevation not given, G. K. MacMillan, 1946:122, as *V. concinnula;* Miller Canyon, 6,000 ft, senior author). Not known fossil. We restrict *V. m. ingersolli* here to Utah, Colorado, Arizona, and New Mexico.

Contrary to Pilsbry's opinion (1948:979), we recognize the name *Pupa ingersolli* as validly defined (according to the Rules) by T. D. A. Cockerell's description of 1891: "allied to *coloradensis,* but 2 mm long, cylindrical, dull brown, with a half whorl more and a double lamella on the parietal wall"; a definition hardly "absurdly inadequate," as Pilsbry claimed. Although *V. concinnula* was placed by Pilsbry (1948:871 and 978) in his "*Vertigo gouldii* Group," we regard it as a variant of *V. modesta* and identify with it *V. modesta insculpta* H. A. Pilsbry (1919). This explains Pilsbry's (1948:989) transferring to *V. m. insculpta* earlier records of *V. concinnula* or *V. modesta parietalis,* from Bill Williams Mtn and Chiricahua Mts, by Pilsbry and Ferriss (1910:144). Synonyms: *Vertigo coloradensis* form *concinnula* T. D. A. Cockerell, 1897 (*V. ingersolli* given as synonym; T.L.: originally only as "Colorado, at higher altitudes"; in 1919 H. A. Pilsbry selected as T.L. Brush Creek, Custer Co., 10,000 ft). *V. modesta insculpta* H. A. Pilsbry, 1919, in 1918-1920: 131, Pl. 10, figs. 12-13 (T.L.: Santa Catalina Mts, Mt Lemmon, supposedly at 9,500 ft [summit is only 9,157 ft]).

The following Recent North American species of *Vertigo* are not known to live at present in Arizona, where they occur now only as Pleistocene fossils, either in situ or washed up in drift. *V. berryi* H. A. Pilsbry, 1919, in Pleistocene of San Pedro Valley at Lehner Mammoth and Murray Springs Sites; in drift of San Pedro Riv at Palominas, Hereford, Charleston, and Fairbank, Cochise Co., and of Santa Cruz Riv at Amado, Santa Cruz Co.; Recent in California and Baja California. *V. elatior* V. Sterki, 1894, in Pleistocene of San Pedro Valley at Lehner Mammoth and Murray Springs Sites; Recent in cold temperate North America, from British Columbia to Atlantic and in New Mexico mountains. *V. ventricosa* (E. S. Morse, 1865), in Pleistocene of San Pedro Valley at Lehner Mammoth and Murray Springs Sites; as sub-Recent fossils at Sylvania Springs in Scotia Canyon, NW slope of Huachuca Mts, 6,150 ft, Cochise Co.; in drift of San Pedro Riv at Hereford and Fairbank, Cochise Co., of Pantano Wash near Vail, Pima Co., and of Little Colorado Riv, 3 mi E of Winslow, Navajo Co.; Recent in SE Canada and NE United States, S to New York, W to Illinois and Missouri. *V. binneyana* V. Sterki, 1890, not recognized thus far in situ in local

Pleistocene; presumably washed-up fossils in drift of San Pedro Riv at Palominas, Hereford, and Benson, Cochise Co.; Recent in British Columbia, Manitoba, Ontario (Hudson Bay), Montana, and New Mexico.

? 125. **Columella columella alticola** (E. Ingersoll, 1875, as *Pupilla alticola*). T.L.: Cunningham Gulch, San Juan Co., Colorado; elevation not given. Not seen by us from Arizona, where it is known only from one Recent Station in Cochise Co.: Huachuca Mts, at Wickersham Rock on Miller Peak, ca 9,200 ft, Pilsbry and Ferriss, 1915a:390 [recorded by error as *C. edentula,* corrected to *C. c. alticola* by Pilsbry, 1926, in 1922-1926:224]. True *Columella edentula* (J. P. R. Draparnaud, 1805; synonym: *Pupa simplex* A. A. Gould, 1840), of Europe and temperate North America, does not occur in Arizona, either Recent or fossil. According to L. Forcart (1959), nominate *Columella c. columella* (G. von Martens, 1830) is a subarctic snail of Scandinavia and Siberia (N of 60° N, below ca 2,200 ft); *C. c. gredleri* (S. Clessin, 1872) is restricted to central European Alps; and *C. c. alticola* is strictly North American, although he did not discuss its exact status. *C. c. alticola* is Recent at 8,000 to 9,200 ft in British Columbia, Alberta, Montana, Wyoming, Colorado, Utah, New Mexico, and Arizona (where it is a relict of a former wider distribution); it is known as a Pleistocene fossil in Illinois, Iowa, and Kansas, but not in Arizona.

Columella hasta (G. D. Hanna, 1911, as *Sphyradium,* from Kansas Pleistocene) was found fossil in Arizona by senior author in Late Pleistocene of San Pedro Valley at Lehner Mammoth Site, Cochise Co.; it is probably a giant subspecies of *C. columella.*

Family Valloniidae

* **Vallonia pulchella** (O. F. Müller, 1778, as *Helix*). T.L.: not given, but presumably in Denmark. Not recorded in print thus far from Arizona, where it is not native but introduced by man, as it is also in Texas, New Mexico, the Rocky Mountain States, Oregon, and California. Seen by us from Maricopa Co. (Phoenix, in lawn of *Dichondra* and *Tradescantia,* May 1968, reported by J. N. Roney in *Arizona Insect Survey* for Nov. 8, 1968), Pima Co. (4 Stations in Tucson: 1249 E Mabel St, A. R. Mead, 1955; 3216 N Jackson Ave, 1967-1970; nursery on Oracle Rd, R. H. Russell and R. L. Reeder, 1970; drift of Rillito Creek at N Campbell Ave, 1963. Ajo, on lawn, M. Lindsley, 1967), Santa Cruz Co. (Nogales, Oct. 1962, Feb. 1965), Gila Co. (drift of small tributary of East Verde Riv, 5 mi NW of Payson, near cottages, 4,500 ft, senior author, 1966), Coconino Co.

(drift of Oak Creek Canyon, 7 mi N of Sedona, 3,500 ft, J. F. Burger, 1965), and Navajo Co. (drift of Little Colorado Riv, 3 mi E of Winslow, R. H. Russell and senior author, 1968). Circumboreal; native in E North America from Newfoundland, Nova Scotia, and Manitoba S to Missouri, Washington, D.C., and Kentucky, and W to E foothills of Rocky Mountains; native also in Europe, North Africa, and Siberia; often adventive, spread by man in both Old and New Worlds. Known as a Pleistocene fossil in both Europe and North America, but not in Arizona.

126. **Vallonia perspectiva** V. Sterki, 1893. T.L., selected by Pilsbry (1948:1034): Woodville, Jackson Co., Alabama. Widespread, living in Arizona highlands at 3,500 to 8,700 ft, mainly in Upper Sonoran and Transition Life-Zones; at lower elevations, only dead in riparian drift. Known Recent in Coconino Co. (Grand Canyon of Colorado Riv; Walnut Creek, 8 mi E of Flagstaff; Padre Canyon, 25 mi E of Flagstaff, 5,800 ft), Yavapai Co. (Mingus Mtn, 6,000 to 6,500 ft), Greenlee Co. (Cosper's Ranch on Blue Riv, 5,060 ft; Eagle Creek at Ole Hagen's Ranch), Gila Co. (Sierra Ancha, 5,000 to 6,500 ft; Pine Creek at Tonto Natural Bridge, 4,500 ft; Pinal Peak, 7,300 to 7,500 ft), Pinal Co. (Galiuro Mts, on N slope, E of Sombrero Peak), Pima Co. (Santa Catalina Mts, SE slope of Marble Peak, at 6,000 to 8,000 ft. Baboquivari Mts, W slope in Baboquivari Canyon, 3,500 to 4,000 ft. Quinlan Mts, Kitt Peak, on N slope, 6,600 to 6,700 ft. Ajo Mts, Arch Canyon, 3,500 ft, westernmost Station of species), Santa Cruz Co. (Santa Rita Mts, 5,200 to 6,200 ft; Patagonia Mts, Camp Washington; Canelo Hills, Lookout Knob, 5,400 ft; foot of Sanford Butte at Sonoita Creek, 3,800 ft), Cochise Co. (Huachuca Mts, widespread at 5,200 to 6,500 ft. Mustang Mts, 5,200 to 5,400 ft; Whetstone Mts, 5,000 ft; Chiricahua Mts, widespread at 5,000 to 7,600 ft; Dragoon Mts, 4,700 to 5,600 ft), and Graham Co. (Pinaleno Mts, Mt Graham, 5,000 to 8,700 ft). Dead in drift in Coconino Co. (Oak Creek), Navajo Co. (Little Colorado Riv), Yavapai Co. (Verde Riv), Maricopa Co. (Sycamore Creek, 1.5 mi N of Sunflower), Pinal Co. (Queen Creek), Pima Co. (Santa Cruz Riv; Pantano Wash; Rillito Creek; San Pedro Riv), Santa Cruz Co, (Santa Cruz Riv; Sonoita Creek), Cochise Co. (San Pedro Riv), Greenlee Co. (Eagle Creek, 6 mi SW of Morenci), and Apache Co. (creek in Buell Park, 12 mi N of Fort Defiance, 7,000 ft, also in alluvium, R. J. Drake, 1949c:28). Fossil in Pleistocene of Cochise Co. in San Pedro Valley, at Lehner Mammoth and Murray Springs Sites, associated with mammoth remains 10,000 to 11,000 yrs old. General distribution discussed in section on Zoogeography. Recent over most of S United States, N to New Jersey, West Virginia, Missouri, and S

Illinois, W to North Dakota and Utah, S to Alabama, trans-Pecos
Texas (E side of North Franklin Mtn at 7,000 ft, A. L. Metcalf and
W. E. Johnson. 1971:90, 31° 54′ N, 106° 29′ W, El Paso Co.; S end of
Guadalupe Mts, Culberson Co.), New Mexico, and Arizona; it reaches
north Mexico in Sonora and Chihuahua.

127. **Vallonia cyclophorella** V. Sterki, 1892. T.L.: Westcliffe,
Custer Co., Colorado. More boreal and more restricted than *V. perspectiva;* in Arizona mainly in N counties and in some S mountains,
at 4,500 to 10,000 ft, in Transition, Canadian, and Hudsonian Life-
Zones; sometimes sympatric with *V. perspectiva.* Known Recent
from Mohave Co. (Mt Trumbull; Hualapai Peak, 8,000 ft), Coconino
Co. (Woody Mtn, 10 mi SW of Flagstaff, 7,500 ft; San Francisco Mtn,
widespread at 8,000 to 10,000 ft; Padre Canyon, 25 mi E of Flag-
staff, 5,800 ft), Yavapai Co. (Mingus Mtn), Gila Co. (Pine Creek at
Tonto Natural Bridge, 4,500 ft), Yuma Co. (drift of Bill Williams
Riv, 30 mi N of Wenden), Navajo Co. (Betatakin Ruin; Holbrook;
drift of Carrizo Creek near Carrizo; drift of Little Colorado Riv, 3 mi
E of Winslow), Apache Co. (Adamana; Reservation Creek; Phelps
Botanical area, 8 mi SW of Greer, 9,200 ft; 19 mi S of Eagar, on State
Hwy 273, 9,200 ft, R. H. Russell, 1970), Greenlee Co. (Cosper's
pasture in Blue Mts and at rim of Blue Mts, ca 10,000 ft), Pima
Co. (Santa Catalina Mts: Mt Lemmon, slopes at 8,000 to 9,000 ft;
S slope of Marble Peak, 7,500 ft; etc), and Cochise Co. (Huachuca
Mts: Wickersham Rock near Miller Peak, ca 9,200 ft; W slope in
Scotia Canyon, at Sylvania Springs, 6,150 ft. Not recorded from
Chiricahua Mts). Fossil in Pleistocene or sub-Recent of Coconino Co.
(Winona Site, 13 mi E of Flagstaff, R. D. Reger and G. L. Batchelder,
1971:191-193), Navajo Co. (Tsegi Canyon, S of Betatakin Ruin,
L. L. Hargrave), and Cochise Co. (San Pedro Valley at Lehner
Mammoth and Murray Springs Sites, associated with mammoth
remains 10,000 to 11,000 yrs old). Recent also in North Dakota,
South Dakota, Wyoming, Montana, Idaho, Washington, Oregon,
Utah, Colorado, Nevada, California, and New Mexico. The few
published Texas records are of Pleistocene fossils, either in situ or
washed up in riparian drift.

Vallonia gracilicosta O. Reinhardt (1883) is Recent in NW
United States, E to North Dakota, and S to Utah, Colorado, and New
Mexico (Sacramento Mts, 7,500 ft, A. L. Metcalf, 1967:46). There is
no reliable evidence that it lives at present in Arizona, where all
records appear to be of fossils, either in situ or in riparian drift. It
is known from in situ Pleistocene fossils in Cochise Co. (San Pedro
Valley at Lehner Mammoth and Murray Springs Sites, associated
with mammoth remains 10,000 to 11,000 yrs old), Coconino Co.

(Winona Site, 13 mi E of Flagstaff, G. L. Batchelder and R. D. Reger, 1969:7, R. D. Reger and G. L. Batchelder, 1971:191-193), and Navajo Co. (Laguna Canyon near Betatakin Ruin, J. H. Ferriss, 1920:14, misspelled *V. gracilicostata*). Dead shells in riparian drift are from Yavapai Co. (Granite Creek, 2 mi N of Prescott, senior author; probably also the shells from drift of Beaver Creek at Montezuma Castle National Monument called *Vallonia sonorana* H. A. Pilsbry, 1915, by A. G. Smith, 1953:8).

Family Cochlicopidae (Cionellidae)

128. **Cochlicopa lubrica** (O. F. Müller, 1774, as *Helix*).[36] T.L. not given with description, but presumably in Denmark. Widespread in Arizona at 4,500 to over 9,000 ft, in Upper Sonoran, Transition, and Canadian Life-Zones. Known Recent in Mohave Co. (Hualapai Peak, 8,000 ft, M. D. Robinson and R. L. Bezy, 1967), Coconino Co. (Grand Canyon, widespread, 5,700 to 6,000 ft. Drift of Oak Creek Canyon, 3,500 ft. Walnut Creek, 8 mi E of Flagstaff, 6,450 ft. Padre Canyon, 25 mi E of Flagstaff, 5,800 ft), Navajo Co. (Show Low Creek Canyon, 0.75 mi S of dam, A. R. Mead. Ta-Biko Canyon near Betatakin Ruin, 7,000 ft. Drift of Little Colorado Riv, 3 mi E of Winslow), Apache Co. (Fish Creek. Chuska Mts, Lukachukai Canyon, 4 mi E of Lukachukai, 7,300 ft), Yavapai Co. (Mingus Mtn, 6,000 to 6,500 ft. Granite Creek, 2 mi N of Prescott, 5,300 ft), Greenlee Co. (Rim of Blue Mts, ca 9,000 ft. Cosper's Ranch on Blue Riv, 5,060 ft), Gila Co. (Pine Creek at Tonto Natural Bridge, 4,500 ft. Foot of Mogollon Rim, 2.5 mi NE of Pine, 6,600 ft. Pinal Peak, 7,300 to 7,500 ft. Sierra Ancha: Parker Creek, 5,080 ft; Reynolds Creek, 5,600 ft; Cherry Creek, 1 mi E of Young, 5,000 ft), Pinal Co. (drift of Queen Creek, 3,700 ft), Pima Co. (Santa Catalina Mts, widespread on N slopes of Mt Lemmon, 6,000 to 8,500 ft; Marble Peak, 8,000 ft. Rincon Mts, Manning Camp, 1 mi S of Spud Rock, 8,000 ft; NE slope of Rincon Peak, 8,200 ft), Maricopa Co. (drift of Sycamore Creek, 1.5 mi N of Sunflower), Cochise Co. (drift of Walnut Creek, 1 mi E of Tomb-

---

[36]*Cochlicopa* A. E. J. de Férussac (1821) and *Cionella* J. G. Jeffreys (1830) are strict synonyms, both having the same type species, *Helix lubrica* O. F. Müller (1774). Most British and other European malacologists accept at present *Cochlicopa* as the older valid name, as was done also until recently by American students, including H. A. Pilsbry in his earlier revision of the genus (1908, in 1907-1908:308-327). Regrettably, Pilsbry later (1948:1045) decided that *Cionella* was the correct name, being influenced by A. S. Kennard's (1942, *Proc. Malacol. Soc. London* 25:113) claim that de Férussac's *Cochlicopa* was an invalid name. H. Watson (1943, *Journ. of Conchology* 22:30-31) has shown, however, that Kennard's interpretation of the Rules of Nomenclature was erroneous in this case. We therefore follow Watson in restoring *Cochlicopa* and Cochlicopidae as the oldest correct generic and family names of *C. lubrica*. It is to be hoped that malacologists on both sides of the Atlantic may now agree on the nomenclature of this interesting Holarctic snail.

stone, 4,500 ft. Huachuca Mts, widespread, 5,200 to 9,200 ft. Dragoon Mts, widespread, 4,700 to 5,600 ft. Chiricahua Mts, widespread, 4,500 to 8,500 ft. Dos Cabezas Mts, Dos Cabezas Peak, 8,000 ft), and Graham Co. (Pinaleno Mts, SE slope of Mt Graham, 5,100 to 6,300 ft). Fossil in Cochise Co. (Late Pleistocene of San Pedro Valley at Lehner Mammoth and Murray Springs Sites, associated with mammoth remains 10,000 to 11,000 yrs old). Holarctic and circumpolar. Recent in Alaska, Canada, and United States: to near sea level in boreal and cold temperate areas, N to beyond Arctic Circle (Point Barrow, Alaska), S to Alabama in the east; at higher elevations in all western and mountain States, except California; in Mexico in NW Chihuahua and S Nuevo Leon, but not known as yet from Sonora. Widespread in Eurasia, from Iceland, N Europe, and Siberia S to NW Africa and E to Japan. Fossil in Pleistocene of both Old and New Worlds.

Pulmonata Limnophila

Family Lymnaeidae[37]

129. **Stagnicola (Stagnicola) elodes nuttalliana** (I. Lea, 1841, *Proc. Amer. Philos. Soc.* 2 [17]:33, as *Lymnea nuttalliana*). T.L.: Oregon, without more precise locality, none selected thus far. Known Recent in Arizona only in Transition, Canadian, and Hudsonian Life-Zones of N sections, in Coconino Co. (Walker Lake, NW end of San Francisco Mtn, 15 mi NW of Flagstaff, 8,250 ft, R. E. C. Stearns, 1891:100, as *Limnaea palustris,* found also by R.H.R., 1969; Hart Prarie, 1 mi SW of Snow Bowl, San Francisco Mtn, 8,730 ft, R.H.R., 1970; Lindbergh Spring on U.S. Hwy 89 Alt, 8 mi S of Flagstaff, 6,500 ft, G. C. C. Bateman, 1967, and R.H.R., 1969; lake on Pinewood Golf Course, 20 mi S of Flagstaff, 4,500 ft, R.H.R., 1969; pond near Interstate Hwy 17, 18 mi S of Flagstaff, 4,300 ft, R.H.R., 1969; pond 5 mi N of Parks on old U.S. Hwy 66, 6,900 ft, R.H.R., 1969), Navajo Co. (Tsegi Canyon, S of Betatakin Ruin, L. L. Hargrave; pond 4 mi E of Clints Well, 50 mi S of Winslow, 6,900 ft, R.H.R., 1969; Cooley Lake, 5 mi SE of McNary, 7,000 ft, R.H.R., 1969), and Apache Co. (pond on State Rd 273, 6 mi W of Greer, 9,000 ft, J. F. Burger and senior author, 1964; Nelson Reservoir, 9 mi S of Springerville, 7,500 ft, M. L. Walton, 1956, and R.H.R., 1969; Hawley Lake, 15 mi E of McNary, 8,500 ft, C. F. Adams, 1965; A-1 Lake, 22 mi W of Eagar, 9,100 ft, R.H.R., 1969; Becker Lake, 2 mi

---

[37]We are under special obligation to Mr. Richard H. Russell for assistance with the taxonomy and nomenclature of Lymnaeidae, as well as for his unpublished records of freshwater mollusks, particularly from the northern sections of the State, here credited to him with his initials (R. H. R.).

N of Springerville, 6,950 ft, R.H.R., 1969; Luna Lake, E of Alpine, 7,800 ft, R.H.R., 1969; Rogers Reservoir, 6 mi E of Nutrioso, 8,560 ft, R.H.R., 1969; canyon 17 mi W of Springerville, 9,050 ft, R.H.R., 1969). Fossil in Late Pliocene, Pleistocene, or sub-Recent of Cochise Co. (Sulphur Springs Valley: Double Adobe Site, 12 mi NW of Douglas, E. Antevs, 1941:66, as *L. palustris nuttalliana*. San Pedro Valley: Sites at Post Ranch, 6 mi, El Paso Pipeline, 7 mi, and California Wash, 9.5 mi S of Benson, R. S. Gray, 1965, and D. W. Taylor, 1966b:94, as *Lymnaea* cf. *elodes;* Choate Ranch 2 Site, W of mouth of Babocomari Creek, ca 20 mi SSW of Benson, E. H. Lindsay, Apr. 1969), Gila Co. (Tonto Creek drainage, 6 mi SW of Payson, D. W. Taylor, 1966b:92, as *Lymnaea* cf. *elodes*). Coconino Co. (Meteor Crater, 20 mi E of Winslow, R. D. Reger and G. L. Batchelder, 1971:191-192, as *Lymnaea palustris* group), Navajo Co. (peat in Laguna Canyon near Betatakin Ruin, J. H. Ferriss, 1920:10, as *Lymnaea palustris* and *L. proxima;* Quaternary near Keet Seel Site, H. G. Richards, 1936:370), and Apache Co. (Fort Defiance, B. Walker, 1915:2, as *Lymnaea proxima*). Both eastern *Stagnicola e. elodes* and western *S. e. nuttalliana* are often regarded as North American subspecies of Old World *Stagnicola palustris* (O. F. Müller, 1774), but it seems advisable to treat *S. palustris* and *S. elodes* as distinct species, since they show significant anatomical differences. *S. elodes* is Recent from the Atlantic to the Pacific and from Alaska to N Arizona and N New Mexico. In S Arizona, S New Mexico, and Texas it is known only as a Late Cenozoic fossil.

130. **Stagnicola (Bakerilymnaea) bulimoides techella** (S. S. Haldeman, 1867, *Amer. Journ. Conch.* 3 [2]:194, Pl. 6, fig. 4, as *Limnea techella*). T.L.: Texas, without precise locality; T.L. not selected thus far. Probably widespread formerly in Arizona, even a century ago; known living now only in Pinal Co. (3.8 mi N of Arizona City, ca 10 mi S of Casa Grande, R.H.R., 1969), Navajo Co. (O'Haco Farm, N side of Winslow, 4,850 ft, R.H.R., 1968-1969; Sitting Bull Trading Post, S side of U.S. Hwy 66, 1.5 mi W of Joseph City, 4,900 ft, R.II.R., 1969; E side of Snowflake, 5,600 ft, R.H.R., 1969), and Apache Co. (3 mi S of St Johns, 5,750 ft, R.H.R., 1969; W side of Springerville, 6,970 ft, R.H.R., 1969). Dead shells, mostly washed-up fossils, in drift of Pima Co. (San Pedro Riv at Redington; Santa Cruz Riv at Tucson and Continental; Rillito Creek at Tucson city limits; Pantano Wash, 2 mi E of Tucson, also 4.5 mi E of Vail), Santa Cruz Co. (Santa Cruz Riv, 5 mi E of Nogales), Cochise Co. (San Pedro Riv at Benson, Fairbank, and Hereford), Pinal Co. (Queen Creek, 2 mi E of Superior), and Maricopa Co. (Salt Riv at Tempe, Pilsbry and Ferriss, 1906:163, fig. 23). Fossil in Late Pliocene to Pleistocene

of San Pedro Valley, Cochise Co. (Choate Ranch 2 Site, W of mouth of Babocomari Creek, E. H. Lindsay, Apr. 1969; Post Ranch Site, 6 mi S of Benson, R. S. Gray, 1965, D. W. Taylor, 1966b:94; Murray Springs Site, associated with mammoth remains 10,000 to 11,000 yrs old); the subfossil *Lymnaea* from Mattie Canyon, W slope of Whetstone Mts, 4,200 ft, Pima Co., figured by R. J. Drake, 1960a: 149, now at Dept. Biol. Sci. Univ. Ariz., is *S. b. techella.* Subspecies *techella* is Recent in south-central United States from Kansas and Missouri S to Alabama in the east and to S California, Colorado, Arizona, New Mexico, and Texas in the west; also in Mexico in Sonora (alive in pond on Mex. Hwy 15, 55 mi S of Nogales, R.H.R., 1968; drift of Rio Sonoyta at Sonoyta, R.H.R., 1968; estuary of Rio Mayo at Huatobampito, R.H.R., 1967) and Tamaulipas (drift of Rio Purificación near Carmen, 24 mi W of Padilla, senior author, 1958). Whether *S. b. techella* is actually separable from *S. b. bulimoides* (I. Lea, 1841) of the Pacific Coast (from Vancouver I to California) is uncertain.

131. **Stagnicola (B.) cockerelli** (H. A. Pilsbry, 1906:130, as *Lymnaea bulimoides cockerelli*). T.L.: Grant, Valencia Co., New Mexico, with original description of March 9, 1906; changed July 24, 1906, to Las Vegas, San Miguel Co., New Mexico, by Pilsbry and Ferriss (1906:162). Recent in Arizona in Maricopa Co. (small pool near Cow Lake, Mesa, C. W. Hibbard and D. W. Taylor, 1960:90), Coconino Co. (Dogtown Wash on Interstate Hwy 40, 3 mi· E of Williams, 6,900 ft, R.H.R., 1969; Spitz Spring, N of Interstate Hwy 40, 1.5 mi W of Parks, 7,000 ft, R.H.R., 1969), Navajo Co. (Holbrook, F. C. Baker, 1911:219; 20 mi N of Winslow, C.W. Hibbard and D. W. Taylor, 1960:90; N side of Winslow at O'Haco Farm, 4,850 ft, R.H.R., 1968-1969; Taylor, 4 mi S of Snowflake, 5,640 ft, R.H.R., 1969; Sitting Bull Trading Post, S of U.S. Hwy 66, 1.5 mi W of Joseph City, 4,900 ft, R.H.R., 1969), and Apache Co. (14 mi W of Ganado, on State Hwy 264, 6,400 ft, R.H.R., 1970). Records from riparian drift of Santa Cruz Riv (Tucson) and San Pedro Riv (Benson) by Pilsbry and Ferriss (1915a:390 and 400) doubtful, as we found only *S. b. techella* in drift there.[38] *S. cockerelli* is Recent in western United States from Montana and South Dakota S to Mexican border, and from east Texas W to California; not definitely known from Mexico; *Lymnaea bulimoides* from Rio Yaqui, 4 mi N of Ciudad Obregón, Sonora, were said by B. A. Branson et al (1964:103) to be "like the form called *L. b. cock-*

---

[38]Pilsbry and Ferriss (1910c:144) report *Lymnaea cockerelli* from a cienega between Chiricahua Mts and Peloncillo Mts, Hidalgo Co., New Mexico [not in Arizona as given], the present San Simon Cienega, 15 mi N of Rodeo. In 1968 R.H.R. and senior author could find only *S. b. techella* there.

erelli"; but all snails seen from Sonora by R.H.R. were *S. b. techella* as listed above. *S. cockerelli* is known from Late Cenozoic in Texas (Late Pleistocene of right bank of Red Riv, 6 mi NW of Denison, Grayson Co., senior author, 1958), New Mexico (F. C. Baker, 1911: 220), Kansas (E. G. Berry and B. B. Miller, 1966, *Malacologia* 4:263), and Oregon-Idaho border (D. W. Taylor, 1966b:72).

    S. cockerelli was originally described as subspecies of *Lymnaea bulimoides,* and retained as such by F. C. Baker (1911:217); Pilsbry and Ferriss (1910:144) cited it as *Lymnaea cockerelli,* and C. W. Hibbard and D. W. Taylor (1960:90) raised it definitely to specific rank, the status accepted by us. R. H. Russell informs us that he found in 1969 thriving colonies of *S. b. techella* and *S. cockerelli* living together (sympatric) in the same pond at two Stations in Navajo Co. (O'Haco Farm and Sitting Bull Trading Post), without transitional specimens or other evidence of interbreeding in nature.

132. **Stagnicola (Hinkleyia) caperata** (T. Say, 1829, *New Harmony Disseminator* 2 [15]: 230, as *Limneus*). T.L.: New Harmony, Posey Co., Indiana. Recent in Arizona in Navajo Co. (irrigation canal 1.5 mi E of Show Low, 6,350 ft, R.H.R., 1969) and Apache Co. (roadside seepages 17 mi E of McNary on State Hwy 73, 9,100 ft, R.H.R., 1969; below dam of Big Lake, 24 mi SSW of Eagar, 8,600 ft, R.H.R., 1970). In Late Pliocene or Pleistocene of Cochise Co. (Double Adobe Site, Sulphur Springs Valley, 12 mi NW of Douglas, E. Antevs, 1946:66. San Pedro Valley: Post Ranch Site, 6 mi, and El Paso Pipeline Site, 7 mi S of Benson, R. S. Gray, 1965, D. W. Taylor, 1966b:94; Murray Springs Site, associated with mammoth remains 10,000 to 11,000 yrs old). Strictly Nearctic; Recent from Alaska, Yukon Territory, Manitoba, Ontario, Quebec, and Maine S to California, Nevada, Utah, Colorado, N Arizona, Nebraska, Iowa, Indiana, and Maryland (after F. C. Baker, 1911:225-234; D. W. Taylor, 1960:53; and D. W. Taylor, H. J. Walter, and J. B. Burch, 1963, *Malacologia* 1 [2]:261-265). Fossil in Pliocene or Pleistocene of Idaho, Nevada, Nebraska, Utah, Kansas, Oklahoma, Indiana, Iowa, New Mexico (Pleistocene, 30 mi NE of Carlsbad, Eddy Co., W. H. Balgemann, Sr., 1963, named by R.H.R., 1969), Arizona, and Texas.

133. **Fossaria obrussa** (T. Say, 1825, *Journ. Acad. Nat. Sci. Philadelphia* Ser. 1, 5 [1]:123, as *Lymneus*). T.L.: Harrowgate near Philadelphia, Pennsylvania. Recent in Arizona in Pima Co. (permanent pond at La Canoa Ranch, 5 mi S of Continental, 2,800 ft, R.H.R. and senior author, 1967; permanent pond at Arivaca, 3,650 ft, R.H.R. and senior author, 1967), Cochise Co. (cienega of Babocomari Creek at Babocomari [now Brophy] Ranch, 4,500 ft, R.H.R.,

1967), Gila Co. (Carrizo Creek at U.S. Hwy 60, 1 mi SW of Carrizo, R.H.R., 1968), Greenlee Co. (Eagle Creek, 6 mi SW of Morenci, 3,550 ft, senior author, 1962), Navajo Co. (irrigation canal, 1.5 mi E of Show Low, 6,350 ft, R.H.R., 1969), and Apache Co. (Little Colorado Riv, at Springerville, 6,900 ft, R.H.R., 1970; head of Canyon del Muerto, Canyon de Chelly National Monument, NE of Chinle, 6,900 ft, R.H.R., 1970). Dead in drift in Yuma Co. (Bill Williams Riv, at Browns Crossing, 30 mi N of Wenden, 1,500 ft, senior author, 1969), Pima Co. (Santa Cruz Riv at Tucson, W. H. Dall, 1897a: 368, as *Limnaea desidiosa,* and Pilsbry and Ferriss, 1915a:400), and Cochise Co. (San Pedro Riv at Benson, senior author). Fossil in Pleistocene of Cochise Co. (Sulphur Springs Valley, 12 mi NW of Douglas, E. Antevs, 1941:67. San Pedro Valley at Lehner Mammoth and Murray Springs Sites, associated with mammoth remains 10,000 to 11,000 yrs old, identified by R.H.R., 1969). Strictly Nearctic; according to F. C. Baker (1911:277-281), Recent from Atlantic to Pacific, N to Mackenzie District and Quebec, SW to New Mexico, Arizona, and north Mexico (not known in Sonora), and SE to Alabama. Fossil in Pliocene, Pleistocene, or sub-Recent of Ontario, Quebec, Maine, New York, Michigan, Wisconsin, Illinois, Iowa, Nebraska, Kansas, Texas, New Mexico, Arizona, Utah, Wyoming, and northern California.

134.   **Fossaria parva** (I. Lea, 1841, *Proc. Amer. Philos. Soc.* 2 [17]: 33, as *Lymnea*); W. G. Binney, 1865a:64, fig. 102 of holotype. T.L.: Cincinnati, Ohio. Known Recent, living in Arizona from Mohave Co. (Pipe Springs at Vermilion Cliffs, Pilsbry and Ferriss, 1911:198), Yavapai Co. (marsh on Verde Riv at Camp Verde, 3,160 ft, R. H. R., 1968; pond at E side of Cornville, 5 mi E of Cottonwood, 3,300 ft, R.H.R., 1969), Gila Co. (pond near Cherry Creek, 1 mi E of Young, 5,100 ft, senior author, 1969), Cochise Co. (Huachuca Mts: Ash Canyon, 5,000 ft, Pilsbry and Ferriss, 1910a:515, F. C. Baker, 1911:251, as *Galba dalli;* Sylvania Springs in Scotia Canyon, NW side of Range, 6,150 ft, R.H.R. and senior author, 1968. Cienega of Babocomari [now Brophy] Ranch, 4,500 ft, R.H.R., 1968), Navajo Co. (O'Haco Ranch at N side of Winslow, 4,850 ft, R.H.R., 1969), and Apache Co. (irrigation canal at W side of Springerville, 6,970 ft, R.H.R., 1969). In drift of Pima Co. (Santa Cruz Riv at Tucson; Rillito Creek at Tucson city limits; Pantano Wash at Vail), Santa Cruz Co. (Santa Cruz Riv at Amado and Calabasas), Cochise Co. (San Pedro Riv at Benson, Pilsbry and Ferriss, 1915a:390, Hereford, B. A. Branson et al, 1966:150, as *Lymnaea humilis,* and Fairbank, senior author), Yavapai Co. (Beaver Creek at Montezuma Castle National Monument, A. G. Smith, 1953:8), and Navajo Co.

(Little Colorado Riv, 3 mi E of Winslow, R.H.R. and senior author, 1968). Fossil in Late Pliocene, Pleistocene, or sub-Recent of Coconino Co. (Winona Site, 13 mi E of Flagstaff, R. D. Reger and G. L. Batchelder, 1971:191 and 193), Cochise Co. (San Pedro Valley: California Wash Site, 9.5 mi S of Benson, R. S. Gray, 1965, D. W. Taylor, 1966b:94, as *Fossaria dalli;* Lehner Mammoth and Murray Springs Sites, associated with mammoth remains 10,000 to 11,000 yrs old), Navajo Co. (subfossil in marl of Laguna Canyon, at Betatakin Ruin, J. H. Ferriss, 1920:10), and Apache Co. (sub-Recent near Fort Defiance, B. Walker, 1915:2, as *Lymnaea* [?] *humilis rustica;* Buell Park, 12 mi N of Fort Defiance, R. J. Drake, 1949c:28 [possibly same Site as Walker's]). *F. parva* is strictly Nearctic; Recent from Alaska and Ontario (James Bay) over most United States (except SE States and Texas), including New Mexico; not known S of Mexican border. Fossil, often cited as *F. dalli*, in Pleistocene of Quebec, Ohio, Illinois, Michigan, Iowa, Nebraska, Kansas, Oklahoma, Colorado, Washington, and Arizona. Synonym: *Lymnaea dalli* F. C. Baker, 1906, *Bull. Illinois State Lab. Nat. Hist.* 17 [6]: 104; 1911:251, Pl. 30, figs. 13-18 (13-16 of cotypes), as *Galba dalli* (T.L.: Lake James, Steuben Co., Indiana, selected in 1907, *Nautilus* 20 [11]:125).

135. **Fossaria modicella** (T. Say, 1825, *Journ. Acad. Nat. Sci. Philadelphia* Ser. 1 5 [1]:122, as *Lymneus*). T.L.: Susquehanna Riv at Owego [not Oswego], Tioga Co., New York. Recent, living in Arizona in Yavapai Co. (E side of Cornville, 5 mi E of Cottonwood, 3,500 ft, R.H.R., 1969), Coconino Co. (Oak Creek, above Sedona, 4,400 ft, M. A. Wetherill, 1940; Circle Bar Draw, 53 mi S of Winslow, 7,000 ft, R.H.R., 1970), Navajo Co. (Show Low Creek, 1 mi S of dam, ca 6 mi S of Show Low, 6,600 ft, A. R. Mead, 1953), and Apache Co. (head of Canyon del Muerto, Canyon de Chelly National Monument, 6,900 ft, R.H.R., 1970; Round Rock Lake, 15 mi WNW of Lukachukai, 5,600 ft, R.H.R., 1970). F. C. Baker (1911:263, fig. 25) included Arizona in Recent range of *F. modicella*, but without definite locality. Fossil in Late Pleistocene of Coconino Co. (Winona Site, 13 mi E of Flagstaff, R. D. Reger and G. L. Batchelder, 1971: 191 and 193) and Cochise Co. (San Pedro Valley at Murray Springs Site, associated with mammoth remains 10,000 to 11,000 yrs old, identified by R.H.R., 1969). Strictly Nearctic; Recent from Vancouver I, Manitoba, E Quebec, and Nova Scotia S to south California, Arizona, New Mexico, Texas, and Alabama; also in N Sonora (10 mi N of Imuris, 3,350 ft, 30° 50′ N, 110° 50′ W, R.H.R., 1969) and Chihuahua (Laguna Toronto, La Boquilla, Distrito Camargo, ca 27° 30′ N, 105° 30′ W, collected by R. J. Drake, Aug. 1947, identified by

R.H.R., 1971). Fossil in Pleistocene of New Jersey, Michigan, Indiana, Illinois, Wisconsin, Iowa, South Dakota, Nevada, New Mexico, Arizona, and Texas (but some of these records questionable).
* **Radix auricularia** (C. von Linné, 1758, *Syst. Nat.* 10th Ed. 2: 774, as *Helix*). T.L.: Europe, without more precise locality, but presumably from Sweden, as the *Fauna Suecica* was cited. Native and widespread in Palearctic Region, Recent as well as fossil in Pleistocene. Not positively known from American Pleistocene; apparently introduced in North America by man in very recent post-Columbian times, being first reported by F. C. Baker (1911:183) as found in 1901 in Illinois (Lincoln Park, Chicago). It since spread rapidly and is now feral (in the open) in S Alaska, Ontario, Vermont, Kentucky, Massachusetts, New York, Pennsylvania, Ohio, Illinois, Colorado, Utah, Arizona, New Mexico (seepage area at S side of Interstate Hwy 10, 1.5 mi E of Deming, 4,300 ft, Luna Co., R.H.R., 1968), Wyoming, Montana, Washington, Oregon, Idaho, and California. First collected in Arizona in 1965 in Apache Co.; now well established (feral) in the northern highlands, but a temporary adventive in the southern counties. Seen from Pima Co. (small nursery pond on Oracle Rd, Tucson, R.H.R., March 1970), Cochise Co. (Willcox, collected by Mary R. Hestand, March 1970), Coconino Co. (Lake Mary, 10 mi SE of Flagstaff, 6,800 ft, R.H.R., 1969; Flagstaff, at 425 E Hutcheson, 6,900 ft, R.H.R., 1969; Ashurst Lake, 18 mi SE of Flagstaff, 7,100 ft, R.H.R., and senior author, 1969; Kinnikinick Lake, 28 mi SE of Flagstaff, ca 7,000 ft, R.H.R. and senior author, 1969), Navajo Co. (Cooley Lake, 5 mi SE of McNary, 7,000 ft, R.H.R., 1969; Rainbow Lake, 10 mi SE of Show Low, 6,700 ft, R.H.R., 1969), and Apache Co. (Becker Lake, 2 mi N of Springerville, 6,950 ft, C. F. Adams, Sept. 1965, R.H.R., Aug. 1969; Crescent Lake, 22 mi SSW of Eager, 8,600 ft, R.H.R., 1970; Big Lake, 24 mi SSW of Eagar, 8,650 ft, R.H.R., 1970; Mexican Hay Lake, 11 mi SW of Eager, 8,600 ft, R.H.R., 1970; Luna Lake, 5 mi E of Alpine, 7,900 ft, R.H.R., 1969; Nelson Reservoir, 9 mi S of Springerville, 7,500 ft, R.H.R., 1969; Rogers Reservoir, 6 mi E of Nutrioso, 8,560 ft, R.H.R., 1969; Earl Park Lake, 16 mi SE of McNary, 8,500 ft, W. J. McConnell, July 1969). Fossils resembling *R. auricularia* are known from North American Oligocene to mid-Pliocene in Idaho, Oregon, and Colorado (D. W. Taylor, 1966b:68-70, Pl. 6, figs. 8-11), but no such shells have been found thus far in American Pleistocene, where they could hardly have escaped notice.
* **Pseudosuccinea columella** (T. Say, 1817, *Journ. Acad. Nat. Sci. Philadelphia* Ser. 1, 1[1]:14, as *Lymneus*). T.L. not given, but probably near Philadelphia, Pennsylvania. Adventive in Arizona; a

thriving colony found in Tucson, Pima Co., in a small nursery pond on Oracle Rd, Nov. 1970 (R. H. Russell, 1971:71); probably imported with water hyacinth (*Eichhornia crassipes*) from SE States, the snails being of the small form (10 to 12 mm high) mentioned by F. C. Baker (1911:170) as peculiar to SE United States. Recent native in North America E of 100th Meridian, from Nova Scotia, Quebec (N to 52° N), and Manitoba S to Florida (S to 27° N) and E Texas, W to Minnesota and E Kansas; also in Mexico and Central and South America (S to Paraguay); reported from Late Cenozoic of New York and Oklahoma.

The following lymneids are known in Arizona from fossils only. *Lymnaea stagnalis appressa* (T. Say, 1821, *Journ. Acad. Nat. Sci. Philadelphia* Ser. 1, 2 [1]:168, as *Lymneus appressus*); the American representative (sibling) of Palearctic *Lymnaea stagnalis* (C. von Linné, 1758), Recent and native in boreal Nearctic Region from Arctic Ocean S to ca 37° N (in Colorado) in the west and to ca 41° N in the east (in Illinois); in Arizona a Late Pleistocene to sub-Recent fossil in Navajo Co. (peat in Laguna Canyon near Betatakin Ruin, J. H. Ferriss, 1920:10; Keet Seel Site, H. G. Richards, 1936:370, as *Lymnaea stagnalis jugularis*; peat in Water Lily Canyon, S of Betatakin Ruin, near the junction of Dogoszki Bito and Tsegi Canyon, L. L. Hargrave, seen by us); we agree with H. B. Baker (1964:153) that the earlier *Lymnaea jugularis* T. Say (1817) is unrecognizable. *Pseudosuccinea dineana* (D. W. Taylor, 1957:659, 3 figs. 1 [1-3], as *Lymnaea* [*Pseudosuccinea*] *dineana*); extinct species of mid-Pliocene (Hemphillian Bidahochi Formation) at White Cone Peak, 50 mi N of Holbrook, Navajo Co.; also collected there by R.H.R., 1969. *Stagnicola reflexa* (T. Say, 1821, *Journ. Acad. Nat. Sci. Philadelphia* Ser. 1, 2 [1]:167, as *Lymneus*); strictly Nearctic, Recent in SE Canada and SE United States, N to ca 69° N (Quebec), S to ca 37° N (Kansas), W to ca 100° W (Manitoba; Nebraska); in Arizona only one (rather dubious) fossil record from Pleistocene in Graham Co. (Ranch 111 Site, 14 mi SE of Safford, P. Seff, 1960: 139). *Stagnicola albiconica* (D. W. Taylor, 1957:657, 3 figs. 1 [4-6], as *Lymnaea* [*Stagnicola*] *albiconica*); extinct species of mid-Pliocene (Hemphillian Bidahochi Formation) at White Cone Peak, 50 mi N of Holbrook, Navajo Co.; also collected there by R.H.R., 1969; first recorded by A. B. Reagan (1929:338, Pl. 14, figs. 6-11; 1932:257-258) as *Limnaea stagnalis* form *appressa*.

Family Physidae

136a. **Physa (Physella) v. virgata** A. A. Gould, 1855a:128. T.L. not selected thus far; described from Gila Riv [presumably in Arizona] and San Diego [presumably in California]; W. G. Binney, 1865a:93, fig. 158 of "authentic specimen" without locality; first selection of lectotype, without locality, by R. I. Johnson, July 1964: 168, Pl. 44, fig. 4 of shell No. 72995, Mus. Comp. Zool., Cambridge, Massachusetts. Recent in Arizona in all counties, at 1,000 to 8,700 ft. Fossil in Pliocene, Pleistocene, and sub-Recent of Gila Co. (6 mi SW of Payson, D. W. Taylor, 1966b:93), Coconino Co. (Winona Site, 13 mi E of Flagstaff; Meteor Crater, 18 mi W of Winslow; both R. D. Reger and G. L. Batchelder, 1971:191-193), Graham Co. (3 mi SE of San Carlos, D. W. Taylor, 1966b:93), and Cochise Co. (San Pedro Valley Sites: Post Ranch, 6 mi, California Wash, 9.5 mi, and El Paso Pipeline, 7 mi S of Benson, R. S. Gray, 1965, D. W. Taylor, 1966b:94; Lehner Mammoth and Murray Springs, associated with mammoth remains 10,000 to 11,000 yrs old). Recent in Nearctic Region in western Canada and United States W of Rocky Mts, S to Arizona, trans-Pecos Texas, and northern Mexico (Sonora, Chihuahua, Coahuila, Hidalgo). Synonyms: *Physa traskii* I. Lea, 1864, *Proc. Acad. Nat. Sci. Philadelphia* 16:115; 1867, *Journ. Acad. Nat. Sci. Philadelphia* Ser. 2, 6:163, Pl. 24, fig. 80 (T.L.: Rio Los Angeles, California). *Physa virgata* mutation *alba* T. D. A. Cockerell, 1902b:138 (T.L.: Salt Riv at Tempe, Maricopa Co.). *Physa humerosa interioris* "Ferriss" H. A. Pilsbry, 1932a:139, Pl. 11, 2 figs. 12 [nomen nudum in J. H. Ferriss, 1920:2 and 4]; (T.L.: West Branch of Navajo Creek, Coconino Co.).

Arizona specimens of *Physa virgata* have been recorded in print as *Physa acuta* J. P. R. Draparnaud, 1805; *Physa heterostropha* T. Say, 1817; *Physa gyrina* T. Say, 1821; *Physa mexicana* R. A. Philippi, 1845; *Physa humerosa* A. A. Gould, 1855; *Physa hawnii* I. Lea, 1864 (by H. C. Yarrow, 1875:940); *Physa anatina* I. Lea, 1864; *Physa ampullacea* W. G. Binney, 1865 (by P. Woolstenhulme, 1942:13); *Physa altonensis* I. Lea, 1864 (by H. C. Yarrow, 1875:939); *Physa mexicana* var. *conoidea* P. Fischer and H. Crosse, 1886; and even *Aplexa hypnorum* (C. von Linné, 1758) by W. H. Dall, 1897a: 369 (D. W. Taylor, 1967b:156, referred Dall's specimens from Santa Cruz Riv at Tucson to *P. virgata*). These names are not necessarily synonyms of *P. virgata,* most of them being misidentifications. Probably, however, *P. virgata* will eventually be recognized as a subspecies, or perhaps a synonym, of one of the earlier species, possibly *Physa gyrina.*

136b. **Physa (P.) virgata bottimeri** W. J. Clench, 1924:12, 2 figs. 4. T.L.: Comanche Spring, Fort Stockton, Pecos Co., Texas, as *Physa bottimeri.* Recent Arizona snails agreeing with original figures and description of *bottimeri* seen by us or recorded from Pima Co.

(pool tank on Redington-Tucson road 12 mi SW of Redington, 4,000 ft, senior author, Oct. 1966; Upper Mendoza Canyon, E side of Coyote Mts, 3,500 ft, A. R. Mead, 1964), Cochise Co. (pond on San Bernardino Ranch at Mexican border, 13 mi E of Douglas, 3,800 ft, senior author, Oct. 1966; this Station is at or near E. A. Mearns' Station 25 of 1892-1893 for U.S. Nat. Mus. lot 130219, recorded by W. H. Dall, 1897a:368 as *Physa mexicana;* D. W. Taylor, 1967b:156, called them *"Physa virgata* in a comprehensive sense," but noted that the series was not typical, most shells being short-spired with expanded body-whorl), and Coconino Co. (Oak Creek below Sedona, with nominate *P. virgata,* M. A. Wetherill, Aug. 1940). We follow D. W. Taylor (1966c:212) in recognizing *bottimeri* as a subspecies, which was thus far reported Recent from trans-Pecos Texas only. It is also Recent in New Mexico (San Simon Cienega, 15 mi N of Rodeo, Hidalgo Co., 3,900 ft, R.H.R. and senior author, 1968). Similar sub-Recent fossils also seen by us from Mexico in Coahuila (3.5 mi S of Artegea, between Saltillo and Matchuala, G. E. Fay, 1956). The correct relationships of *P. v. virgata, P. virgata bottimeri,* and *P. humerosa* remain obscure at present.

? 137. **Physa (Physella) humerosa** A. A. Gould, 1855a:128; 1856: 331, Pl. 11, figs. 6-9. T.L.: described from Colorado Desert and Pecos Riv. D. W. Taylor (1966c:210) selected as T.L. Colorado Desert, California (without more precise locality), where the dead syntypes were collected by W. P. Blake, Nov. 17, 1853, on the dry shore of a former lake, ca 13 mi SE of Deep Well, as described in his Report (W. P. Blake, 1856 and 1858:97); his diary and map place this Station at or near the present-day Point of Rocks, near Salton View, Riverside Co. We saw no Recent Arizona shells definitely referable to the species, which is included here only for D. W. Taylor's record (1966c:212) of two shells (lot 29119 U.S. Nat. Mus.), "fresh and retaining periostracum, labeled Gila, Arizona," collected by Dr. O. Loew in 1873. The supposed *P. humerosa* recorded by Pilsbry and Ferriss (1911:198) from Coconino Co. (Indian Gardens on Bright Angel Trail, Grand Canyon of Colorado Riv), should be revised or collected again; they might have been what we are calling *P. virgata bottimeri.* Senior author refers to *P. humerosa* Late Pleistocene fossils from the San Pedro Valley at Murray Springs Site, Cochise Co. (one shell in Unit F_1 of C.V. Haynes, Jr., associated with mammoth remains 10,000 to 11,000 yrs old; another in Unit G_1, ca 5,000 yrs old, possibly washed in from an older Unit); also from sub-Recent alluvium at The Narrows 10 mi N of Benson, R.H.R., 1970. D. W. Taylor restricts Recent *P. humerosa* to SE California, SW Arizona, and N Baja California.

Family Planorbidae

138. **Gyraulus (Torquis) parvus** (T. Say, 1817, in *W. Nicholson's British Encyclopedia* 1st Amer. Ed., 2, Art. Conchology, signature Xx, 9th unnumbered page, Pl. 1, fig. 5, as *Planorbis*). T. L.: Delaware Riv, Pennsylvania. Widespread Recent in Arizona in Coconino Co. (Walker Lake of San Francisco Mtn, 8,250 ft, R.E.C. Stearns, 1891:103, and R.H.R., 1969; Spitz Spring, 9 mi E of Williams, 7,000 ft, R.H.R., 1969; pond 20 mi S of Flagstaff, R.H.R., 1969; Lindbergh Spring on Alternate U.S. Hwy 89, 8 mi S of Flagstaff, 6,500 ft, R.H.R., 1969; Ashurst Lake, 18 mi SE of Flagstaff, 7,114 ft, R.H.R., 1969; Chevelon Creek, 16 mi NW of Heber, 5,200 ft, R.H.R., 1969), Navajo Co. (N side of Winslow, 4,800 ft, R.H.R., 1968-1969; 8 mi S of Show Low, 6,600 ft, R.H.R., 1969; Cholla Lake, 10 mi E of Holbrook, 5,000 ft, R.H.R., 1969; Clear Creek Reservoir, 6 mi SE of Winslow, 4,850 ft, R.H.R., 1969; Rainbow Lake, 10 mi SE of Show Low, 6,700 ft, R.H.R., 1969), Apache Co. (East Fork of Black Riv, 3 mi N of Buffalo Crossing, 7,700 ft, senior author, 1964; A-1 Lake, 14 mi E of McNary, 8,600 ft, R.H.R., 1969; Rogers Reservoir, 6 mi E of Nutrioso, 8,560 ft, R.H.R., 1969; Luna Lake, 5 mi E of Alpine, 7,900 ft, R.H.R., 1969; Wilson Reservoir, 9 mi S of Springerville, 7,500 ft, R.H.R., 1969), Greenlee Co. (Eagle Creek, 6 mi SW of Morenci, 3,550 ft, senior author, 1962), Cochise Co. (Parker Canyon Lake, 8 mi SE of Canelo, 5,500 ft, senior author, 1967. Chiricahua Mts: Cave Creek at Herb Martyr Dam, 5,600 ft, and Rucker Canyon, 5,900 to 6,000 ft, both senior author), Santa Cruz Co. (Pena Blanca Lake, Pajaritos Mts, 3,900 ft, senior author), Yavapai Co. (E side of Cornville, 5 mi E of Cottonwood, 3,300 ft, R.H.R., 1969), and Gila Co. (Sierra Ancha, Reynolds Creek, 5,400 ft, senior author, 1964; Seneca Lake near Seneca, 3,000 ft, senior author, 1968; Carrizo Creek, 1 mi SW of Carrizo, senior author, 1968). Dead drift shells, some washed-up fossils, in Pima Co. (Santa Cruz Riv at Tucson and Marana; Rillito Creek at Tucson city limits; Pantano Wash, 1 mi E of Tucson; Tanque Verde Creek, E of Tucson; San Pedro Riv at Redington), Santa Cruz Co. (Santa Cruz Riv at Calabasas and 7 mi NW of Nogales), Cochise Co. (San Pedro Riv at The Narrows, Benson, Hereford, Charleston, and Palominas), Pinal Co. (San Pedro Riv at Mammoth; Gila Riv at Sacaton), Maricopa Co. (New River, 29 mi N of Phoenix; Canyon Lake), Gila Co. (Salt Riv at U.S. Hwy 60, B. A. Branson et al, 1966:150, as *Gyraulus similaris*), Yavapai Co. (Beaver Creek at Montezuma Castle National Monument, A. G. Smith, 1953a:8, as *Planorbis vermicularis;* Agua Fria Riv at junction of Badger Spring Creek, 6 mi NE of Bumblebee), Coconino Co. (Oak Creek, 7 mi N of Sedona), and Navajo Co.

(Little Colorado Riv, 3 mi E of Winslow). Fossil in Late Pliocene, Pleistocene, and sub-Recent of Coconino Co. (Walnut Creek Site, 5 mi W of Winona, H. S. Colton, 1929:94; Meteor Crater Site, 18 mi W of Winslow, R. D. Reger and G. L. Batchelder; 1971:191-193), Apache Co. (Fort Defiance, B. Walker, 1915:2), and Cochise Co. (Sulphur Springs Valley Site, 9 mi NW of Douglas, S. Antevs, 1941:37. San Pedro Valley Sites: Post Ranch, 6 mi, California Wash, 9.5 mi, and El Paso Pipeline, 7 mi S of Benson, R. S. Gray, 1965, D. W. Taylor, 1966b:94; Lehner Mammoth and Murray Springs, associated with mammoth remains 10,000 to 11,000 yrs old). Strictly Nearctic, Recent in S Canada and most of United States, N to Alaska and Yukon, S to Florida, and in Mexico to N Sonora (Rio Nacozari, 7 mi S of Nacozari, B. A. Branson, 1964:103). Synonyms: *Planorbis vermicularis* A. A. Gould, 1847, *Proc. Boston Soc. Nat. Hist.* 2:212 (T.L.: "interior of Oregon at Drayton" [probably Dayton, Yamhill Co., Oregon]). *Planorbis similaris* F. C. Baker, 1917, *Bull. Amer. Mus. Nat. Hist.* 41:529 (T.L.: Smartweed Lake near Tolland, Gilpin Co., 8,850 ft, Colorado).

? 139. **Gyraulus (T.) circumstriatus** (G. W. Tryon, 1866a:113, Pl. 10, figs. 6-8, as *Planorbis* subgenus *Gyraulus*). T.L.: Weatogue, Hartford Co., Connecticut. Not seen by us from Arizona, either Recent or fossil. Included tentatively for published records from Yavapai Co. (Pecks Lake, 3 mi E of Clarkdale and Verde Riv, ca 3,300 ft, collected presumably alive by E. H. Ashmun, southernmost Station of species, C. W. Hibbard and D. W. Taylor, 1960:97) and Apache Co. (dead shells in riparian drift of Zuni Riv near St Johns, B. A. Branson et al, 1966:150).[39] According to C. W. Hibbard and D. W. Taylor (1960), Recent in S Canada and N and central United States, N to British Columbia, Alberta, N Ontario, and Labrador, S to West Virginia in the east and to N Arizona in the west; not known from California or Sonora. Recorded from Late Pliocene of Cochise Co. (California Wash Site, ca 9.5 mi S of Benson, R. S. Gray, 1965, D. W. Taylor, 1966b:94, as *Torquis circumstriatus*), and from Late Pleistocene of Coconino Co. (Meteor Crater, 18 mi E of Winslow, R. D. Reger and G. L. Batchelder, 1971:192-193); also known fossil from Pleistocene of Kansas and from sub-Recent alluvium in New Mexico (Rio Grande floodplain in Dona Ana Co., A. L. Metcalf, 1967:31 and 37).

* **Biomphalaria havanensis** (L. Pfeiffer, 1839, *Arch. f. Naturgeschichte* 5 (1):354, as *Planorbis*); specific name and synonyms have

---

[39]*G. circumstriatus* could not be found in Pecks Lake in 1969 by R.H.R. and senior author. The drift shells of Zuni Riv were probably washed-up fossils.

been combined also with *Taphius, Tropicorbis, Armigerus, Obstructio,* and *Planorbina,* but *Biomphalaria* H. B. Preston (1910) has been given official preference (1965, Opinion 735, *Bull. Zool. Nomencl.* 22 [2]:94-99); combinations with *Segmentina* and *Planorbula* were erroneous. T.L.: Cuba, without more precise locality; presumably near Havana. Not native in Arizona, where it is not known fossil; only an occasional, perhaps temporary adventive, in artificial bodies of water stocked with introduced fish or sometimes transported by water birds. The one published record is of dead shells in riparian drift of San Pedro Riv at Benson, Cochise Co. (Pilsbry and Ferriss, 1915a:390, as *Planorbis liebmanni*). Living populations are known at three Stations in Santa Cruz Co. (Pena Blanca Lake, Pajaritos Mts, 3,900 ft, senior author and R.H.R., 1961-1967; small permanent pond 3.5 mi due W of Pena Blanca Lake, near Ruby road, ca 4,000 ft, R.H.R., 1967) and Pima Co. (pond at Quitobaquito Springs, Organ Pipe Cactus National Monument, 1,110 ft, senior author, 1969); probably to be found also in some other artificial lakes in the State. Originally a tropical and subtropical snail, widespread native in Mexico, Central America, and Antilles; native in S half of Texas (N to Travis Co.; farther N, as at Waco, Dallas, etc, probably now only an adventive in artificial bodies of water); in Sonora thus far only in riparian drift of Rio Mayo at Navajoa and near its mouth at Huatobampito, R.H.R., 1967, probably washed down from some impounded water ("presa"). A common Pleistocene fossil in Texas and Louisiana. As investigated by C. S. Richards (1963) for other species of *Biomphalaria,* populations of *B. havanensis,* particularly those with young snails, consist normally of toothless shells and others with internal calcareous lamellae or teeth, usually at some distance within the aperture on the parietal and outer walls. As a rule, the armature is formed at an early growth stage, but not always at the same age of the snail nor in all individuals, and often before a growth diapause; eventually, when growth resumes, the lamellae are resorbed by the snail, so that they are not found in the early whorls of the adult, and only exceptionally near the aperture. What initiates the production and later resorption of the lamellae, as well as their possible purpose, is conjectural. Perhaps the lamellae merely store surplus calcium temporarily for later use. It seems most unlikely that they protect the snail against parasites or predators, since many young and most adults dispense with them. The presence of an armature in only part of the population has caused some confusion in taxonomy, particularly when dealing with small lots or lots consisting only of adult snails. A similar internal armature is a normal feature of *Planorbula,* so that *Planorbula armigera* (T. Say, 1818) and *B. havanensis* are easily confused; furthermore, toothed and toothless individuals, even in the same population, have also been referred to different species or genera. *Planorbula* and *Biomphalaria* are anatomically distinct, and F. C. Baker (1945:80 and 85) pointed out some differences also in their armature. He failed to recognize, however, that toothed and toothless individuals of *B. havanensis* are cospecific and often occur in the same population, so that he called the toothless *havanensis* (placing them in subgenus *Tropicorbis,* sensu stricto) and the toothed *obstructus* (in subgenus *Obstructio*). The lamellae vary in number (1 to 6), shape, size, and arrangement, again within the same population. For instance, a set of 356 living snails, from the pond 3.5 mi W of Pena Blanca Lake, consisted mostly of young snails (8.5 mm or less in diam., while adults may reach 11 mm); 247 were toothless (snails 2.8 to 8.5, but mostly over 7 mm in diam.) and 109 toothed (snails 3.4 to 6.8 mm in diam.; only five over 5 mm; 102 of these had six, and seven had four or five teeth). The ratio one to two for toothed to toothless snails in this lot is abnormally high due to the excess of young snails. As immatures are often ignored by collectors, many museum lots are mainly of adults, hence with few or no toothed shells. The shell of *havanensis* varies also in shape, relative height and diameter, size for same number of whorls, number of whorls and maximum size of final adult, thickness of aperture, sculpture, etc. Hence the many synonyms in current use; only three of local importance cited here: *Planorbis obstructus* A. Morelet, 1849, *Test. Noviss. Ins. Cubanae Amer. Centr.* 1:17 (T.L.: Carmen I off Campeche, Mexico), for toothed snails. *Planorbis decipiens* C. B. Adams, 1849, *Contrib. to Conchology* 3:43 (T.L.: Jamaica, without precise locality); syntypes seen by senior author at Mus. Comp. Zool., Cambridge, Massachusetts, from Westmoreland and Caymanas, Jamaica, some toothed, others not, are *Biomphalaria havanensis;* the trivial name *decipiens,* in the combination *Taphius decipiens,* was applied by error to Arizona *Drepanotrema (Antillorbis) aeruginosum* (A. Morelet) by B.A. Branson et al, 1964:103 and 1966:150. *Planorbis liebmanni* W. B. R. H. Dunker, 1850, *Syst. Conch.-Cab.* 1 Abt. 17:59, Pl. 10, figs. 32-34 (T.L.: Vera Cruz, Mexico). In view of the great variability of the shell, the specific distinctiveness from *havanensis* of the supposedly extinct Pleistocene *Biomphalaria* is questionable: *B. goodrichi* (A. B. Leonard and D. S. Franzen, 1944, *Univ. Kansas Sci. Bull.* 30 [1]:25, Pl. 5, fig. 22, as *Helisoma;* Beaver Co., Oklahoma); and *B. kansasensis* E. G. Berry and B. B. Miller, 1966, *Malacologia* 4 (1):266, figs. 3-5; Meade Co., Kansas. If correctly placed in *Biom-*

phalaria, these fossils extend the former range of the genus to ca
37° N, some 500 mi N of the Recent range.

140. **Helisoma (Pierosoma) tenue** (W. B. R. H. Dunker, 1850,
Syst. Conch.-Cab. 1, Abt. 17:45, Pl. 10, figs. 14-19, as *Planorbis).*
T.L.: near Mexico City. One of the most widely distributed snails
of Arizona; known Recent from all counties except Mohave, Yavapai, Apache, and Greenlee, where no doubt it lives also. Usually
at low elevations, below 6,000 ft; but it occurs at 8,500 ft in Pinaleno Mts on Mt Graham at Riggs Flat, at 7,000 ft on Santa Catalina Mts in Rose Canyon Lake, and at 8,700 ft on San Francisco
Mtn in a tank of NW slope in Hart Prairie near Snow Bowl (R.H.R.,
1970). Recent in most of Mexico (Baja California, Sonora, Chihuahua,
etc), Arizona, New Mexico, S California, and trans-Pecos Texas.
Sometimes placed in genus *Planorbella.* Fossil in Late Pleistocene
of Coconino Co. (Winona Site, 13 mi E of Flagstaff; Meteor Crater
Site, 18 mi W of Winslow; both R. D. Reger and G. L. Batchelder,
1971:191-193, as *Planorbella tenuis).* Synonyms: *Planorbis sinuo-*
sus A. Bonnet, 1864, *Rev. Mag. Zool.* (2) 16:280, Pl. 22, figs. 31-32
(T.L.: Rio Grande in New Mexico). *Planorbis tumens* P. P. Carpenter, 1857, *Cat. Mazatlan Shells Brit. Mus.*:181 (T.L.: Mazatlan,
Sinaloa, Mexico). Arizona specimens also recorded as *Planorbis*
trivolvis T. Say, 1817, *P. caribaeus* A. d'Orbigny, 1841, or *P. tumi-*
dus W. B. R. H. Dunker, 1850, names not necessarily synonyms
of *H. tenue.* The *Helisoma t. trivolvis* recorded by J. P. Woolstenhulme (1942:12) from Soldier Lake and Ashurst Lake, Coconino
Co., were probably *H. tenue,* found in Ashurst Lake by R.H.R. and
senior author, 1969.

? * **Helisoma (Pierosoma) subcrenatum** (P. P. Carpenter, 1857,
Proc. Zool. Soc. London 22 for 1856:220, as *Planorbis subcrenatus).*
T.L.: Oregon, without precise locality; holotype figured by W. G.
Binney (1865a:103, 2 figs. 176) with Washoe [? = Wasco, Sherman Co.]
as T.L. Recent in Apache Co.: living in Tsaile Lake, at head of
Canyon del Muerto, Canyon de Chelly National Monument, ca
7,000 ft, R.H.R., Aug. 1970; whether this is a Relict of a former wider
pre-Columbian distribution or a post-Columbian introduction by
man cannot be decided at present. Known as a Pliocene fossil in
Navajo Co.: Hemphillian Bidahochi Formation at White Cone
Peak, 50 mi N of Holbrook, D. W. Taylor, 1957:660, collected also
by R.H.R., 1969. Recent shells of Apache Co. show no appreciable
differences from Recent shells of Oregon or the fossils of Navajo
Co. Known Recent from Alaska, British Columbia, Northwest Territories, Alberta, Manitoba, Washington, Utah, Oregon, Idaho,
Montana, California, and Colorado. Synonym: *Planorbis binneyi*

G. W. Tryon, 1867, *Amer. Journ. Conch.* 3 (2):197 (new name for *Planorbis corpulentus* S. S. Haldeman, 1844, *Monogr. Lymniades* 7:19, Pl. 3, figs. 7-9; not *P. corpulentus* T. Say, 1824).

141. **Drepanotrema (Antillorbis) aeruginosum** (A. Morelet, 1851, *Test. Noviss. Ins. Cubanae Amer. Centr.* 2:15, as *Planorbis*). T.L.: Lake Yzabal, Guatemala. Recent in south-central Arizona, mostly found as dead, but often fresh, shells; known from Pima Co. (Santa Cruz Riv at Tucson and 0.5 mi W of Continental; San Pedro Riv at Redington; Rillito Creek at Tucson city limits; Cienega Creek, 3.5 mi NE of Empire Ranch; irrigation ditch on Hwy 86, 3 mi W of Robles Junction), Santa Cruz Co. (small creek in Pajaritos Mts, 2.5 mi W of Pena Blanca Lake, 4,000 ft; wash at NE foot of Tumacacori Peak; Santa Cruz Riv at Amado, 3,050 ft, and 2.5 mi NE of Lochiel, 4,600 ft), Cochise Co. (cienega of Babocomari Creek at Babocomari [now Brophy] Ranch, 4,500 ft; San Pedro Riv at The Narrows, Benson, Hereford, and Palominas), and Pinal Co. (Gila Riv at Sacaton). H. W. Harry and B. Hubendick (1964:30-33, figs. 90-92 and 114-117) referred *Planorbis aeruginosus* to subgenus *Antillorbis;* we definitely unite with it *H. lineatus sonorensis, P. filocinctus,* and *P. arizonensis,* cited below, which these authors only regarded as possible synonyms; Arizona shells agree perfectly with their drawings of Puerto Rico *D. aeruginosum.* As here understood, the species lives in Antilles, Guatemala, Mexico (including Sonora), S Texas, and S Arizona. Not known fossil. Synonyms: *Planorbis salleanus* W. B. R. H. Dunker, 1855, *Proc. Zool. Soc. London* 21:53 (T.L.: Santo Domingo). *Planorbis circumlineatus,* R. J. Shuttleworth, 1854, *Mitth. Naturf. Ges. Bern:*96 (T.L.: Humacao, Puerto Rico). *Helicodiscus lineatus sonorensis* J. G. Cooper, 1893:343, Pl. 14, figs. 10a-d (T.L.: San Miguel, Sonora). *Planorbis filocinctus* H. A. Pilsbry and J. H. Ferriss, 1906:165, Pl. 9, figs. 1-3 (T.L.: drift of San Pedro Riv at Benson, Cochise Co.); not *Planorbis filocinctus* C. L. F. von Sandberger, 1875. *Planorbis arizonensis* H. A. Pilsbry and J. H. Ferriss, 1915c:390, footnote (new name for *P. filocinctus* H. A. Pilsbry and J. H. Ferriss, 1906; same T.L.). *Planorbis (Gyraulus) santacruzensis* L. Germain, 1923, *Rec. Indian Mus.* 21:138, figs. 18-21 (T.L.: St Croix I, Virgin Is). B. A. Branson et al, 1964:103 and 1966:150, reported *D. aeruginosum,* from drift of San Pedro Riv at Hereford, by error as *Taphius decipiens; Planorbis decipiens* C. B. Adams, 1849, is a synonym of *Biomphalaria havanensis* (L. Pfeiffer, 1839).

? *Promenetus (Phreatomenetus) carus* (H. A. Pilsbry and J. H. Ferriss, 1906:164, Pl. 9, figs. 4-5, as *Planorbis*). T.L.: canyon of Pecos Riv, ca 1 mi above RR high bridge, ca 42 mi NW of Del Rio,

Val Verde Co., Texas. Doubtfully Recent in Arizona; only published record from dead shells in drift of San Pedro Riv at Hereford, Cochise Co., by B. A. Branson et al, 1966:150. Possibly washed-up fossils were found by senior author in drift of San Pedro Riv at Palominas and Benson, Cochise Co., but the species is not known in situ from Pleistocene thus far. Definitely Recent in trans-Pecos and central Texas only.

The following planorbids are known thus far in Arizona only from Pliocene, Pleistocene, or sub-Recent fossils in situ or washed-up in drift. *Gyraulus deflectus* (T. Say, 1824, in W. H. Keating, *Narrative S. H. Long's Exped.* 2, Appendix:261, Pl. 15, fig. 8, as *Planorbis;* T.L.: "Northwest Territory" [= Minnesota]); recorded by Pilsbry and Ferriss, 1911:198, from Fredonia, Coconino Co., without details; and by B. Walker, 1915:2, as a post-Pleistocene fossil from Fort Defiance, Apache Co.; not seen by us. *Omalodiscus pattersoni* (F. C. Baker, 1938, *Nautilus* 51, Pt. 4:129, as *Planorbis;* T.L.: 6 mi N of Ainsworth, Brown Co., Nebraska; A. B. Leonard, 1957, *Illinois State Geol. Surv. Rept. Invest.* 201:12, Pl. 2, figs. 4-6 of holotype); extinct species known from Pliocene of Late Pleistocene in Ohio, Oklahoma, Texas, Kansas, Nebraska, Utah, Washington, Wyoming, and Idaho (D. W. Taylor, 1958, *Journ. of Paleontology* 32:1149-1153); senior author refers to it a washed-up shell, presumably a Pleistocene fossil, from riparian drift of San Pedro Riv at Fairbank, Cochise Co. *Promenetus kansasensis* (F. C. Baker, 1938, *Nautilus* 51 (4):129, as *Menetus;* from Upper Pliocene in Meade Co., Kansas, and Brown Co., Nebraska); extinct species recorded from Late Pliocene (Blancan) in Cochise Co. (San Pedro Valley: El Paso Pipeline Site, ca 7 mi S of Benson, R. S. Gray, 1965, D. W. Taylor, 1966b:94, as *Promenetus exacuous kansasensis*). *Promenetus (Phreatomenetus) umbilicatellus* (T. D. A. Cockerell, 1887, *Conchol. Exchange* 2 (5):68, as *Planorbis;* T.L.: Recent in Manitoba); in Arizona in Pliocene of Navajo Co. (Hemphillian Bidahochi Formation at White Cone Peak, 50 mi N of Holbrook, D. W. Taylor, 1957:660; collected also by R.H.R., 1969); in Pleistocene of Gila Co. (Tonto Creek drainage, 6 mi SW of Payson, D. W. Taylor, 1966:92) and Cochise Co. (San Pedro Valley: Post Ranch Site, 3 mi S of Benson, R. S. Gray, 1965, D. W. Taylor, 1966b:94); washed-up fossils in riparian drift of Pima Co. (Santa Cruz Riv at Tucson and Amado) and Cochise Co. (San Pedro Riv at Hereford and Palominas). R. D. Reger and G. L. Batchelder's (1971:193) record from "modern Walnut Creek drainage" in Coconino Co. was probably also of washed-up fossils, not of live snails. *Armiger crista* (C. von Linné, 1758, *Syst. Nat.* 10th Ed. 2:709, as *Nautilus;*

T.L.: Germany, without precise locality); not known Recent in Arizona; in Late Pleistocene of Coconino Co. (Winona, 13 mi E of Flagstaff, and Meteor Crater, 18 mi W of Winslow, R. D. Reger and G. L. Batchelder, 1971:191-193); washed-up fossils in riparian drift of Little Colorado Riv, 3 mi E of Winslow, 4,810 ft, Navajo Co., R.H.R. and senior author, 1968; Recent in Eurasia and, in North America, in Alaska (N to Fort Yukon, 66° 34' N, 145° 16' W), southern Canada, and northern United States (according to D. W. Taylor, 1960:58, S to 41° N, sporadically in W California and central Utah); also in Pleistocene of Texas.

Family Ancylidae (including Ferrissiidae)

142. **Laevapex (Ferrissia) californica** (J. Rowell, May 1863:21, 3 figs. 5, as *Gundlachia*). T.L.: "Feather Riv at Marysville," Yuba Co., California; holotype at Univ. Michigan Mus. Zool. No. 102011, collected by J. Rowell at T.L., selected by P.F. Basch (1965, *Bull. Mus. Comp. Zool.* 129 [8]:435). The only freshwater limpet known Recent in Arizona, probably more widespread than reported thus far. First found by J. H. Ferriss in 1913 at one Station in Sabino Canyon, Santa Catalina Mts, Pima Co., "at Alkali Spring, Lowell U.S. Ranger Station, on dead leaves of *Platanus* [misspelled *"Plantinus"*] *wrightii*," and recorded by Pilsbry and Ferriss (1919a:305) as follows: "Very few examples are in the *Gundlachia* stage; none were found in the septate stage; many have the narrow, high, oblique shape of septates, but without septum. Many of them reached the normal size of septates, then had a resting stage during which the shell became blackened, resuming growth subsequently along the margins, forming a narrow, oblique shell, somewhat like *A. parallelus* in outline. Other examples become wider, about as in *A. rivularis,* in the second period of growth. The early stages are similar in all, having the usual *Ferrissia* sculpture. Those individuals in the *Gundlachia* stage do not appear specifically separable from *G. californica.*" The limpet was not collected again at this Station until Oct. 1968, when Floyd G. Werner found it breeding profusely in an indoor aquarium stocked with plant debris from the Creek, ca 0.5 mi N of the Lower Sabino Canyon dam; he also observed in the aquarium a limpet attached to the elytra of a ditiscid beetle kept with the snails. It was found in Dec. 1968 at the same Station and elsewhere in Sabino Canyon at 2,800 to 3,000 ft, by R.H.R. Later also in Pima Co., in NW section of Tucson, within the city limits on Oracle Rd, in two small private ponds, at ca 2,300 ft, by R.H.R., Sept. 1969; and in Navajo Co., in a small pond at W side of State Hwy 173 near Lakeside, ca 8 mi S of Show Low, at 6,600 ft,

by R.H.R., Sept. 9, 1969, and in Clear Creek Reservoir, 5.5 mi SE of Winslow, at 4,850 ft, by R.H.R., Aug. 27, 1969; at all later Stations on leaves and stalks of waterlilies. It is known fossil from Cochise Co., in San Pedro Valley, in Late Pleistocene of Murray Springs Site, associated with mammoth remains 10,000 to 11,000 yrs old; and in sub-Recent alluvium of San Pedro River banks at The Narrows, ca 10 mi N of Benson, collected by R.H.R. and T. R. Van Devender, Nov. 1970. In New Mexico, A. L. Metcalf (1967:37) reports it, as *Ferrissia fragilis,* from Rio Grande Valley, living, as well as sub-fossil in valley alluvium. *L. californica* is widely distributed in North America. Synonyms: *Ancylus fragilis* G. W. Tryon, June-July 1863:149, Pl. 1, fig. 15 (T.L.: Laguna Honda, California; so far as could be traced, the description was published in June or early July, while that of *G. californica* appeared not later than May, according to information kindly supplied by Allyn G. Smith; he also informed us that the T.L. of *fragilis* is the small reservoir still called Laguna Honda in San Francisco Co.). J. G. Cooper (1890, *Proc. California Acad. Sci.* Ser. 2 3:83) first suggested that *Ancylus fragilis* was the non-septate form or stage of *Gundlachia californica. Gundlachia meekiana* W. Stimpson, Dec. 1863, *Proc. Boston Soc. Nat. Hist.* 9:250, 3 figs. 2 (T.L.: pond near Potomac Riv, Washington, D.C.).

Laevapex parallela (S. S. Haldeman, Jan. 1841, *Monogr. Amer. Lymniades,* 2:unnumbered 3rd p. of wrapper, as *Ancylus;* T.L.: "Vermont," without precise locality) is known in Arizona only from Late Pleistocene fossils in Coconino Co.: Walnut Creek near Winona, 13 mi E of Flagstaff (H. S. Colton, 1929:94; G. L. Batchelder and R. D. Reger, 1969:7); Meteor Crater, 20 mi W of Winslow (R. D. Reger and G. L. Batchelder, 1971:191-193, as *Ferrissia parallela*).

Prosobranchia

Family Viviparidae

* **Viviparus (Cipangopaludina) chinensis malleatus** (L. Reeve, 1863, *Conch. Icon.* 14 *Paludina:*Pl. 5, figs. 5a-b, with letterpress, as *Paludina malleata*). T.L.: Japan, without more precise locality. Native in East Asia; introduced in North America and permanently established in the open in several United States localities. Thus far in Arizona only as a temporary introduction. In 1962 alive in Tucson in a small pond on campus of Univ Arizona, as recorded by W. J. Clench and S. L. H. Fuller (1965:404); it lived also for some time on campus in open-air fish tanks; later (1967-1970) it disappeared from the Tucson area.

Family Valvatidae

143. **Valvata (Valvata) humeralis californica** H. A. Pilsbry, 1908:82. T.L.: Bear Lake, San Bernardino Co., California. Recent in Arizona in Apache Co.: dead, fresh shells in drift of East Fork of Black Riv, 3.5 mi NE of Buffalo Crossing, 7,700 ft, senior author, Aug. 1964; many alive in Big Lake, 24 mi SSW of Eagar, 8,650 ft, R.H.R., Aug. 1970. Subspecies also Recent in Washington, Oregon, Idaho, Montana, Wyoming, California, Utah, and New Mexico (Grant, Valencia Co., Pilsbry, 1908:82).

D. W. Taylor (1957:657) reports *Valvata humeralis* T. Say (1829, *New Harmony Disseminator* 2[16]:244; T.L.: near Mexico City) from mid-Pliocene of White Cone Peak, 50 mi N of Holbrook, Navajo Co.; collected also by R.H.R., 1969. G. L. Batchelder and R. D. Reger (1969:7) and R. D. Reger and G. L. Batchelder (1971:192-193) found it in Late Pleistocene of Coconino Co. at Winona Site, 13 mi E of Flagstaff, and at Meteor Crater Site, 20 mi W of Winslow. The subspecies at these Sites was not determined.

Valvata (Tropidina) tricarinata (T. Say, 1817, *Journ. Acad. Nat. Sci. Philadelphia* 1 (1):13, as *Cyclostoma*) is not known Recent in Arizona, but perhaps occurs there as a Pleistocene fossil.

Family Hydrobiidae (Truncatellidae)

144. **Tryonia protea** (A. A. Gould, Oct. 1855a:129, as *Amnicola*). T.L.: "Gran Jornada," Colorado Desert, California; the dead, fossil types collected by Thomas H. Webb and Wm. P. Blake in Riverside Co., near Salton View of present-day maps, as explained for the types of *Physa humerosa*. Collected alive in Arizona by senior author, Apr. 24, 1962, in an irrigation ditch N of Buckeye, Maricopa Co. Previously reported from the State only for dead shells in riparian drift of Santa Cruz Riv at Tucson, Pima Co., by Pilsbry and Ferriss (1915b:400), as *Paludestrina protea*. Synonym: *Melania exigua* T. A. Conrad, Feb. 1855:269 (T.L.: Colorado Desert, California); not *Melania exigua* A. Morelet, 1851.

145. **Tryonia imitator** (H. A. Pilsbry, 1899b:124, as *Paludestrina*). T.L.: Santa Cruz, Santa Cruz Co., California. Known Recent in Arizona at Quitobaquito Springs, Organ Pipe Cactus National Monument, Pima Co., collected by senior author, Apr. 18, 1963.

146. **Fontelicella** sp. One or several, possibly undescribed species known Recent in Arizona from Cochise Co. (Huachuca Mts: Ash and Tanner [now Garden] Canyons, Pilsbry and Ferriss, 1910a:516, as *Paludestrina stearnsiana;* Sylvania Spring in Scotia Canyon, NW side of Range, 5,200 ft, collected by J. T. Marshall, Jr., June 17, 1951, and by senior author and R.H.R., Jan. 30, 1968; spring in

Huachuca Canyon, 5,250 ft, collected by senior author, June 3, 1967; Tombstone Reservoir in Miller Canyon, 6,000 ft, collected by C. Lehner, 1970) and Santa Cruz Co. (Monkey Spring, 5.5 mi SW of Sonoita, 4,400 ft, collected by senior author and others, 1961-1967).

Amnicola longinqua A. A. Gould, 1855a:129, described from fossils of Colorado Desert of California, is placed tentatively in *Fontelicella* by W. O. Gregg and D. W. Taylor (1965:108).

Other hydrobiids, of doubtful or unidentified genera and species, perhaps mostly washed-up Pleistocene fossils, are found dead in Arizona in drift. A. G. Smith (1953:8) records as *Amnicola palomasensis* (H. A. Pilsbry, 1895, originally described from Chihuahua, Mexico, as *Bythinella*) drift shells from Beaver Creek, at Montezuma Castle National Monument, Yavapai Co. Drift shells from San Pedro Riv at Benson, Cochise Co., were listed as *Amnicola* sp. by Pilsbry and Ferriss (1915b:390). Unidentified hydrobiids were found by senior author in drift of San Pedro Riv at Fairbank, Cochise Co.; and, in Pima Co., on Rillito Creek, at N Campbell Ave near Tucson city limits, and in Pantano Wash, between Pantano and Tucson. In Sylvania Spring of Scotia Canyon, at NW side of Huachuca Mts, some unidentified hydrobiids live with *Fontelicella*.

PELECYPODA

Family Sphaeriidae

147. **Sphaerium transversum** (T. Say, 1829, *New Harmony Disseminator* 2 (23):356, as *Cyclas*). T.L.: Kentucky, without more precise locality. Known Recent in Arizona at three localities in Santa Cruz Co.: Potrero Creek, 4 mi NW of Nogales, W of U.S. Hwy 89, 3,700 ft (D. W. Taylor, 1967a:202); large, permanent pond at Arivaca, 3,650 ft, collected by senior author and R.H.R., July and Nov. 1967; small, permanent pond, 3.5 mi NE of Arivaca, ca 1 mi W of Arivaca Rd, ca 3,700 ft, R.H.R., Nov. 1967. Senior author found dead valves in drift of San Pedro Riv at Hereford, Cochise Co. Sometimes placed in a subgenus (or genus) *Musculium*. Arizona not included by H. G. Herrington (1962) in Recent native range over most of temperate and subtropical North America, N to Northwest Territories, S to central Mexico (Sonora: 55 mi S of Nogales on Mex. Hwy 15, R.H.R., 1968; Chihuahua: Presa Chihuahua, ca 7 mi S of the city of Chihuahua, 5,800 ft, R.H.R., 1971; Coahuila; Tabasco; Tamaulipas); introduced by man in Europe.

? 148. **Sphaerium striatinum** (J. B. P. A. de Lamarck, 1818, *Hist. Nat. Anim. Sans Vert.* 5:550 [printed 560 by error], as *Cyclas*). T.L.: Lake George, New York. Published Recent, living records from Arizona are questionable. *Sphaerium nobile* (A. A. Gould, 1855), a synonym of *striatinum* for H. B. Herrington (1962), was described from a clam collected by T. H. Webb allegedly "near San Pedro, California," but it could not have come from California. Senior author is inclined to accept San Pedro Riv, Arizona, as the T.L. of *nobile*. D. W. Taylor (1967b:154) disagrees. Although he found a paratype of *nobile* (U. S. Nat. Mus. No. 11592) to be *striatinum*, he claims that the latter does not occur in Arizona. The only similar species in the State, he says, is *S. triangulare* (T. Say), which he regards as specifically distinct from *striatinum*, whereas H. B. Herrington (1962) synonymizes them. W. H. Dall (1897a:370) recorded as *Sphaerium solidulum* (T. Prime) clams collected by E. A. Mearns in San Bernardino Riv, reputedly in Arizona, but possibly from across the border in Sonora. D. W. Taylor (1967b:154), who saw these clams (U.S. National Museum No. 130236), refers them to what he calls *S. triangulare* (T. Say), citing as synonym the MS nomen nudum *Sphaerium eminens* given by V. Sterki to Lot No. 130236. Pilsbry and Ferriss (1915b:390) referred a dead valve from drift of San Pedro Riv at Benson, Cochise Co., to *Sphaerium triangulare,* after comparing it with T. Say's type; this drift shell may have been a washed-up fossil. E. Antevs (1941:47) reports as *Sphaerium aureum* (T. Prime), another synonym of *S. striatinum,* Pleistocene fossils from a Sulphur Springs Valley Site, 12 mi SW of Douglas, Cochise Co. Mid-Pliocene fossils from Bidahochi Formation (Hemphillian) of White Cone Peak, 50 mi N of Holbrook, Navajo Co., called *S. striatinum* by D. W. Taylor (1957:656), were also collected there by R.H.R., 1969. We have seen Recent *S. striatinum* from New Mexico, collected by A. L. Metcalf (in litt., 1970) in Cimarron Riv, Union Co., and in Rio Grande, 1 mi N of Rinconada, Rio Arriba Co. According to H. B. Herrington (1962:28), Recent *S. striatinum* ranges in North America from Northwest Territories S to Mexico and from Atlantic to Pacific. Synonyms, for H. B. Herrington (1962:27): *Cyclas triangularis* T. Say, 1829, *New Harmony Disseminator* 2 (23):356 (T.L.: Mexico, without more definite locality). *Cyclas solidula* T. Prime, March 1852 (not 1851, as sometimes cited), *Proc. Boston Soc. Nat. Hist.* 4 (for 1851-1854):158 (T.L.: Ohio, without more definite locality). *Cyclas aurea* T. Prime, March 1852 (not 1851), ibidem 4 (for 1851-1854):159 (T.L.: "Lake Superior," without more precise locality. *Cyclas nobilis* A. A. Gould, 1855a:229 (T.L. discussed before).

Sphaerium simile (T. Say, 1817, in *W. Nicholson's British Encyclopedia,* 1st American Ed. 2, Article "Conchology," signature

Yy:14th unnumbered page, Pl. 1, fig. 9, as *Cyclas similis*); T.L.: Delaware Riv, Pennsylvania. Not positively known Recent in Arizona. Recorded by H. Hannibal (1912a:21) from Colorado Riv at Yuma, without mention of date or collector, but probably no longer living there, in view of the many changes the river has undergone since. We have seen fossils of it from post-Pleistocene or sub-Recent alluvium in the banks of San Pedro Riv at The Narrows 10 mi N of Benson, Cochise Co., collected first by E. L. Montgomery, though not mentioned by him in 1963, and in Nov. 1970 by R.H.R. and T. R. Van Devender. According to H. B. Herrington (1962:69), *S. simile* is Recent in North America from Maine, Quebec, Ontario (N to James Bay), Saskatchewan, and Alberta S to Virginia, Ohio, Iowa, Minnesota, Montana, Wyoming, and (?) Washington; known from Pleistocene of Kansas.

149. **Pisidium (Rivulina) casertanum** (G. S. Poli, 1795, *Test. Utrinque Siciliae* 1 Ordo II:65; Pl. 16, fig. 1; as *Cardium*). T.L.: Caserta, Italy. Widespread Recent in Arizona, in permanent, usually stagnant water, but surviving alive for some weeks in damp mud; known from Coconino Co. (Oak Creek Canyon, 7 mi N of Sedona, 3,500 ft, senior author; Deer Lake, 47 mi E of Strawberry, 7,500 ft, R.H.R.; pond at E side of Interstate Hwy 17, 25 mi S of Flagstaff, 4,200 ft, R.H.R.; Pinewood Golf Course, 20 mi S of Flagstaff, 4,500 ft, R.H.R.; Spitz Spring, 9 mi E of Williams, 7,000 ft, R.H.R.; Lindbergh Spring on Alt. U.S. Hwy 89, 8 mi S of Flagstaff, 6,500 ft, R.H.R.), Navajo Co. (N side of Winslow at O'Haco Farm, 4,800 ft, R.H.R.; 1.5 mi S of Show Low, 6,350 ft, R.H.R.), Apache Co. (East Fork of Black Riv, 3 mi NE of Buffalo Crossing, 7,700 ft, senior author; below Lyman Lake Dam, 13 mi S of St Johns, ca 6,000 ft, R.H.R.; Nutrioso Riv at Nutrioso, 7,680 ft, R.H.R.; small canyon 17 mi W of Springerville, off State Hwy 73, 9,050 ft, alive and also dead on caddis fly [Trichoptera] cases, R.H.R.), Yavapai Co. (pond at E side of Cornville, 5 mi E of Cottonwood, 3,300 ft, R.H.R.; Granite Creek, 2 mi N of Prescott, 5,300 ft, senior author), Gila Co. (East Verde Riv, 5 mi NW of Payson, 4,500 ft, senior author), and Cochise Co. (Babocomari Creek at Bobocomari [now Brophy] Ranch, 4,500 ft, A. R. Mead et al. Huachuca Mts: Carr Canyon at Reef, 7,000 ft; Tanner [now Garden] Canyon, 6,000 to 7,000 ft; Ash Canyon, 4,500 ft; mouth of Hunter Canyon, 4,500 ft; all 4 Stations by Pilsbry and Ferriss, 1906:173, and 1910a:516, as *P. abditum huachucanum;* spring in Huachuca Canyon, 5,250 ft, senior author; Sylvania Spring in Scotia Canyon, W side of Range, 6,150 ft, senior author. Chiricahua Mts: Turkey Creek West, 6,000 ft, J. F. Burger; spring branch at head and at box of Rucker Canyon, Pilsbry and Ferriss, 1910b:144; Cave

Creek at Herb Martyr Dam, 5,600 ft, senior author). Recent also in New Mexico and trans-Pecos Texas; not known from Sonora. Dead shells, mostly washed-up fossils, in drift of Pima Co. (Pantano Wash near Vail, senior author) and Yavapai Co. (Beaver Creek at Montezuma Castle National Monument, A. G. Smith, 1953:8, as *P. abditum huachucanum*). Fossil in Late Cenozoic of Coconino Co. (Walnut Creek Site near Winona, 13 mi E of Flagstaff, H. S. Colton, 1929:94; Meteor Crater Site, 18 mi W of Winslow, R. D. Reger and G. L. Batchelder, 1971:192-193), Pima Co. (Mattie Canyon, W slope of Whetstone Mts, R. J. Drake, clams seen by senior author), Cochise Co. (San Pedro Valley: El Paso Pipeline Site, 7 mi S of Benson, R. S. Gray, 1965, D. W. Taylor, 1966b:94; Lehner Mammoth and Murray Springs Sites, associated with mammoth remains 10,000 to 11,000 yrs old), and Apache Co. (Peridot Creek in Buell Park, 12 mi N of Fort Defiance, R. J. Drake, 1949:28, as *P. abditum*); also of Nebraska, Kansas, Oklahoma, Texas, and New Mexico (in Picacho alluvium of Rio Grande, Dona Ana Co., A. L. Metcalf, 1967:16 and 35). Cosmopolitan in Recent fauna: in America from Alaska to Patagonia; in Old World in Eurasia, Australia, and New Zealand, in Africa S to Rhodesia; also in Pleistocene of Old World. Synonyms (after H. B. Herrington, 1962:33): *Pisidium abditum* S. S. Haldeman, 1841, *Proc. Acad. Nat. Sci. Philadelphia* 1, (4):53 (T.L.: Lancaster Co., Pennsylvania). *P. abditum huachucanum* H. A. Pilsbry and J. H. Ferriss, 1906:173; not figured (T.L.: Huachuca Mts: Carr Canyon at Reef, ca 7,000 ft, Cochise Co.).

150. **Pisidium (R.) compressum** T. Prime, March 1852 (not 1851 as sometimes cited), *Proc. Boston Soc. Nat. Hist.* 4 (for 1851-1854):164. T.L.: Fresh Pond in Cambridge, Massachusetts. Possibly a Recent record from Arizona by W. H. Dall (1897a:370): San Bernardino Riv, "Arizona," coll. by E. A. Mearns; specimens checked at U.S. Nat. Mus. by D. W. Taylor (1969b:156); but the Station may have been S of Mexican border in Sonora. Seen Recent by us from Coconino Co. (Oak Creek Canyon, 7 mi N of Sedona, 3,500 ft, senior author, 1964), Navajo Co. (Show Low Lake, 4 mi S of Show Low, 6,500 ft, R.H.R., 1968), and Apache Co. (East Fork of Black Riv at Diamond Rock Forest Camp, 7 mi N of Buffalo Crossing, 7,900 ft, J. F. Burger, 1969). Riparian drift shells, possibly washed-up fossils, in Cochise Co. (San Pedro Riv at Benson, Pilsbry and Ferriss, 1915a:390 and 400). Fossil in Late Cenozoic of Coconino Co. (Late Pleistocene at Winona Site, 13 mi E of Flagstaff, R. D. Reger and G. L. Batchelder, 1971:191 and 193), Navajo Co. (Mid Pliocene: Hemphillian Bidahochi Formation at White Cone Peak, 50 mi N of Holbrook, D. W. Taylor, 1957:657, and R.H.R., 1969), and

Cochise Co. (Late Pleistocene of San Pedro Valley at Murray Springs Site, associated with mammoth remains 10,000 to 11,000 yrs old; post-Pleistocene alluvium of San Pedro Riv at The Narrows, 10 mi N of Benson, R.H.R. and T. R. Van Devender, 1970). According to H. B. Herrington (1962:35), Recent in North America, N to Alaska and Northwest Territories (Great Slave Lake), S to Mexico in Chihuahua (Lake Palomas, Mimbres Valley, just S of U.S. border, W. H. Dall, 1897a:370), Coahuila, and Tamaulipas. Also in Late Cenozoic of Nebraska, Kansas, Oklahoma, Texas, and New Mexico (Pleistocene of Baxter Ranch Site, 5 mi N of Portales, Roosevelt Co., senior author, 1958).

? 151. **Pisidium (R.) nitidum** L. Jenyns, 1832, *Trans. Cambridge Philos. Soc.* 4 (2): 304, Pl. 20, figs. 7-8. Described from Surrey and Cambridgeshire, England. Tentatively in Recent Arizona fauna for a record from riparian drift of Santa Cruz Riv at Tucson, Pima Co., by Pilsbry and Ferriss (1915a:400, as *P. pauperculum*). Fossil in Late Pleistocene of Coconino Co. (Winona Site, 13 mi E of Flagstaff, R. D. Reger and G. L. Batchelder, 1971:191 and 193). Not seen by us from the State. H. B. Herrington (1962:46) cites it Recent from Arizona and New Mexico, without precise localities, and gives the general range as Europe and North America, from Newfoundland, Northwest Territories, and British Columbia southward over most of the States (except the southern) to Mexico. In Late Cenozoic also in Nebraska, Kansas, Oklahoma, and Texas. Synonym (for H. B. Herrington, 1962:45):*Pisidium pauperculum* V. Sterki, 1896, *Nautilus* 10 (6):64 (described from Massachusetts, New York, New Jersey, Pennsylvania, Michigan, Wisconsin, and Minnesota; no T.L. selected thus·far).

152. **Pisidium (R.) walkeri** V. Sterki, 1895, *Nautilus* 9 (7):75. Described from Michigan, Pennsylvania, New York, and Minnesota; no T.L. selected thus far. Arizona included in Recent range, without precise locality, by H. B. Herrington (1962:51). Seen Recent from Apache Co. (Hawley Lake, 15 mi SE of McNary, 8,500 ft, G. F. Adams, 1965; NW side of St Johns, 5,725 ft, R.H.R., 1970; Big Lake, 24 mi SSW of Eagar, 8.650 ft, R.H.R., 1970), and Gila Co. (pond near Cherry Creek, 1 mi E of Young, 5,100 ft, senior author, 1969). In Late Pleistocene of Cochise Co. (San Pedro Valley at Murray Springs Site, associated with mammoth remains 10,000 to 11,000 yrs old). According to H. B. Herrington (1962:51), Recent in Nearctic Region from Maine, Ontario, Saskatchewan, Alberta, and Northwest Territories (N to Great Slave Lake, ca 61° 30′ N), SE to Virginia, Missouri, and Michigan, NW to Wisconsin, Iowa, South Dakota, and Montana, and SW to Arizona. In Pleistocene of Quebec,

Ontario, Kansas, New Mexico (Baxter Ranch Site, 5 mi N of Portales, Roosevelt Co., senior author, 1958), and Arizona.

? 153. **Pisidium (Neopisidium) punctatum** V. Sterki, 1895, *Nautilus* 8 (9):99, Pl. 2, figs. 7-12 and 14. T.L.: Tuscarawas Riv, Tuscarawas Co., Ohio. Accepted tentatively for the Recent Arizona fauna because H. B. Herrington (1962:48) included the State in the Recent range of *P. punctiferum* (R. J. L. Guppy, 1867), with which he synonymized *P. punctatum*. It now appears that these are two distinct species; most probably H. B. Herrington's supposed *punctiferum* from Arizona were *punctatum,* since this is reported fossil from the State. Recorded from Late Pleistocene in Coconino Co. (Winona Site, 13 mi E of Flagstaff, R. D. Reger and G. L. Batchelder, 1971:191 and 193). We have not seen it, either Recent or fossil.

? 154. **Pisidium (? Neopisidium) insigne** W. M. Gabb, 1888, *Amer. Journ. Conch.* 4 (2):69, Pl. 2, fig. 2. T.L.: Spring at Fort Tejon, Kern Co., California. Included tentatively in Arizona fauna for H. B. Herrington's (1962:43) mentioning the State in the general range (without precise locality), given as North America from S Ontario and British Columbia SE to New York and Michigan, SW to California, Nevada, Arizona, and New Mexico. We have not seen it.

Pisidium subtruncatum A. W. Malm, 1855, *Göteborgs K. Vet.-o. Vitterh.-Samh. Handl.* 3:92, figs. (T.L.: Sweden). Not known Recent in Arizona. R. D. Reger and G. L. Batchelder (1971:191 and 193) list it from Late Pleistocene of Coconino Co. (Winona Site, 13 mi E of Flagstaff). Recent in Europe and in North America N to Northwest Territories (Great Slave Lake), S to Ohio, South Dakota, Colorado, and California.

We wish to acknowledge our indebtness to Mr. P. N. D'Eliscu for assistance with identification of our local Sphaeriidae.

Family Cyrenidae

* **Corbicula manilensis** (R. A. Philippi, Nov. 1844 [not 1841, as sometimes cited], *Zeitschr. f. Malakozool.:*162, as *Cyrena*); sometimes misspelled *manillensis*. T.L.: Manila, Philippine Is. The Asian freshwater clam, native of the Far East (Japan, Korea, China, Taiwan, Philippines, and Indonesia). Introduced to North America, either deliberately or by accident; first noticed there in 1938 on N bank of Columbia Riv, Washington, by John Q. Burch (1944, *Minutes Conch. Club South. California* 38:18); now well established, feral in much of warm temperate United States and NW Mexico, from Pacific to Atlantic, N to 46° 15′ N in the west and to 40° N in the east. Common in Arizona in the Colorado, Gila, and Verde river

drainages, in Mohave Co. (Lake Mead near Temple Bar), Yuma
Co. (Yuma; Roll; 4 mi SW of Parker; Martinez Lake, 15 mi NE of
Yuma), Yavapai Co. (Verde Riv at mouth of Tangle Creek, 25 mi E
of Rock Springs, C. Leeke, 1962; Verde Riv at Edge Farm, 2 mi W
of Camp Verde, 3,300 ft, G. Huffner, 1969), and Maricopa Co. (near
Papago Park in Phoenix, first State record, June 1956, D. S. and H.
A. Dundee, 1958:51, as *C. fluminea;* Mesa; 8 mi S of Chandler;
Theba; Buckeye; Apache Lake); also in SE New Mexico and trans-
Pecos Texas (irrigation canals from Rio Grande in Mesilla Valley
and near El Paso, A. L. Metcalf, 1966:16), in SE California (Salton
Sea Basin, R. O. Fox, 1970:145; Westmorland, Imperial Co., R.H.R.,
1970), northern Baja California (0.5 mi N of Cierro Prieto, S of
U. S. border, R. O. Fox, 1970:145; on Mex. Hwy 4, 2 mi W of Car-
ranza, and on Mex. Hwy 5, 16.4 mi S of U.S. border, R.H.R., 1970),
and in NW Sonora (2 mi S of San Luis, R.H.R., 1970). We follow
the current use of *C. manilensis* for the introduced Asian freshwater
clam, as explained recently by R. M. Sinclair (1971). We are, however, not fully satisfied that it should supersede the older name
Corbicula fluminea (O. F. Müller, 1774, *Verm. Terr. Fluv. Hist.*
2:206, as *Tellina;* for dead valves from river sand of China). Synonym: *Corbicula leana* T. Prime, June 1864 (not 1884), *Ann. Lyc.
Nat. Hist. New York* 6:68, fig. 14 (T.L.: Japan, without precise locality); name used for Arizona clams by G. E. Gunning and R. D.
Suttkuss, 1966, *Nautilus* 79 (4):113-116.

Family Unionidae

* **Anodonta (Pyganodon) corpulenta** W. Cooper, 1834, *Appendix
to Narrative of Expedition through Missouri to Itasca Lake:* 154.
T.L.: "Upper Missouri Riv," without more precise locality. Native
in Missouri and Mississippi drainages; introduced in Arizona, probably with fish carrying the glochidia. Established thus far only in
Coconino Co.: Lake Mary, 10 mi SE of Flagstaff, 6,800 ft, first
taken by A. R. Phillips, Aug. 26, 1951, later by A. G. Smith, Apr.
11, 1955 (as reported by D. W. Taylor, 1966a:198, Pl. 28, figs. 1
and 3), M. L. Walton, March 1956, and R.H.R., Sept. 1969; Lake
Kinnikinick, 28 mi SE of Flagstaff, 7,000 ft, R.H.R. and senior
author, Aug. 1969. *A. corpulenta* should probably be ranked as a
subspecies of *Anodonta grandis* (T. Say, 1829).

155. **Anodonta (Anodonta) californiensis** I. Lea, 1852:286,
Pl. 25, 3 figs. 47. T.L.: "Rio Colorado, California"; two or three
syntypes collected by J. L. Le Conte, probably in the lower section
of the river at or near the California-Arizona-Mexico boundary.
R. E. C. Stearns (1879:145, with fig., and 151) suggested that the T.L.

was "presumably in the neighborhood of Fort Yuma," where J. L. Le Conte was stationed as Army Surgeon when Fort Yuma was on the California side of the river. C. T. Simpson (in W. H. Dall, 1897:372) synonymized his *Anodonta mearnsi* with *Anodonta dejecta* J. Lewis, itself now a synonym of *A. californiensis,* and referred to it clams collected by E. A. Mearns in Colorado Riv at Station 68 (of 1894) near the Mexican border. According to D. W. Taylor (1967b:157), the clams from Station 68 near Boundary Monument No. 204 (U. S. National Museum Lot 130171) are from a laguna of Colorado Riv in Yuma Co. Whether the clam still lives in Colorado Riv is doubtful in view of the alterations the river has undergone in recent years. The later and contemporary history of *A. californiensis* in Arizona offers instructive evidence of the impact on the indigenous biota of the steady deterioration by man of the surfacial water of the State. In this connection it should be noted that, like many other Unionidae, *A. californiensis* can survive as a species only in association with certain species of fish. The glochidium, or swimming stage that hatches from the egg of the clam, must attach to the fins (in *Anodonta*; not to the gills as in some other unionids) of a host fish, which it parasitizes while transforming into the free-living, eventually stationary clam. When the proper (specific) host fish disappear from the habitat, the clam also will become extinct. This explains why *A. californiensis* is now near extinction in Arizona, while perhaps only a century ago it may have been widespread in the State.

Some 30 years after the original description, R. E. C. Stearns (1883:1019, fig. 3) reported that J. F. James had collected *A. californiensis* in June 1880 in "the Little Santa Cruz Riv, just outside Tucson." That the clam was thriving at that time in the Tucson area was proved dramatically in May 1968, when in the course of an urban renewal project Jim Ayres excavated in downtown Tucson (at Alameda St and Main Ave) a Chinese trash dump dating from 1880-1885. Many valves of *A. californiensis* were dug up, mostly well preserved and some of them with periostracum. Their numbers, distance from the river, and association with kitchen utensils suggest that they may have been eaten. In 1893 E. A. Mearns collected in Santa Cruz Riv at Tucson (his Station 46) clams listed as *Anodonta dejecta* (our *A. californiensis*) by C. T. Simpson (in W. H. Dall, 1897:372) and D. W. Taylor (1967b:156, for Lots 130172 and 130180, U. S. National Museum). The clam probably died out at Tucson before 1915, when Pilsbry and Ferriss (1915a:400) cited *A. dejecta* "fossil and Recent, fragmentary" in drift of Santa Cruz Riv; they later (1919a:329) recorded the same specimens as "subfossil" in the river's banks. There is also evidence that *A. californiensis* lived in relatively recent, but pre-Columbian times in Gila Riv (valves dug up at Indian Sites by E. W. Haury, 1950:362, as *"Anodonta cynea beriniana"* [sic; wrong identification], in Ventana Cave, 12 mi NW of Quijotoa, Pima Co., and clams seen by us from Snaketown near Sacaton, Pinal Co.) and in San Pedro Riv (subfossil valves in alluvium at The Narrows, 3,340 ft, 10 mi N of Benson, Cochise Co., associated with *Pisidium compressum* and *Sphaerium simile,* R.H.R. and T. R. Van Devender, Nov. 1970).

C. T. Simpson (1893) described *Anodonta mearnsiana* from clams collected alive by E. A. Mearns in 1892 at his Station 25, San Bernardino Ranch, near Boundary Monument 77, at Mexican border in Rio Yaqui drainage, but whether this was in Arizona or Sonora is uncertain. In 1966, senior author was unable to find either living or dead clams in the now impounded creek at San Bernardino Ranch, 13 mi E of Douglas, Cochise Co. Possibly, however, the clam may yet live farther downstream in the more natural waters of Rio San Bernardino in Sonora. C. T. Simpson later (in W. H. Dall, 1897:372) regarded his *A. mearnsiana* as not separable from the earlier *Anodonta dejecta.* J. Lewis (1875), itself now recognized as a synonym of *A. californiensis.* The clam found by E. H. Ashmun in Oak Creek Canyon, Coconino Co., referred by H. Hannibal by error to *Anodonta cygnea impura* (1912a:19, with *A. californiensis* cited as synonym), was probably *A. californiensis,* but it has not been seen there since. Pilsbry and Ferriss (1919a:329) reported *californiensis* as *A. dejecta* from Black Riv, at S border of Apache Co. It was found again in that general area Oct. 19, 1956, by W. O. Gregg and M. L. Walton, at Buffalo Crossing, 7,200 ft, Greenlee Co. (clams seen by us). In addition, M. L. Walton informs us (in litt., 1968) that he collected the same clam with W. O. Gregg in 1956 in Little Colorado Riv near Springerville, at ca 7,000 ft, Apache Co., and in Chevelon Creek, a tributary of Little Colorado Riv, ca 30 mi S of Winslow, Navajo Co.; in 1970 R.H.R. could find only a few dead valves on Chevelon Creek, 2 mi below Chevelon Dam, at 6,400 ft. W. O. Gregg (1947, 67:17) first suggested that *A. dejecta* was perhaps a subspecies only, or even a mere form of *A. californiensis.* After comparing Recent clams from Buffalo Crossing and subfossil valves from Tucson with Recent clams from Los Angeles Riv, California (sent by M. L. Walton), we are unable to separate the *A. dejecta* (= *mearnsiana*) of Arizona even as a subspecies from *A. californiensis.*

A. californiensis is restricted to the Pacific drainage, where its main range is in the Southwestern Molluscan Province. At present it is definitely known alive in Oregon, California, Utah, Arizona (where it is now on the verge of extinction), and Chihuahua (living clams, not separable from Recent or subfossil *californiensis* of California and Arizona, in Arroyo Rincon de la Concha, tributary of Rio Papigochic [Rio Yaqui drainage], 1.5 mi NW and 1 mi NE of Yepomera, ca 29° N, 108° W, T. R. Van Devender and R. H. R., 1971). The range of the species thus extends inland to over 600 mi from the Pacific. No doubt it will eventually be found in Sonora also.

Synonyms: *Anodonta dejecta* J. Lewis, 1875a:26, and 1875b: 952 (T.L.: originally given by error as "Arkansas Riv and its tributaries," but known only from the Pacific drainage. *Anodonta mearnsiana* C. T. Simpson, 1893:134; syntypes figured by C. T. Simpson in W. H. Dall, 1897a:Pl. 32, figs. 4-5 (T.L.: "San Bernardino Ranch," near Mexican border, 13 mi E of Douglas, Cochise Co); collected by E. A. Mearns in 1892 at his Station 25 near Boundary Monument 77; probably extinct there now; but, as mentioned before, the T.L. might have been across the border, in Sonora, rather than in Arizona. *A. californiensis* is sometimes placed in a distinct subgenus *Arnoldina* H. Hannibal, 1912, a name here regarded as a synonym of *Anodonta,* sensu stricto.

BIBLIOGRAPHY

Publications on Recent and Late Cenozoic inland mollusks of Southwestern Molluscan Province, namely arid and semiarid southern California, northeast Baja California, northern Sonora, northern Chihuahua, Arizona, New Mexico, and trans-Pecos Texas (see Section on Zoogeography and map, Fig. 1).

Antevs, E.
1941   Age of the Cochise Culture Stages. Medallion Papers, (Globe, Arizona), 29:31-81 (fossil mollusks of Sulphur Spring Valley: 37 and 66-67, named by H. G. Richards).

1959   Geological age of the Lehner Mammoth Site. American Antiquity 25 (1):31-34.

Arnaud, P. H., Jr.
1970   The Sefton Foundation Orca Expedition to the Gulf of California, March-April, 1953. Occas. Pap. California Acad. Sci. 86:1-37 (*Micrarionta,* etc, on some islands; no species named).

Ashmun, E. H.
1899   Collecting in Arizona and New Mexico. Nautilus 13 (2):13-17.

Baily, J. L., Jr.
1953   *Oreohelix* in San Diego County, California. Nautilus 67 (1):20-22.

1956   Observations on the recently extinct mollusk fauna of Panamint Lake. Nautilus 69 (3):100-103.

Baily, J. L., Jr., and R. I. Baily
1940   A new urocoptid mollusc from the State of Sonora, Mexico. Nautilus 53 (3):94 and 95; Pl. 12.

1951   Further observations on the Mollusca of the relict lakes in the Great Basin. Nautilus 63 (2):46-53; (3):85-93; Pl. 4 (*Amnicola longinqua:* 51, Pl. 4, 3 figs. 7, of Colorado Desert).

Baker, F. C.
1911   The Lymnaeidae of North and Middle America, Recent and fossil. Chicago Acad. Sci. Special Publ. 3:i-xvi + 1-539; 55 Pls.

1945   The molluscan family Planorbidae. (Urbana, Illinois):i-xxxvi + 1-530; 141 Pls.

Baker, H. B.
1929   Pseudohyaline American landsnails. Proc. Acad. Nat. Sci.
            Philadelphia 81:251-266; Pls. 8-10.
1930a   Mexican mollusks collected for Dr. Bryant Walker in
            1926. Part II. Auriculidae, Orthurethra, Heterurethra
            and Aulacopoda. Occas. Pap. Mus. Zool. Univ. Michigan
            220:1-46; Pls. 7-10.
1930b   The North American *Retinellae.* Proc. Acad. Nat. Sci.
            Philadelphia 82:193-219; Pls. 9-14.
1958   Land snail dispersal. Nautilus 71 (4):141-148.
1962   Type land snails in the Academy of Natural Sciences of
(May)   Philadelphia. I. North America North of Mexico. Proc.
            Acad. Nat. Sci. Philadelphia 114:1-21.
1963   Type land snails in the Academy of Natural Sciences of
(Nov.)   Philadelphia. II. Land Pulmonata, exclusive of North
            America North of Mexico. Proc. Acad. Nat. Sci. Philadel-
            phia 115:191-259.
1964   Type land snails in the Academy of Natural Sciences of
(Oct.)   Philadelphia. III. Limnophile and thalassophile Pulmon-
            ata. IV. Land and fresh-water Prosobranchia. Proc.
            Acad. Nat. Sci. Philadelphia 116:149-193.

Barnes, W. C.
1935   Arizona place names. Univ. Arizona Bull. 6 (1):1-503 (Gen-
            eral Bull. No. 2).

Bartsch, P.
1903   A new landshell from California. Proc. Biol. Soc. Wash-
            ington 16:103-104.
1904a   *Sonorella wolcottiana*—A correction. Proc. Biol. Soc.
            Washington 17:101.
1904b   Notes on the genus *Sonorella* with descriptions of new
(Oct.)   species. Smithson. Misc. Coll. 47 (for 1905) (2) No. 1481:
            187-200; Pls. 27-33.
1906   The urocoptid mollusks from the mainland of America
(Sept.)   in the collection of the United States National Museum.
            Proc. U.S. Nat. Mus. 31 (for 1907) (1483):109-160; Pls. 3-5.
1943   Notes on Mexican urocoptid mollusks. Journ. Washington
            (D.C.) Acad. Sci. 33 (2):54-59.
1947   Notes on some Mexican urocoptid mollusks, with the
            description of new species. Journ. Washington (D.C.)
            Acad. Sci. 37 (8):284-288.
1950   A new terrestrial mollusk from Mexico. Journ. Washing-
            ton (D.C.) Acad. Sci. 40 (8):265.

Batchelder, G. L., and R. D. Reger
1969   Additional molluscan evidence of a Late Pleistocene lake
            near Winona, Coconino County, Arizona. The Echo, West-
            ern Soc. Malacologists, Inc., 1st Meeting, 1968:7 (privately
            printed for members only; received in Tucson, March
            1969). [Abstract in 1968, Journ. Arizona Acad. Sci. 5
            Suppl. (Proc. 12th Ann. Meet.):9, as "New evidence for a
            Late Pleistocene or early Recent lake near Winona,
            Coconino Co., Arizona"].

Berry, S. S.
1917 A new *Sonorella* from Sonora. Nautilus 31 (1):14-15.
1922 Notes on the mollusks of the Colorado Desert. I. Proc. Acad. Nat. Sci. Philadelphia 74:69-100; Pls. 8-10.
1926 (Nov.) Two new helicoid snails from the Mohave Desert. Ann. Mag. Nat. Hist. Ser. 9 18:490-493.
1928a New helicoid snails from the Mohave Desert. Ann. Mag. Nat. Hist. Ser. 10 1:274-279 and 618-622.
1928b A new land snail from Lower California with notes on other species. Journ. Entom. Zool. 20 (4):73-83; 2 Pls.
1929 Three new snails from the hills of California. Nautilus 43 (2):39-40. Correction, 1930 43 (4):138.
1930a Snail notes from the California Desert. Nautilus 43 (3):73-75.
1930b Preliminary notices of two new snails from the Colorado Desert. Ann. Mag. Nat. Hist. Ser. 10 5:543-545.
1930c New helicoid snails from the Mohave Desert, IV. Ann. Mag. Nat. Hist. Ser. 10 6:187-193.
1931a The genus *Oreohelix* in California, Nautilus 45 (3):73-75.
1931b New helicoid snails from the Mohave Desert. 5. The genus *Oreohelix* in southern California and Nevada. Ann. Mag. Nat. Hist. Ser. 10 8:115-120.
1933 Snails and other Mollusks. *In* E. C. Jaeger, "The California Deserts: . . . ." 67-73.
1943 On the generic relationships of certain California xerophile snails. Trans. San Diego Soc. Nat. Hist. 10 (1):1-24; Pls. 1-2.
1947a A surprising molluscan discovery in Death Valley. Leaflets in Malacology 1 (2):5-8 (*Assiminea infima* n. sp.).
1947b On the generic relationships of certain Lower Californian helicoid snails. Leaflets in Malacology 1 (3):9-12.
1948 Snails of the Sierra Ancha, Arizona. American Midland Naturalist 39 (1):151-159.
1953 Two Californian mountain snails of the genus *Helminthoglypta*—A problem in the relationship of species. Trans. San Diego Soc. Nat. Hist. 11:329-344; 2 Pls.
1964 Notices of new Eastern Pacific Mollusca. VI. Leaflets in Malacology 1 (24):147-154 (*Melampus mousleyi* n. sp. of Cholla Bay).

Binney, W. G.
1865a Land and fresh water shells of North America. II. Pulmonata Limnophila and Thalassophila. Smithson. Misc. Coll. 7 (2) No. 143:i-ix + 1-161 (reprinted in Sterkiana No. 18, June 1965:21-50; No. 19, Sept. 1965:3-34; No. 21, March 1966:2-40; figs. on 2 Pls. in No. 23, Sept. 1966, and 2 Pls. in No. 24, Dec. 1966).
1865b Descriptions of new species of North American land and freshwater shells. Amer. Journ. Conch. 1 (1):49-51; Pl. 7.
1865c Land and fresh water shells of North America. III. Ampullariidae, Valvatidae, Viviparidae, fresh water Rissoidae, Cyclophoridae, Truncatellidae, fresh water Neritidae,

Helicinidae. Smithson. Misc. Coll. 7 (144):i-viii + 1-120 (reprinted in Sterkiana No. 22, June 1966:3-30; No. 23, Sept. 1966:25-48; No. 24, Dec. 1966:7-27; figs. on 3 Pls. in No. 24, Dec. 1966).

1883   A supplement to the fifth volume of the terrestrial air-breathing mollusks of the United States and adjacent territories. Bull. Mus. Comp. Zool. 11 (8):135-166; Pls. 1-4.

1885   A manual of American land shells. Bull. U.S. Nat. Mus. 28:1-528.

1886   A second supplement to the fifth volume of the terrestrial air-breathing mollusks of the United States and adjacent territories. Bull. Mus. Comp. Zool. 13 (2):23-48; Pls. 1-2.

1892   A fourth supplement to the fifth volume of the Terrestrial air-breathing mollusks of the United States and adjacent territories. Bull. Mus. Comp. Zool. 22 (4):163-204; Pls. 1-4.

Binney, W. G., and T. Bland

1869   Land and fresh water shells of North America. I. Pulmonata Geophila. Smithson. Misc. Coll. 8 (3) (194):1-316.

Blair, W. F.

1950   The Biotic Provinces of Texas. Texas Journ. Sci. 2 (1):93-117.

Blake, W. P.

1855   Preliminary geological report. *In* R. S. Williamson, Report of explorations for RR routes from Mississippi River to Pacific Ocean (U.S. House of Representatives, Doc. 129):174-228; Appendix, Art. 2:22-28 (A. A. Gould's List of Shells, published separately, not as part of Congress. Doc. 129).

1856   Report of a geological reconnaissance in California, etc. In R. S. Williamson, Report of explorations in California . . . for RR routes, etc (in 5 vols.) 5 Pt. 2 Appendix, Art. 3:330-336; Pl. 11 (A. A. Gould's Catalogue of Recent Shells).

1858   Report of a geological reconnaissance in California . . . for RR routes, etc (New York; [separate edition of vol. 5 of the Official Report of 1856] Appendix, Art. 3:330-336; Pl. 11 (A. A. Gould's Catalogue of Recent Shells).

Blanchard, R. C.

1913   The geology of the western Buckskin Mountains, Yuma Co., Arizona. Contrib. Columbia Univ. Dept. Geol. 26 (1):1-80 (fossil mollusks:39, named by W. H. Dall).

Bland, T.

1881   Description of a new species of *Triodopsis,* from New Mexico. Ann. New York Acad. Sci. 2 (for 1882) (1)ff115-116.

Boss, K. J., J. Rosewater, and F. A. Ruhoff

1968   The zoological taxa of William Healey Dall. U.S. Nat. Mus. Bull. 287:1-427.

Branson, B. A.

1963   A note on molluscan zoogeography: distribution of *Gyrau-*

lus arizonensis (Pilsbry and Ferriss). Southwestern Natur. 9 (1):51-52.

Branson, B. A., C. L. McCoy, Jr., and M. E. Sisk
[cited as B. A. Branson et al].
1964   Notes on Sonoran gastropods. Southwestern Natur. 9 (2): 103-104.

Branson, B. A., M. E. Sisk, and C. L. McCoy, Jr.
[cited as B. A. Branson et al].
1966   Observations on and distribution of some western and southwestern mollusks. Veliger 9 (2):145-151.

Bryant, F. W.
1900a   Description of a new California land snail. Nautilus 13 (11):122 (reprinted in West American Scientist 11:31, 1900).

1900b   *Epiphragmophora harperi* n. sp. Nautilus 13(12):143-144 (reprinted in West American Scientist 11:30, 1900).

Bryson, R. A.
1957   The annual march of precipitation in Arizona, New Mexico and northwestern Mexico. Techn. Repts. Meteor. Climat. Arid Regions (Tucson, Arizona) 6:1-24.

Burch, J. B. and F. G. Thompson
1957   Three new Mexican land snails of the genus *Humboldtiana*. Occas. Pap. Mus. Zool. Univ. Michigan 590:1-11; Pls. 1-5.

Buxton, P. A.
1923   Animal life in deserts. (London):i-ix + 1-176.

Chalfant, W. O.
1930   Death Valley. The facts (Stanford, California):i-ix + 1-155; 16 Pls. (revised edition, 1936: i-ix + 1-160; 18 Pls.)

Chase, E. M.
1946   George Willett. Nautilus 59 (3):101-104; portrait.

Cheatum, E. P.
1935   Gastropods of the Davis Mountains vicinity in West Texas. Nautilus 48 (4):112-116; Pl. 5.

1971   A new species of *Ashmunella* from the Davis Mountains in West Texas. Nautilus 84 (3):107-109.

Chevallier, H.
1965   Les mollusques de l'Expédition du Mexique. Journ. de Conchyliologie 105:4-39.

Clapp, G. H.
1907   *Epiphragmophora (Micrarionta) hutsoni* n. sp. Nautilus 20 (12):130-137; Pl. 9.

1908   New land shells from Arizona and New Mexico. Nautilus 22 (8):76-78; Pl. 7 (explained in Pt. 9, 1909:96).

1913   *Gundlachia hjalmarsoni* Pfr. in the Rio Grande, Texas. Nautilus 27 (7):77-78.

Clarke, A. H., Jr.
1960   Catalogue and illustrations of mollusks described by Wesley Newcomb. Bulletins Amer. Paleontol. 41 (188):133-160; Pl. 17 (holotype of *Helix rowelli*:154).

Clench, W. J.
1924   A new species of *Physa* from Texas. Nautilus 38 (1): 12-13.

Clench, W. J., and S. L. H. Fuller
1965   The genus *Viviparus* in North America. Occas. Papers Moll. Dept. Moll. Mus. Comp. Zool. 2 (32):385-412; Pls. 65-69 (*V. chinensis malleatus* in Arizona:404).

Clench, W. J., and W. B. Miller
1966   A new species of *Ashmunella* from west Texas. Breviora, Mus. Comp. Zool. 244:1-6.

Clench, W. J., and H. A. Rehder
1930   A new *Humboldtiana* from Texas. Nautilus 44 (1):10-13; Pl. 2.

Clench, W. J., and R. D. Turner
1962   New names introduced by H. A. Pilsbry in the Mollusca
(June)   and Crustacea. Acad. Nat. Sci. Philadelphia Special Publ. 4:1-218.

Cockerell, T. D. A.
1896a   *Agriolimax campestris* in the Pecos Valley, New Mexico. Nautilus 10 (3):35.

1896b   Land Mollusca from the rejectamenta of the Rio Grande, New Mexico. Nautilus 10 (4):41-43.

1897a   *Vertigo coloradensis* and its allies. Nautilus 10 (12):134-135.

1897b   A new form of *Pupa*. Nautilus 10 (12):143.

1898   New *Polygyras* from White Mountain, New Mexico. Nautilus 12 (7):76-77.

1899a   *Bifidaria armifera* var. nov. *ruidosensis*. Nautilus 13 (3):36.

1899b   *Polygyra triodontoides* in New Mexico. Nautilus 13 (7):84.

1899c   Notes on the *indentata* group of *Vitrea*. Nautilus 12 (10):120.

1899d   Another new snail from New Mexico. Nautilus 12 (11):131.

1900   Note on *Ashmunella hyporhyssa* (Ckll.). Nautilus 14 (6):72.

1901   An evolving *Ashmunella*. Nautilus 15 (3):35.

1902a   *Unio popeii* Lea in New Mexico. Nautilus 16 (6):69-70.

1902b   *Physa virgata* mut. *alba* nov. Journ. of Malacology 5 (4):138.

1903a   A new fossil *Ashmunella*. Nautilus 16 (9):105.

1903b   *Ashmunella thomsoniana cooperae*. Nautilus 17 (3):36.

1905a   A new *Oreohelix*. Nautilus 18 (10):113-114.

1905b   A fossil form of *Oreohelix yavapai* Pilsbry. Nautilus 19 (4):46-47.

1905c   The snails of New Mexico and Arizona. Nautilus 19 (6): 68-71.

1906   The fossil Mollusca of Florissant, Colorado. Bull. Amer. Mus. Nat. Hist. 22 (27):459-462 ("*Holospira* [*H. leidyi*]" in Eocene of New Mexico: 459, footnote).

1912   *Ashmunella thomsoniana*. Nautilus 26 (6):69-70.

1914   Tertiary Mollusca from New Mexico and Wyoming. Bull.
         Amer. Mus. Nat. Hist. 33 (6):101-107; Pls. 8-10.
1927   A large form of *Oreohelix yavapai* in Grand Canyon.
         Nautilus 40 (3):101.
1930   An apparently extinct *Euglandina* from Texas. Proc.
         Colorado Mus. Nat. Hist. 9 (5):52-53.
1945   The Colorado Desert of California: its origin and biota.
         Trans. Kansas Acad. Sci. 48:1-39; 4 Pls. (mollusks:14,
         15-16, 29, 38).
1946   *Helix aspersa* in New Mexico. Nautilus 59 (4):144.

Cockerell, T. D. A., and Mary Cooper
1902   Notes on *Ashmunella*. Nautilus 15 (10):109-110.

Cole, G. A., and Whiteside, M. C.
1965   An ecological reconnaissance of Quitobaquito Spring, Ari-
         zona. Journ. Arizona Acad. Sci. 3 (3):159-163 (*Physa* sp.:
         162).

Colton, H. S.
1929   Fossil fresh water shells from Winona, Coconino County,
         Arizona. Nautilus 42 (3):93-94 (sub-Recent fossils).
1958   Precipitation about the San Francisco Peaks, Arizona.
         Mus. North Arizona Techn. Series 2:1-18.

Connolly, M.
1931   Contributions to a knowledge of the fauna of South-
         West Africa. IX. The non-marine Mollusca of South-
         West Africa. Ann. South African Mus. 29 (2):277-336;
         Pl. 3 + map.

Conrad, T. A.
1855   Description of a new species of *Melania*. Proc. Acad.
(Feb.)   Nat. Sci. Philadelphia 7:269.

Cooper, J. G.
1892   Catalogue of the land and fresh-water Mollusca of Lower
         California. Zoe 3:12-25.
1893   On land and fresh-water Mollusca of Lower California.
         Proc. California Acad. Sci. (2) 3:338-444; Pls. 13-14.

Crandall, O. A.
1901   The American *Physae*. Nautilus 15 (3):25-30; (4):42-45;
         (5):54-58; (6):69-71; Pls. 1-2.

Crosse, H., and P. Fischer
1870   Etude sur la mâchoire et l'armature linguale des Cylin
         drellidae et de quelques genres voisins sous le rapport
         conchyliologique. Journ. de Conchyliologie 18:5-27; Pls.
         3-5.
1892a   Note sur le genre *Holospira*, Martens, et sur la distri-
         bution géographique des espèces dont il se compose.
         Journ. de Conchyliologie 40:256-279; Pl. 5.
1892b   Note sur le *Neritina picta*, Sowerby. Journ. de Conchy-
         liologie 40:292-293.

Dall, W. H.
1885   Notes on some Floridian land and fresh-water shells; with
(July)   a revision of the Auriculacea of the eastern United
         States. Proc. U.S. Nat. Mus. 8 (for 1885) (17):255-289; Pls.
         17-18.

1893   Land shells of the genus *Bulimulus* in Lower California,
(Nov.)   with descriptions of several new species. Proc. U.S. Nat.
              Mus. 16 (for 1893) (958):639-647; Pls. 71-72.
1895a   On a new species of *Holospira* from Texas. Nautilus 8
              (10):112.
1895b   Synopsis of the subdivisions of *Holospira* and some re-
              lated genera. Nautilus 9 (5):50-51.
1896   Diagnoses of new mollusks from the Survey of the Mexi-
(Apr.)   can Boundary. Proc. U.S. Nat. Mus. 18 (for 1895) (1033):
              1-6.
1897a   Report on the mollusks collected by the International
(Jan.)   Boundary Commission of the United States and Mexico,
              1892-1894. Proc. U.S. Nat. Mus. 19 (for 1896-1897) (1111):
              333-379; Pls. 31-33 (Unionidae by C. T. Simpson:370-374).
1897b   On a new form of *Polygyra* from New Mexico. Nautilus
              11 (1):2-3.
1897c   A new *Holospira* from Texas. Nautilus, 11, Pt. 4:38.
1897d   New land shells from Mexico and New Mexico. Nautilus
              11 (6):61-62.
1898   A new *Polygyra* from New Mexico. Nautilus 12 (7):75.
1902   Illustrations and descriptions of new, unfigured or imper-
(March)   fectly known shells, chiefly American, in the United
              States National Museum. Proc. U.S. Nat. Mus 24 (for
              1902) (1264):499-566; Pls. 27-40.
1911   Notes on *Gundlachia* and *Ancylus*. American Naturalist
              45:175-189.

Damon, P. E., C. V. Haynes, Jr., and A. Long
1964   Arizona Radiocarbon dates, V. Radiocarbon 6:91-107 (Leh-
              ner Site:100).

Damon, P. E., and A. Long
1962   Arizona Radiocarbon dates, III. Radiocarbon 4:239-249
              (Lehner Site:243; Murray Springs Site:245).

Daniels, L. E.
1911   Notes on *Oreohelix*. Nautilus 25 (2):18-19.
1912   Abnormal shells. Nautilus 26 (4):38-42; Pl. 5.

Dees, Lola T.
1970   Edible land snails in the United States. U.S. Dept. Interior
              Fish and Wildlife Service Resource Publ. 91:1-8 (only *Helix
              aspersa* mentioned for Arizona).

Dice, L. R.
1943   The Biotic Provinces of North America. (Ann Arbor,
              Michigan):i-viii + 1-78; 1 map.

Drake, R. J.
1948a   Mollusca of the eastern basin of the Chaco River, New
              Mexico. 1. Nautilus 62 (1):5-8.
1948b   Mexican localities of shells illustrated in F. C. Baker's
              "The Molluscan Family Planorbidae." Minutes Conchol.
              Club South. California (76):8-10.
1949a   Mollusca of the eastern basin of the Chaco River, New
              Mexico. 2. Nautilus 62 (3):94-97 (essentially a duplica-
              tion of 1948a).

1949b   A new species of *Oreohelix,* subgenus *Radiocentrum* from southeastern Chihuahua. Nautilus 62 (4):109-112; Pl. 8.

1949c   Mollusk shells found in alluvium at Buell Park, Apache Co., Arizona. Plateau 22 (2):26-31.

1951a   Three new species of *Coelostemma* from southeastern Chihuahua. Rev. Soc. Malac. "Carlos de la Torre" Habana 8 (1):39-43; Pl. 6.

1951b   A new species of *Polygyra* from west Texas. Rev. Soc. Malac. "Carlos de la Torre" Habana 8 (1):44-46.

1952a   *Haplocion yucatanensis* (Bartsch) from Coahuila. Rev. Soc. Malac. "Carlos de la Torre" Habana 8:131-132.

1952b   Change in genus for *Polygyra pasonis.* Rev. Soc. Malac. "Carlos de la Torre" Habana 9 (1):30.

1953a   Studies of the species (and distributions) of the non-marine mollusk fauna of Sonora. Amer. Philos. Soc. Yearbook for 1952:154-157.

1953b   *Amnicola brandi,* a new species of snail from northwestern Chihuahua. Journ. Washington (D.C.) Acad. Sci. 43 (1):26-28.

1954   A study of Arizona Pleistocene and Recent mollusks; a tool for local geochronological research. El Palacio 61 (11):355-367.

1956   Further study of the species of the nonmarine mollusk fauna of Sonora and the preparation of a monograph of the species of land shells in particular of Sonora. Amer. Philos. Soc. Yearbook for 1955:131-132.

1957   *Micrarionta* from northwestern Sonora. Bull. South. California Acad. Sci. 56 (2):76-80; Pls. 14-15.

1960a   Nonmarine molluscan remains from Recent sediments in
(Jan.)   Matty Canyon, Pima County, Arizona. Bull. South. California Acad. Sci. 58 (for 1959) (3):146-154; Pls. 43-44.

1960b   Nonmarine molluscan remains from an archaeological site at La Playa, northern Sonora, Mexico. Bull. South. California Acad. Sci. 59 (3):133-137; Pls. 42-43.

1961a   Nonmarine molluscan distribution in the Gulf of California region of Mexico. Amer. Philos. Soc. Yearbook for 1960:288-290.

1961b   Nonmarine molluscs from the La Playa Site, Sonora, Mexico. 2. Bull. South. California Acad. Sci. 60 (3):127-129.

1962a   Nonmarine molluscs from Recent sediments near Vernon, Apache County, Arizona. Bull. South. California Acad. Sci. 61 (1):25-28.

1962b   G. W. Horn's gastropod locality in Arizona. Bull. South. California Acad. Sci. 61 (1):44.

Dundee, D. S., and H. A. Dundee
1958   Extension of known ranges of 4 mollusks. Nautilus 72 (2):51-53 (*Corbicula fluminea* and *Sonorella coltoniana* in Arizona).

Edson, H. M.
1912   Two new land shells from California. Nautilus 26 (3):37.

Edson, H. M., and H. Hannibal
1911   A census of the land and fresh-water mollusks of south-

western California. Bull. South. California Acad. Sci. 10:47-64; Pl. 1.

Emerson, W. K., and M. K. Jacobson
1964   Terrestrial mollusks of the Belvedere Expedition to the Gulf of California. Trans. San Diego Soc. Nat. Hist. 13 (16):313-332.

Ferriss, J. H.
1904   Southwestern shells. Nautilus 18 (5):49-54.
1910   A collecting excursion north of the Grand Canyon of the Colorado. Nautilus 23 (9):109-112.
1913   (Correspondence from Arizona). Nautilus 27 (5):60.
1914a   Camps in the Catalinas and White Mountains of Arizona, with description of a new species. Nautilus 27 (10): 109-112.
1914b   Notes. Nautilus 27 (12):134-136 (odor emitted by *Sonorella*).
1915   Our New Mexican expedition of 1914. Nautilus 28 (10): 109-113.
1917   A shell hunt in the Black Range, with description of a new *Oreohelix*. Nautilus 30 (9):99-103.
1918   Camping in the Sierras and the Desert. Nautilus 32 (1):3-9; Pl. 1.
1919a   My journey to the Blue and White Mountains, Arizona. Nautilus 32 (3):81-86.
1919b   Along the Mexican border. Nautilus 33 (2):37-44.
1920   The Navajo Nation. Nautilus 34(1):1-14, and (4):109-111; 1 Pl.
1924   On the Rio Grande. Nautilus 38 (2):37-43.
1925   A new Texan *Bulimulus*. Nautilus 39(1):24-25.

Feth, J. H., and J. D. Hem
1963   Reconnaissance of headwater springs in the Gila River drainage basin, Arizona. U.S. Geol. Surv. Water-Supply Paper 1619H:H1-H54 (*Physa* in limestone, San Carlos Indian Reservation:H14).

Fischer, P., and H. Crosse
1870-1894 Mission Scientifique au Mexique et dans l'Amérique Centrale. 7me Partie. Etudes sur les mollusques terrestres et fluviatiles du Mexique et du Guatémala. (Paris), Vol. 1, Pt. 1, (1870):1-152, 6 Pls.; Pt. 2, (1872):153-304, 6 Pls.; Pts. 3-4, (1873):305-464, 8 Pls.; Pt. 5, (1875):465-546, 4 Pls.; Pt. 6, (1877):547-624, 4 Pls.; Pt. 7, (1878):625-702, 3 Pls. Vol. 2, Pt. 8, (1880):1-80, 5 Pls.; Pt. 9, (1886):81-128, 6 Pls.; Pt. 10, (1888):129-176, 4 Pls.; Pt. 11, (1890):177-256, 2 Pls.; Pt. 12, (1891):257-312, 4 Pls.; Pt. 13, (1892):313-392, 2 Pls.; Pt. 14, (1893):393-488, 4 Pls.; Pts. 15-16, (1894):489-656, 8 Pls.; Pt. 17, (1902):657-731, 7 Pls.

Forcart, L.
1955   Die nordischen Arten der Gattung *Vitrina*. Archiv f. Molluskenk. 84 (4-6):155-166; Pl. 12.
1959   Die palaearktischen Arten des Genus *Columella*. Verh. Naturf. Ges. Basel 70 (1):7-18.

Fox, R. O.
1970   *Corbicula* in Baja California. Nautilus 83 (4):145.
Frierson, L. S.
1927   A classified and annotated check list of North American
            Naiades. (Baylor University, Waco, Texas):1-111.
Frömming, E.
1954   Biologie der mitteleuropäischen Landgastropoden. (Ber-
            lin):1-404.
Gabb, W. M.
1865   Descriptions of three new species of Mexican land shells.
            Amer. Journ. Conch. 1 (3):208-209; Pl. 19.
1866   Descriptions of three new species of land shells from Ari-
            zona. Amer. Journ. Conch. 2 (4):330-331; Pl. 21.
1868   Description of new species of land shells from Lower Cali-
(Jan. 2)   fornia. Amer. Journ. Conch. 3 (3):235-238; Pl. 16.
Gerhard, P., and H. E. Gulick
1967   Lower California guidebook. (Glendale, California), 4th
            Edition (revised):1-243; 13 Pls., 17 maps (earlier editions
            in 1956, 1958, and 1962).
Gilbertson, L. H.
1965   The biology of the snail *Sonorella odorata* Pilsbry and
            Ferriss. [MS Master Sci. Dissertation, Dept. Zoology,
            Univ. of Arizona, Tucson:i-viii + 1-49].
1969   Notes on the biology of the snail *Sonorella odorata* in
            Arizona. Nautilus 83 (1):29-33.
Goldman, E. A., and R. T. Moore
1945   The biotic provinces of Mexico. Journ. Mammalogy 26:
            347-360.
Gould, A. A.
1855a   New species of land and fresh-water shells from western
            North America. Proc. Boston Soc. Nat. Hist. 5 (for 1854-
            1856):128-130 and 228-229 (publ. Oct. 1855 according to G.
            W. Tryon, 1863, Proc. Acad. Nat. Sci. Philadelphia
            15:148).
1855b   Catalogue of the Recent shells, with descriptions of the
            new species. *In* W. P. Blake, "Preliminary Geological Re-
            port, etc" (as listed for W. P. Blake), Appendix, Art. 2:22-
            28.
1856   (Same title as 1855b). *In* W. P. Blake, in R. S. William-
            son, "Report of Explorations in California, etc" (as listed
            for W. P. Blake) 5 Pt. 2 Appendix, Art. 3:330-336; Pl. 11.
1858   (Same title as 1855b) *In* W. P. Blake, "Report of a Geo-
            logical Reconnaissance in California, etc" (as listed for
            W. P. Blake), (New York) Appendix, Art. 3:330-336; Pl. 11.
1862   Otia Conchologica: descriptions of shells and mollusks,
            from 1839 to 1862. (Boston):i-iv + 1-256.
Granger, B. H.
1960   Will C. Barnes' Arizona Place Names (revised and en-
            larged). (Tucson):i-xix + 1-519.
Gray, R. S.
1965   Late Cenozoic sediments in the San Pedro Valley near

St. David, Arizona. [MS Ph. D. Dissertation, Dept. Geology, Univ. of Arizona, Tucson:i-xii + 1-198 (fossil mollusks named by D. W. Taylor:85)] (Dissertation Abstracts 26 (2):984, Aug. 1965).

1967 Petrography of the Upper Cenozoic non-marine sediments in the San Pedro Valley, Arizona. Journ. Sedim. Petrology 37:774-789 (mollusks mentioned but not named: 786).

Green, C. R., and W. D. Sellers
1964 Arizona climate. (Tucson, Arizona):i-viii + 1-503.

Gregg, W. O.
1947 The fresh water Mollusca of California; including a few forms found in adjoining areas. Minutes Conch. Club South. California 67 (March):3-21; 69 (May):3-18; 70 (June):14-20.

1949a A new and unusual helicoid snail from Los Angeles County, California. Bull. South. California Acad. Sci. 47 (for 1948) (3):100-102; Pl. 23.

1949b Helicoid snails of the desert regions of California. Ann. Rept. Amer. Malac. Un. 15 (for 1949):24 (reprinted in Sterkiana 9:36, 1963).

1950 Collecting land snails in southeastern Arizona. Ann. Rept. Amer. Malac. Un. 16 (for 1950:24-25 (reprinted in Sterkiana 9:35, 1963).

1951 A new *Sonorella* from the Chiricahua Mountains, Arizona. Bull. South. California Acad. Sci. 50 (3):156-158.

1953 Two new land snails from Arizona. Bull. South. California Acad. Sci. 52 (2):71-75.

1960 (Jan. 1) Derivation of the Helminthoglyptinae with particular reference to the desert forms. Ann. Rept. Amer. Malac. Un. 25 (for 1959):45-46.

Gregg, W. O., and W. B. Miller
1969 A new *Sonorella* from Phoenix, Arizona. Nautilus 82 (3): 90-93.

Gregg, W. O., and D. W. Taylor
1965 *Fontelicella,* a new genus of west American freshwater snails. Malacologia 3 (1):103-110.

Gregory, H. E.
1917 Geology of the Navajo country: a reconnaissance of parts of Arizona, New Mexico and Utah. U.S. Geol. Surv. Profess. Paper 93:1-181; Pls. 1-34 (*Physa* in White Cone Peak fossil fauna).

Hand, E. E.
1922 *Sonorellae* and scenery. Nautilus 35 (4):123-129.

Hanna, G. D.
1923 Expedition of the California Academy of Sciences to the Gulf of California in 1921. Land and freshwater mollusks. Proc. California Acad. Sci. (4) 12:483-527; Pls. 7-11.

1966 Introduced mollusks of western North America. Occ. Papers California Acad. Sci. 48:1-108; 4 Pls.

Hanna, G. D., and A. G. Smith
1968   The Diguet-Mabille land and freshwater mollusks of Baja California. Proc. California Acad. Sci. (4) 30 (18):381-399.
1971   Note on *Planorbis mysarus* Mabille, 1895. The Veliger 13 (4):369-370.

Hannibal, H.
1912a   The aquatic molluscs of southern California. Bull. South.
(Jan.)   California Acad. Sci. 11 (1):18-46.
1912b   A synopsis of the Recent and Tertiary freshwater Mollusca of the Californian Province, based upon an ontogenetic classification. Proc. Malacol. Soc. London 10 (2, for June): 112-165, and (3, for Oct.):167-211; Pls. 5-8.

Harry, H. W., and B. Hubendick
1964   The freshwater pulmonate Mollusca of Puerto Rico. Göteborgs K. Vet.- o. Vitterh.-Samh. Handl. (6) Sect. B, 9 (5):1-77; 16 Pls. (synonymy of *Helicodiscus lineatus sonorensis* suggested:33).

Hastings, J. R.
1964a   Climatological data for Baja California. Univ. Arizona, Inst. Atmosph. Physics Techn. Repts. Meteor. Climat. Arid Regions 14:i-ix + 1-132.
1964b   Climatological data for Sonora and northern Sinaloa. Univ. Arizona, Inst. Atmosph. Physics Techn. Repts. Meteor. Climat. Arid Regions 15: i-viii + 1-52.

Hastings, J. R., and R. R. Humphrey
1969   Climatological data and statistics for Baja California. Univ. Arizona, Inst. Atmosph. Physics Techn. Repts. Meteor. Climat. Arid Regions 18:i-iv + 1-95.

Hastings, J. R., and R. M. Turner
1965   Seasonal precipitation regimes in Baja California, Mexico. Geografiska Annaler (Stockholm) 47 Ser. A 4:204-223.

Haury, E. W.
1950   The stratigraphy and archaeology of Ventana Cave, Arizona. (Albuquerque, New Mexico):i-ix + 1-599; 66 Pls. (*Anodonta* and *Sonorella* at pre-Columbian Indian Site:362; the specific identifications questionable).

Haury, E. W., E. B. Sayles, and W. W. Wasley
1959   The Lehner Mammoth Site, southeastern Arizona. American Antiquity 25 (1):2-30 and (references) 39-42.

Haynes, C. V., Jr.
1968   Preliminary report on the Late Quaternary geology of the San Pedro Valley, Arizona. *In* Arizona Geological Society, Southern Arizona Guidebook, III (S. R. Titley, Editor):79-96.
1969   The earliest Americans. Science 166 (3906):709-715.
1970   Geochronology of man-mammoth sites and their bearing on the origin of the Llano complex. *In* Pleistocene and Recent environments of the Central Great Plains (W. Dort, Jr. and J. K. Jones, Editors), Dept. Geol. Univ. Kansas Special Publ 3:77-92.

Haynes, C. V., Jr., and E. T. Hemmings
1968   Mammoth-bone shaft wrench from Murray Springs, Arizona. Science 159 (3811):186-187.

Heindl, L. A., and J. F. Lance
1960   Physiographic and structural subdivisions of Arizona. Arizona Geol. Soc. Digest 3:12-18.

Hemmings, E. T.
1970   Early man in the San Pedro Valley, Arizona. [MS Ph. D. Dissertation, Dept. Anthropology, Univ. Arizona, Tucson:i-xv + 1-236] (Dissertation Abstracts 31 (2):1029B, 1970).

Henderson, J.
1912   Mollusca from northern New Mexico. Nautilus 26 (7): 80-81.

1914   A new *Sonorella* from the Grand Canyon, Arizona. Nautilus 27 (11):122-124.

1917   A new Pleistocene mollusk locality in New Mexico. Nautilus 30 (12):134-135.

1918   On the North American genus *Oreohelix.* Proc. Malacol. Soc. London 13 (1-2):21-24.

1928   Molluscan Provinces in the western United States. Nautilus 41 (3):85-91.

1931   Molluscan Provinces in the western United States. Univ. Colorado Studies 18 (4):177-186.

1933   *Lampsilis* at old New Mexican camp sites. Nautilus 46 (3):107.

1935   Fossil non-marine Mollusca of North America. Geol. Soc. America Special Papers 3:i-iv + 1-313.

1939   The Mollusca of New Mexico and Arizona. *In* D. D. Brand and F. E. Harvey, "So Live the Works of Man" (Albuquerque, New Mexico):187-194.

Herre, A. W. C. T.
1923   Lichens, impossible plants. Scientific Monthly 16 (2): 130-140.

Herrington, H. B.
1950   Some wrong identifications of Sphaeriidae. Nautilus 63 (4):115-119.

1962   A revision of the Sphaeriidae of North America. Misc. Public. Mus. Zool. Univ. Michigan 118:1-74; Pls. 1-7.

Hibbard, C. W., and D. W. Taylor
1960   Two Late Pleistocene faunas from southwestern Kansas. Contrib. Mus. Paleont. Univ. Michigan 16 (1):1-223; 16 Pls.

Hoff, C.C.
1961   Place names used by Pilsbry in his Monograph of the Land Mollusca of North America. Bull. New Mexico Acad. Sci. 2:50-57.

1962   Some terrestrial Gastropoda from New Mexico. Southwestern Natur. 7 (1):51-63.

Hopkins, D. M. (Editor)
1967   The Bering land bridge. (Stanford, California):i-xiii + 1-495.

Howard, E. B.
1935   Evidence of Early Man in North America. The Museum Journal Univ. Mus. Univ. Pennsylvania 24:61-175.

Hubendick, B.
1951   Recent Lymnaeidae. Their variation, morphology, taxonomy, nomenclature, and distribution. Kongl. Svenska Vet. Ak. Handl. (Stockholm) (4) 3:1-223.

1964   Studies in Ancylidae. The subgroups. Göteborgs K. Vet.-o. Vitterh.-Samh. Handl. (6) Section B 9, (6):1-72.

Hunt, C. B.
1956   Cenozoic geology of the Colorado Plateau. U.S. Geol. Surv. Profess. Paper 279:1-99 (White Cone Peak fossils: 29, after A. B. Reagan, 1932).

Ingram, W. M.
1946a   Mollusk food of the beetle, *Scaphinotus interruptus* (Men.). Bull. South. California Acad. Sci. 45 (1):34-36.

1946b   A check list of the helicoid snails of California from Henry A. Pilsbry's Monograph. Bull. South. California Acad. Sci. 45(2):61-93.

Ingram, W. M., L. Keup, and C. Henderson
1964   Asiatic clams at Parker, Arizona. Nautilus 77 (4):121-125.

Jaeger, E. C.
1933   The California Deserts. A visitor's handbook. (Stanford, California):i-x + 1-207 (revised editions, 1938 and 1955: 1-209).

1957   The North American Deserts. (Stanford, California):i-x + 1-308.

James, J. F.
1882   The Colorado Desert. Pop. Sci. Monthly 20:384-390.

Jameson, D. A.
1969   Rainfall patterns on vegetation zones in northern Arizona. Plateau 41 (3):105-111.

Jenkins, O. P.
1923   Verde River lake beds near Clarkdale, Arizona. Amer. Journ. Sci. (5) 5 (25), Art. 6:65-81 (poor freshwater gastropods in Verde Formation:76-77).

C. W. J. (Johnson, C. W.)
1905   Edward H. Ashmun. Nautilus 18 (11):121-122.

Johnson, R. I.
1964   The Recent Mollusca of Augustus Addison Gould. U.S. (July)   Nat. Mus. Bull. 239:1-182; Pls. 1-45.

Jones, D. T.
1935   Burrowing of snails. Nautilus 43(4):140-142.

1940   Recent collections of Utah Mollusca, with extralimital records from certain Utah cabinets. Proc. Utah Acad. Sci. Arts Lett. 17:33-45 (mollusks of TaBiko Canyon, Navajo Co., collected by M. Wetherill).

Kaltenbach, H.
1934   Die individuelle ökologische und geographische Varia-
            bilität der Wüstenschnecken *Eremina desertorum, has-*
            selquisti und *zitteli.* Archiv f. Naturgesch. (N.F.) 3(3):
            383-404 (land snails in Egyptian Desert).
1943   Beitrag zur Kenntnis der Wüstenschnecken *Eremina de-*
(Jan. 31)   *sertorum* und *hasselquisti* mit ihren individuellen, ökolo-
            gischen und geographischen Rassen. Archiv f. Naturgesch.
            (N.F.) 11 (for 1942) (4):350-386 (land snails in Egyptian
            Desert).

Karlin, E. J.
1961   Ecological relationship between vegetation and the dis-
            tribution of land snails in Montana, Colorado and New
            Mexico. American Midland Natur. 65(1):60-66.

Keen, A. Myra
1958   Sea shells of tropical West America. (Stanford, Cali-
            fornia):i-viii + 1-624; 10 Pls. (*Cerithidea* and Ellobiidae
            of Gulf of California).

King, P. B.
1948   Geology of southern Guadalupe Mountains, Texas. U.S.
            Geol. Surv. Prof. Paper 215:i-vi + 1-183; 23 Pls. (mol-
            lusks:145).

Knechtel, M. M.
1936   Geological relations of the Gila Conglomerate in south-
            eastern Arizona. Amer. Journ. Sci. (5) 31:81-92 (fossil
            mollusks in Graham Co.:88).
1938   Geology and ground-water resources of the valley of Gila
            River and San Simon Creek, Graham Co., Arizona. U.S.
            Geol. Surv. Water-Supply Paper 796F:181-222 (fossil mol-
            lusks:199).

Kottlowski, F. E., M. E. Cooley, and H. V. Ruhe
1965   Quaternary geology of the Southwest. In H. E. Wright
            and D. G. Frey, "The Quaternary of the United States"
            :287-298.

Künkel, K.
1916   Zur   Biologie der Lungenschnecken. (Heidelberg):1-440.
Lance, J. F.
1959   Faunal remains from the Lehner Mammoth Site. Ameri-
            can Antiquity 25 (1):35-39.
1960   Stratigraphic and structural position of Cenozoic fossil
            localities in Arizona. Arizona Geol. Soc. Digest 3:155-159
            (location of 35 Cenozoic Sites; mollusks not mentioned).

Lea, I.
1852   Descriptions of new species of the family Unionidae.
            Trans. Amer. Philos. Soc. (N.S.) 10 (for 1853), Art. 18:
            253-294; Pls. 12-29 (*Anodonta californiensis*:286).

Lehner, R. E.
1958   Geology of the Clarkdale Quadrangle, Arizona. Bull. U.S.
            Geol. Surv. 1021N:511-592; Pls. 45-47 (mollusks from Verde
            Formation:561, after R. H. Mahard, 1949).

Leonard, A. B., and J. C. Frye
   1962   Pleistocene molluscan faunas and physiographic history
             of Pecos Valley in Texas. Univ. Texas Bur. Econ. Geol.
             Rept. Invest. 45:1-42.

Leonard, A. B., and Tong-Yun Ho
   1960a   A new species of *Callipyrgula* from the Pleistocene of
             Texas. Nautilus 73 (3):110-113; Pl. 11.

   1960b   New *Callipyrgula* from Pleistocene of Texas and notes
             on *Cochliopa riograndensis*. Nautilus 73 (4):125-129; Pls.
             12-13.

Lewis, J.
   1875a   (*Anodonta dejecta*). Field and Forest 1 (2-3):26-27.

   1875b   Description of a new species of *Anodonta*. *In* H. C. Yar-
             row, 1875 (as cited below):952.

Lindroth, C. H.
   1957   The faunal connections between Europe and North Amer-
             ica. (New York and Stockholm):1-344.

Long, A.
   1965   Smithsonian Institution radiocarbon measurements, II.
             Radiocarbon 7:245-256 (shells of Willcox Playa sediment:
             254).

   1966   Late Pleistocene and Recent chronologies of Playa lakes
             in Arizona and New Mexico. [MS Ph. D. Dissertation,
             Dept. Geol., Univ. Arizona, Tucson:i-xi + 1-141 (*Sphaer-
             ium, Gyraulus, Lymnaea, Succinea, Vertigo* in sediment
             at W side of Willcox Playa, the shells radiocarbon dated
             ca 10,000 yrs old:16, 29, 86, 122-123)].

Long, G. E., and D. W. Taylor
   1970   Estuarine mollusks of the Cholla Bay, Sonora, Mexico.
             The Echo, Abstr. Proc. 2nd Ann. Meet. Western Soc.
             Malacol. (1969):17-18 (*Assiminea californica, Melampus
             mousleyi, Marinula rhoadsi, Cerithidea mazatlanica* at
             Cholla Bay).

Lowe, C. H.
   1955   The eastern limit of the Sonoran Desert in the United
             States. Ecology 36:343-345.

   1961   Biotic communities in the sub-Mogollon region of the
             inland Southwest. Journ. Arizona Acad. Sci. 2:40-49.

   1968   Fauna of desert environments. *In* W. G. McGinnies, B. J.
             Goldman, and P. Paylore, "Deserts of the World." An ap-
             praisal of research into their physical and biological en-
             vironments. (Tucson, Arizona):567-645.

Lowe, C. H. (Editor)
   1964   The vertebrates of Arizona. (Tucson, Arizona):i-vii +
             1-259 (new printings, without changes, 1965, 1967).

Lowe, H. N.
   1934   On the Sonoran side of the Gulf. Nautilus 48 (1):1-4,
             and (2):43-46.

Mabille, J.
   1895   Mollusques de la Basse Californie, recueillis par M. Diguet.
             Bull. Soc. Philomath. Paris (Ser. 8) 7 (2):54-76.

MacMillan, G. K.
1946   Notes on some southwestern Pupillidae. Nautilus 59 (4): 121-124.

Mahard, R. H.
1949   Late Cenozoic chronology of the upper Verde Valley, Arizona. Journ. Denison Univ. Sci. Labor. 41 Art. 7:97-128; 12 Pls. (Pleistocene *Sphaerium, Lymnaea, Physa, Gyraulus, Pupilla?*, named to genera only by Teng-Chien Yen: 119, from lake beds of Verde Formation, 1 mi N of Clarkdale Smelter, N side of Verde Riv).

Mallery, T. D.
1936   Rainfall records for the Sonoran desert. Ecology 17 (1): 110-121; (2):212-215.

Marshall, W. B.
1929   Three new land shells of the genus *Oreohelix* from Ari-
(Oct.)   zona. Proc. U.S. Nat. Mus. 76 (for 1930) (2802):1-3; Pl. 1.

Martens, E. von
1890-   Biologia Centrali-Americana. Terrestrial and Fluviatile
1901   Mollusca. (London):i-xxviii + 1-706; Pls. 1-44 (1890:1-40; 1891:41-96; 1892:97-176; 1893:177-248; 1897:249-288; 1898: 289-368; 1899:369-472; 1900:473-616; 1901:617-706 + i-xxviii).

Martin, P. S.
1963   The last 10,000 years; a fossil pollen record of the American Southwest. (Tucson, Arizona):1-87.

Martin, P. S., and P. J. Mehringer, Jr.
1965   Pleistocene pollen analysis and biogeography of the Southwest. *In* H. E. Wright and D. G. Frey, "The Quaternary of the United States" (Princeton, New Jersey):433-451.

Mead, A. R.
1951   Recent invasion of Arizona by the European brown snail. Ann. Rept. Amer. Malac. Un. 17 (for 1951):21-22.
1952a   Two new records of foreign mollusks in Arizona. Journ. Colorado-Wyoming Acad. Sci. 4 (4):90.
1952b   Foreign mollusks in Arizona. Ann. Rept. Amer. Malac. Un. 18 (for 1952):30 (reprinted in Sterkiana, 1963, 9:27).
1953   Additional introductions of foreign snails into Arizona. Ann. Rept. Amer. Malac. Un. 19 (for 1953):11-12 (reprinted in Sterkiana, 1963, 9:27).
1959   The appearance of the giant African snail in Arizona. Proc. Hawaiian Ent. Soc. 17 (for 1958):85-86.
1971a   Helicid land mollusks introduced into North America. The Biologist 53 (3):104-111 (abstract in Ann. Rept. Amer. Malac. Un. No. 36 for 1970, Febr. 18, 1971:55).
1971b   Status of *Achatina* and *Rumina* in the United States. The Biologist 53 (3):112-117 (abstract in Ann. Rept. Amer. Malac. Un. No. 36 for 1970, Febr. 18, 1971:56).

Mearns, E. A.
1907   Mammals of the Mexican Boundary of the United States; a descriptive catalogue of the species of mammals occurring in that region, with a general summary of the

natural history and a list of trees. Bull. U.S. Nat. Mus. 56:i-xv + 1-530; Pls. 1-13 (location of Arizona stations of mollusks).

Mehringer, P. J., Jr., and C. V. Haynes, Jr.
1965   The pollen evidence for the environment of Early Man and extinct mammals at the Lehner Mammoth Site, southeastern Arizona. American Antiquity 31 (1):17-23.

Mehringer, P. J., Jr., P. S. Martin, and C. V. Haynes, Jr.
1967   Murray Springs, a mid-Postglacial pollen record from southern Arizona. Amer. Journ. Sci. 265:786-797.

Meinzer, O. E.
1922   Map of the Pleistocene lakes of the Basin and Range Province and its significance. Bull. Geol. Soc. America 33 (3):541-552.

Melton, M. A.
1965   The geomorphic and paleoclimatic significance of alluvial deposits in southern Arizona. Journ. of Geology 73:1-38; Pls. 1-3.

Merriam, C. H.
1890   Results of a biological survey of the San Francisco Mountain region and desert of the Little Colorado in Arizona. North American Fauna (U.S. Dept. Agric.) 3:1-128.

Metcalf, A. L.
1966   *Corbicula manilensis* in the Mesilla Valley of Texas and New Mexico. Nautilus 80(1):16-20.

1967   Late Quaternary mollusks of the Rio Grande Valley, Caballo Dam, New Mexico to El Paso, Texas. Univ. Texas at El Paso, Science Ser. 1:1-62 (mollusks in alluvium of trans-Pecos Texas, New Mexico, and Greenlee Co., Arizona).

1968   Mollusca of El Paso County, westernmost Texas. Ann. Rept. Amer. Malac. Un. Bull. 35:32-33.

1969   Quaternary surfaces, sediments, and mollusks: southern Mesilla Valley, New Mexico, and Texas. New Mexico Geol. Soc. Field Conference Guidebook 20:158-164.

1970a   Field journal of Henry A. Pilsbry pertaining to New Mexico and trans-Pecos Texas. Sterkiana No. 39:23-37.

1970b   Late Pleistocene (Woodfordian) gastropods from Dry Cave, Eddy County, New Mexico. Texas Journ. Sci. 22 (1):41-46.

1972   Journals of Henry A. Pilsbry pertaining to Arizona. Sterkiana No. 45:21-31.

Metcalf, A. L., and P. A. Hurley
1971   A new *Ashmunella* (Polygyridae) from Doña Ana County, New Mexico. Nautilus 84 (4):120-127.

Metcalf, A. L., and W. E. Johnson
1971   Gastropods of the Franklin Mountains, El Paso County,
(July)   Texas. Southwestern Natur. 16 (1):85-109.

Metzger, D. G.
1931   Basin sediments near Cibola, Arizona. Special Papers Geol. Soc. America 73:51-52 (unnamed fossil mollusks).

Miles, C. D.
1961a   The occurrence of head-warts on the land snail *Rumina*

decollata (L.) from Arizona (U.S.A.). Journ. de Conchyliologie 101:179-182; Pl. 2.

1961b Regeneration and lesions in pulmonate gastropods. [MS Ph. D. Dissertation, Dept. Zoology, Univ. Arizona, Tucson:i-ix + 1-206]. Dissertation Abstracts, 22:1757 (experiments with *Rumina decollata, Helix aspersa* and *Sonorella odorata* in Arizona).

Miles, C. D., and A. R. Mead

1960 New *Pallifera* (*Pancalyptus*) from Arizona. Nautilus 74
(Oct.) (2):75-78 (Pl. in Pt. 3, Jan. 1961).

1961 Rediscovery of Pilsbry's *Philomycus* (*Pallifera*) *arizonen-*
(Jan. 1) *sis.* Ann. Rept. Amer. Malac. Un. 27 (for 1960):25.

1963 Type material of the slug *Pallifera pilsbryi.* Nautilus 76 (3):112-113.

Miller, W. B.

1965 Preliminary observations on *Sonorella* in Arizona and Mexico. Ann. Rept. Amer. Malac. Un. 32 (for 1965):50-51.

1966 Three new *Sonorella* from southwest Arizona. Nautilus
(Oct.) 80 (2):46-52; Pls. 1-2.

1967a Two new *Sonorella* from Sonora, Mexico. Nautilus 80
(Apr.) (4):114-119.

1967b Two new *Sonorella* from Sonora, Mexico, with notes on
(July) southern limits of genus. Nautilus 81 (1):1-6; Pl. 1.

1967c Two new *Sonorella* from Rincon Mountains of Arizona.
(Oct.) Nautilus 81 (2):54-61.

1968a Anatomical revision of the genus *Sonorella.* [MS 1967 Ph. D. Dissertation at Dept. Biological Sciences, Univ. Arizona, Tucson:i-xiii + 1-293; 107 figs.] (Dissertation Abstracts, 28:3530 B, 1968).

1968b New *Sonorella* from Arizona, Nautilus 82 (2):50-52 and 59-63.

1969 A new *Sonorella* from the Salt River Mountains of Phoenix, Arizona. Nautilus 82 (3):87-89.

1970 A new species of *Helminthoglypta* from the Mojave Desert. The Veliger 12 (3):275-278; Pl. 41.

Minckley, W. L.

1969 Aquatic biota of the Sonoita Creek basin, Santa Cruz County, Arizona. Nature Conservancy Ecological Studies, Leaflet No. 15:1-8 (*Physa virgata* and *Helisoma* [as *Planorbella*] *tenue*:6).

Montgomery, E. L.

1963 The geology and ground water investigation of the Tres Alamos Site of the San Pedro River, Cochise County, Arizona. [MS Master Sci. Dissertation, Dept. Geology, Univ. Arizona, Tucson:i-viii + 1-61] (The Narrows Fossil Site described, but mollusks not mentioned).

Morrison, J. P. E.

1940 A new species of *Fluminicola* with notes on "Colorado Desert" shells, and on the genus *Clappia.* Nautilus 53 (4):124-127.

1943 *Oreohelix* east of the Mississippi. Nautilus 56 (3):104.

1956   How many *Syncera* species are living in Death Valley?
              Ann. Rept. Amer. Malac. Un. 22 (for 1955):29.

Nelson, E. W.
1922   Lower California and its natural resources. Mem. Nat.
              Ac. Sci. (Washington, D.C.) 16 1st Mem.:1-194; 34 Pls.;
              1 map (Reprint, 1966, Riverside, California:i-xvi + 1-194;
              34 Pls.; 1 map).

Newcomb, W.
1865   Descriptions of new species of land shells. Proc. California
(Jan.)   Ac. Sci. (1) 3 (for 1863-1867):179-182.

Nichol, A. A.
1937   The natural vegetation of Arizona. Agric. Exper. Sta.,
              Tucson, Techn. Bull. 68:179-222; Map (2nd Edition, re-
              vised by W. S. Phillips, 1952, Techn. Bull. 127:187-230;
              map).

Noel, Martha S.
1954   Animal ecology of a New Mexico springbrook. Hydro-
              biologia 6:120-135; (*Physa integra* and *Amnicola neomexi-
              cana* near Roswell, Chaves Co.).

Orcutt, C. H.
1889a   Recent and sub-fossil shells of the Colorado Desert. West
              American Scientist 6:92-93.
1889b   (Bowers' new *Helix* at Indio). West American Scientist
              6:96.
1890   West American notes. Nautilus 4 (6):67-68.
1891   Contributions to West American Mollusca, 1. West Ameri-
              can Scientist 7:222-224.
1900   Notes and news. West American Scientist 11:28 and 70.

Pallary, P.
1924   Faune malacologique du Sinai. Journ. de Conchyliologie
              68 (3):181-217; Pls. 10-12.

Parodiz, J. J.
1954   A new species of *Humboldtiana* from Texas. Nautilus 67
              (4):107-108; Pl. 9.

Pedersen, E. P., and C. F. Royse
1970   Late Cenozoic geology of the Payson Basin, Gila County,
              Arizona. Journ. Arizona Acad. Sci. 6, (2):168-178 (un-
              named gastropods at fossil Site on Rye Creek, ca 10 mi
              S of Payson).

Péwé, T. L., and R. G. Updike
1970   Guidebook to the geology of the San Francisco Peaks,
              Arizona. Plateau 43 (2):43-102 (fossil mollusks mentioned:
              60).

Pilsbry, H. A.
1889   Nomenclature and check-list of North American land
              shells. Proc. Acad. Nat. Sci. Philadelphia 41:191-210.
1890a   Two new species of United States land shells. Nautilus
              4 (1):3-4.
1890b   New and little-known American mollusks, No. 3. Proc.
              Acad. Nat. Sci. Philadelphia 42:296-302; Pl. 5.
1890c   Note on a southern *Pupa*. Proc. Acad. Nat. Sci. Phila-
              delphia 42:44-45; Pl. 1.

1891   On *Helix harfordiana* and other shells. Nautilus 5 (4):
            39-42; Pl. 2.

1893-   Manual of Conchology (2) 9, Guide to the study of the
1895   helices. (Philadelphia):i-xlviii + 1-366 + 1-126; Pls. 1-71;
            (1893:1-48; 1894:49-160; 1895:161-366 + i-xlviii + index:1-
            126).

1894   New forms of western helices. Nautilus 8 (7):81-82.

1897-   A classified catalogue of American land shells with lo-
1898a   calities. Nautilus 11 (1897):45-48, 59-60, 71-72, 83-84, 90-
            96; (1898):105-108, 117-120, 127-132, 138-144.

1897-   Manual of Conchology (2) 11, American Bulimulidae:
1898b   *Bulimulus, Neopetraeus, Oxychona* and South American
            Drymaeus. (Philadelphia):1-339; Pls. 1-51 (1897:1-144;
            1898:145-339).

1898c   Descriptions of new species and varieties of American
            Zonitidae and Endodontidae. Nautilus 12 (8):85-87.

1899a   Remarks on the American species of *Conulus.* Nautilus
            12 (10):113-117.

1899b   Catalog of the Amnicolidae of the western United
            States. Nautilus 12 (11):121-127.

1899c   (New*Mexican helicoid land snail . . . . . named Ashmun-*
(Feb.)   *ella*). Science (N.S.) 9:182 (with brief definition, but with-
            out mention of a species).

1899d   Confirmation of the generic characters of *Ashmunella.*
(Sept.)   Proc. Acad. Nat. Sci. Philadelphia 51:379.

1900a   *Sonorella,* a new genus of helices. Proc. Acad. Nat. Sci.
            Philadelphia, 52:556-560; Pl. 21.

1900b   New records of New Mexican snails. Nautilus 14 (7):
            82-83.

1900c   Note on *Thysanophora hornii* Gabb. Nautilus 14 (9):98-
            99.

1900d   Notes on the anatomy of the helicoid genus *Ashmun-
            ella.* Proc. Acad. Nat. Sci. Philadelphia 52:107-109.

1900e   New species and subspecies of American land snails.
            Nautilus 13 (10):114-115.

1900f   Land shells from rejectamenta of the Rio Grande at
            Mesilla, New Mexico, and of the Gallinas River at Las
            Vegas, New Mexico. Nautilus 14 (1):9-10.

1901   *Holospira minima* v. Martens. Nautilus 14 (10):118-119
            (unsigned article).

1902a   New American land shells. Nautilus 16 (3):30-33.

1902b   *Vertigo coloradensis* and *V. ingersolli.* Nautilus 16
            (5):58-59.

1902c   *Ashmunella levettei* (Bld.). Nautilus 16 (5):59-60.

1902d   Southwestern land snails. Proc. Acad. Nat. Sci. Phila-
(Oct.)   delphia 54:510-512.

1902-   Manual of Conchology (2) 15, Urocoptidae. (Philadel-
1903   phia):i-viii + 1-323; Pls. 1-66 (1902:1-128; 1903:129-323
            + i-viii).

1904a   (J. H. Ferriss' collecting in Arizona). Nautilus 17 (11):131
(March)   (first mention in print of the name *Oreohelix,* with brief

description in footnote, "for the Rocky Mountain helices of the *H. strigosa* group").

1904b   A new Lower California *Sonorella*. Nautilus 18 (5):59.

1905   Mollusca of the southwestern States. I. Urocoptidae;
(May)   Helicidae of Arizona and New Mexico. Proc. Acad. Nat. Sci. Philadelphia 57:211-290; Pls. 11-27 (reviewed by T. D. A. Cockerell, 1905b:68-71).

1906   Shells of Grant, Valencia Co., New Mexico. Nautilus
(March)   19 (11):130.

1907   On the soft anatomy of *E. (Micrarionta) hutsoni*. Nautilus 20 (12):138-139; Pl. 9.

1907-   Manual of Conchology (2) 19, Oleacinidae, Ferussaci-
1908   dae. (Philadelphia):i-xxvii + 1-366; Pls. 1-52 (1907:1-192; 1908:193-366 + i-xxvii).

1908   *Valvata humeralis californica*. Nautilus 22 (8):82.

1910   A new species of *Marinula* from near the head of the
(May)   Gulf of California. Proc. Acad. Nat. Sci. Philadelphia 62:148.

1912   Two new American land shells collected by Messrs. Hebard and Rehn. Nautilus 26 (8):88-90.

1913   Notes on some Lower Californian helices. Proc. Acad. Nat. Sci. Philadelphia 65:380-393.

1914   Shells of Duran, New Mexico. Nautilus 28 (4):37-38; Pl. 2.

1915a   A new subspecies of *Oreohelix cooperi*. Nautilus 29 (4):48.

1915b   Mollusca of the southwestern States. VI. The Hacheta Grande, Florida and Peloncillo Mountains, New Mexico. Proc. Acad. Nat. Sci. Philadelphia 67:323-350; Pls. 5-6.

1916a   Helices of Lower California and Sinaloa. Nautilus 29 (9):97-102; Pls. 2-3.

1916b   A new Californian land snail. Nautilus 29 (9):104-105; Pl. 4.

1916c   New species of *Amnicola* from New Mexico and Utah. Nautilus 29 (10):111-112; Pl. 5 (Pl. in vol. 30, Pt. 12, 1917).

1916d   On some ill-understood *Oreohelices*. Nautilus 29 (12):139-142.

1916e   Notes on the anatomy of *Oreohelix*, with a catalogue of the species. Proc. Acad. Nat. Sci. Philadelphia 68:340-359; Pls. 19-22.

1916-   Manual of Conchology (2) 24, Pupillidae (Gastrocoptinae).
1918   (Philadelphia):i-xii + 1-380; Pls. 1-49 (1916:1-112; 1917: 113-256; 1918:257-380 + i-xii).

1917a   *Philomycus* in Arizona. Nautilus 30 (10):119.

1917b   A new *Holospira* from Chihuahua. Nautilus 30 (11):124-125; Pl. 4.

1917c   Notes on the anatomy of *Oreohelix*. II. Proc. Acad. Nat. Sci. Philadelphia 69:42-46.

1918a   On the generic position of *Sonorella wolcottiana* Bartsch. Proc. Acad. Nat. Sci. Philadelphia 70:139-140.

1918-   Manual of Conchology (2) 25, Pupillidae (Gastrocoptinae,
1920   Vertigininae). (Philadelphia):i-v + 1-404; Pls. 1-34 (1918: 1-64; 1919:65-224; 1920:225-404 + i-v).

1919a   A new Californian *Micrarionta*. Nautilus 33 (2):53.

1919b   Shells from the Chiricahuan Mountains, Arizona. Nautilus 33 (2):70.

1920-   Manual of Conchology (2) 26, Pupillidae (Vertigininae, 1921   Pupillinae) (Philadelphia):i-iv + 1-254; Pls. 1-24 (1920:1-64; 1921:65-254 + i-iv).

1921   Land shells from Palm Canyon, California and the Grand Canyon. Nautilus 35 (2):48.

1922-   Manual of Conchology (2) 27, Pupillidae (Orculinae, 1926   Pagodulinae, Acanthinulinae, etc.). (Philadelphia):i-v + 1-369; Pls. 1-32 (1922:1-80; 1923:81-128; 1924:129-176; 1926:177-369 + i-v).

1925   R[*adiodiscus*] *orizabensis*. Nautilus 39 (1):28.

1926a   Costa Rican land shells collected by A. A. Olsson. Proc. Acad. Nat. Sci. Philadelphia 78:127-133; Pl. 11 (adult *Radiodiscus millecostatus* from Miller Canyon, Huachuca Mts:132, 2 figs. 4A).

1926b   James H. Ferriss. Nautilus 40 (1):1-6; portrait.

1926c   The land mollusks of the Republic of Panama and the Canal Zone. Proc. Acad. Nat. Sci. Philadelphia 78:57-126; Pls. 9-10 (*Thysanophora hornii* in Arizona:113).

1927a   Expedition to Guadalupe Island, Mexico, in 1922. Land and freshwater mollusks. Proc. California Acad. Sci. (4) 16 (7):159-203; Pls. 6-12.

1927b   The structure and affinities of *Humboldtiana* and related helicid genera of Mexico and Texas. Proc. Acad. Nat. Sci. Philadelphia 79:165-192; Pls. 11-14.

1927-   Manual of Conchology (2) 28, Geographical distribution 1935   of Pupillidae, etc. (Philadelphia):1-226; Pls. 1-31 (1927: 1-48; 1928:49-96; 1934:97-160; 1935:161-226).

1928a   Mexican mollusks. Proc. Acad. Nat. Sci. Philadelphia 80:115-117.

1928b   Helices from California and Texas and a zonitid from Virginia. Nautilus 41 (3):81-83.

1931   Landshells collected by H. N. Lowe in western Mexico. Nautilus 44 (3):81-84 (with notes by H. N. Lowe).

1932a   *Physa humerosa interioris* "Ferriss, 1920." Nautilus 45 (4):139; Pl. 11.

1932b   A New Mexican *Ashmunella*. Nautilus 46 (1):19.

1934   Notes on the anatomy of *Oreohelix*. III. With descrip-
(March)   tions of new species and subspecies. Proc. Acad. Nat. Sci. Philadelphia 85 (for 1933):383-410; Pls. 14-15.

1935a   *Ashmunella metamorphosa* Pilsbry. Nautilus, 49 (2):67-68.

1935b   Western and southwestern Amnicolidae and a new *Humboldtiana*. Nautilus 48 (3):91-94 (see: W. O. Gregg, 1941, Nautilus 54 [4]:118).

1935c   Descriptions of Middle American land and freshwater Mollusca. Proc. Acad. Nat. Sci. Philadelphia 87:1-6; Pl. 1.

1936a   Land shells from Texas and New Mexico. Nautilus 49 (3):100-102.

1936b   The eastern limit of *Sonorella.* Nautilus 49 (4):109-110.

1939-   Land Mollusca of North America (North of Mexico).
1948   Acad. Nat. Sci. Philadelphia, Monograph No. 3. Vol. 1, Pt. 1 (1939):i-xvii + 1-573 + i-ix; Pt. 2 (1940):i-iv + 575-994 + i-ix; Vol. 2, Pt. 1 (1946):i-vi + 1-520, 1 Pl.; Pt. 2 (1948):i-xlvii + 521-1113 (cited as "Land Moll. North America").

1948b   Inland mollusks of northern Mexico. I. The genera *Humboldtiana, Sonorella, Oreohelix* and *Ashmunella.* Proc. Acad. Nat. Sci. Philadelphia 100:185-203; Pls. 12-14.

1950   *Pseudosubulina,* a genus new to the United States. Nautilus 64 (1):55-56.

1952   A *Holospira* new to the United States. Nautilus 66 (2):69-70 (with Pl. 6 in 66 [3], 1953).

1953   Inland Mollusca of northern Mexico. II. Urocoptidae, Pupillidae, Strobilopsidae, Valloniidae and Cionellidae. Proc. Acad. Nat. Sci. Philadelphia 105:133-167; Pls. 3-10.

1954   *Holospira riograndensis.* Nautilus 68 (1):34.

1956   Inland Mollusca of northern Mexico. III. Polygyridae and Potadominae. Proc. Acad. Nat. Sci. Philadelphia 108:19-40; Pls. 2-4.

Pilsbry, H. A., and E. P. Cheatum

1951   Land snails from the Guadalupe Range, Texas. Nautilus 64 (1):87-90; Pl. 4 (Pl. in 64 [2], 1951).

Pilsbry, H. A., and T. D. A. Cockerell

1899a   A new genus of helices. Nautilus 12 (9):107 (first valid
(Jan.)   publication of *Ashmunella,* with a type species; signed "H.A.P. and T.D.A.C.").

1899b   *Ashmunella,* a new genus of Helices. Proc. Acad. Nat.
(June)   Sci. Philadelphia, 51:188-194.

1899c   Another new *Ashmunella.* Nautilus 13 (5):49-50.
(Sept.)

1900   Records of Mollusca from New Mexico. Nautilus 14 (8):85-86.

Pilsbry, H. A., and J. H. Ferriss

1906   Mollusca of the southwestern States. II. Proc. Acad. Nat. Sci. Philadelphia, 58:123-175 (123-160, June 20; 161-175, July 24); Pls. 5-11 (reviewed by T. D. A. Cockerell, 1910, Nautilus 24 (5):70-72).

1907   Notes on some New Mexican *Ashmunellas.* Nautilus 20 (12):133-135.

1908   A new *Micrarionta* from Arizona. Nautilus 21 (12):134-136; Pl. 11.

1909   Mollusks from around Albuquerque, New Mexico. Nautilus 22 (10):103-104.

1910a   Mollusca of the southwestern States. III. The Huachuca
(Jan.)   Mountains, Arizona. Proc. Acad. Nat. Sci. Philadelphia 61 (for 1909):494-516; Pls. 19-22.

1910b A new *Sonorella* from the Rincon Mountains, Arizona.
(Jan.)   Proc. Acad. Nat. Sci. Philadelphia 61 (for 1909):517-518;
         Pl. 22.
1910c   Mollusca of the southwestern States. IV. The Chirica-
(Apr.) hua Mountains, Arizona. Proc. Acad. Nat. Sci. Phila-
        delphia 62:44-147; Pls. 1-14.
1911   Mollusca of the southwestern States. V. The Grand Can-
(May)   yon and northern Arizona. Proc. Acad. Nat. Sci. Phila-
        delphia 63:174-199; Pls. 12-14.
1915a   Mollusca of the southwestern States. VII. The Dragoon,
(July) Mule, Santa Rita, Baboquivari and Tucson Ranges, Ari-
        zona. Proc. Acad. Nat. Sci. Philadelphia 67:363-418; Pls.
        8-15.
1915b   The New Mexican Expedition of 1914 — *Ashmunella*.
        Nautilus 29 (2):13-16; Pls. 1-2; (3):29-35; and (4):41-43 (*A.
        pilsbryana*:16; misspelled *"pilsbryi"* on Pl. 2).
1917   Mollusca of the southwestern States. VIII. The Black
(Apr.) Range, New Mexico. Proc. Acad. Nat. Sci. Philadelphia
        69:83-107; Pls. 7-10.
1919a   Mollusca of the southwestern States. IX. The Santa
(Feb.) Catalina, Rincon, Tortillita and Galiuro Mountains. X.
        The mountains of the Gila headwaters. Proc. Acad. Nat.
        Sci. Philadelphia 70 (for 1918):282-333; Pls. 3-7.
1919b   New land snails collected by the Ferriss and Hinkley
        Expedition of 1919. Nautilus 33 (1):19-21.
1923   Mollusca of the southwestern States. XI. From the Tuc-
(Apr.) son Range to Ajo, and mountain ranges between the
        San Pedro and Santa Cruz Rivers, Arizona. Proc. Acad.
        Nat. Sci. Philadelphia 75:47-104; Pls. 1-8.

Pilsbry, H. A., and S. C. Field
1931   Description of a new desert helicid snail. Nautilus 45
        (1):20-21; Pl. 7 (Pl. in 44 [4], 1931).
Pilsbry, H. A., and C. W. Johnson
1898   Classified catalogue of the land shells of America North
        of Mexico. (Philadelphia):1-35 (reprint of H. A. Pilsbry,
        1897-1898).
Pilsbry, H. A., and H. N. Lowe
1934   Some desert helices of the genus *Micrarionta*. Nautilus
        48 (2):67-68.
Pilsbry, H. A., and E. G. Vanatta
1900   A partial revision of the *Pupae* of the United States.
        Proc. Acad. Nat. Sci. Philadelphia 52:582-611; Pls. 22-23.
1923   *Ashmunella hebardi,* a new snail from the Hacheta
        Grande Mountains, New Mexico. Nautilus 36 (4):119-120.
Rascop, A. M.
1960   The biology of *Rumina decollata* (Linnaeus). [MS Master
        Sci. Dissertation at Dept. Zoology, Univ. of Arizona,
        Tucson].
Rawls, H. C.
1969   Concerning the type locality of *Micrarionta rowelli hut-
        soni* (Clapp). Nautilus 82 (3):83-87.

1971   New localities for *Micrarionta rowelli mccoiana* Willett. Sterkiana 43:19.

Reagan, A. B.

1929   Pleistocene mollusks from Hopi Buttes. Pan-American Geologist, 51 (5):337-338; Pl. 14 (fossils of White Cone Butte, Navajo Co.).

1932   The Tertiary-Pleistocene of the Navajo country in Arizona, with a description of some of its included fossils. Trans. Kansas Acad. Sci. 35:253-259; Pl. 1 (fossil mollusks of White Cone Butte, Navajo Co.:257-258).

Reger, R. D., and G. L. Batchelder

1971   Late Pleistocene mollusks and a minimum age of Meteor Crater, Arizona. Journ. Arizona Acad. Sci. 6 (3):190-195.

Rehder, H.

1932   *Polygyra kiowaensis*. Nautilus 45 (4):141.

Richards, C. S.

1963   Apertural lamellae, epiphragms and estivation of planorbid mollusks. Amer. Journ. Trop. Med. Hyg. 12 (2):254-263.

Richards, H. G.

1936   Mollusks associated with Early Man in the Southwest. American Naturalist 70 (729):369-371 (fossils from Clovis Site, Curry Co., New Mexico, and Keet Seel Site, Navajo Co., Arizona:369-370).

Ross, P. C.

1922   Geology of the Lower Gila River region, Arizona. U.S. Geol. Surv. Prof. Paper 129:183-197 (fossil mollusks:189).

1923   The Lower Gila River. U.S. Geol. Surv. Water-Supply Paper 498:1-23 (fossil mollusks:23).

Rowell, J.

1863   (Description of a new California mollusc, *Gundlachia*
(May)   *californica*). Proc. California Acad. Sci. (1) 3 (for 1863-1867):21-22.

Russell, R. H.

1970   Zoogeography of Late Cenozoic Mollusca from the San Pedro Valley, southeastern Arizona. Journ. Arizona Acad. Sci. 6 Suppl. (Abstracts for 14th Meet., Apr. 17-18):22-23.

1971a   The appearance of *Pseudosuccinea columella* (Say) in Arizona. Nautilus 85 (2):71.

1971b   Late Pleistocene molluscan history of the San Pedro Valley, southeastern Arizona. The Echo, Abstracts Proc. 4th Ann. Meet. Western Soc. Malacol. (for 1971):29.

Sabels, D. E.

1962   Mogollon Rim volcanism and geochronology. New Mexico Geol. Soc., 13th Field Conference, Mogollon 1962 Guidebook:100-106 (mollusks in Lower Bidahochi Beds).

Seff, P.

1960   Preliminary report on the 111 Ranch beds, Graham Co., Arizona. Arizona Geol. Soc. Digest 3:137-139 (fossil mollusks:139).

1962   Stratigraphic geology and depositional environments of the 111 Ranch area, Graham Co., Arizona. [MS Ph. D. Dissertation at Dept. Geology, Univ. Arizona, Tucson: i-v + 1-171 (fossil mollusks:83-95, 1 Pl.)] (Dissertation Abstracts, 23 (4):1328, Oct. 1962).

Shields, L. M., and L. J. Gardner (Editors)
1961   Bioecology of the arid and semiarid lands of the Southwest. New Mexico Highlands Univ. Bull., Las Vegas:1-69 (papers by R. B. Cowles, R. A. Darrow, J. E. Fletcher, L. J. Gardner, E. B. Kurtz, Jr., P. S. Martin, and L. M. Shields).

Showers, L. I.
1959   The snail, *Rumina decollata* (L.), in Arizona. Ann. Rept. Amer. Malac. Un. 25 (for 1958):38 (title only; no localities).

Shreve, F.
1951   Vegetation of the Sonoran Desert. Carnegie Inst. Washington Public. 591:i-xii + 1-192; 27 maps; 37 Pls.

Simpson, C. T.
1893   A new *Anodonta*. Nautilus 6 (12):134-135.
1894   Types of *Anodonta dejecta* rediscovered. Nautilus 8 (4): 52-53.
1900   Synopsis of the Naiades, or pearly freshwater mussels.
(Oct.)   Proc. U.S. Nat. Mus. 22 (for 1900), (1205):501-1040; Pl. 18.
1914   A descriptive catalogue of the Naiades, or pearly freshwater mussels. (Detroit, Michigan):i-xi + 1-1540.

Sinclair, R. M.
1971   Annotated bibliography on the exotic bivalve *Corbicula* in North America, 1900-1971. Sterkiana 43:11-18.

Skinner, M. F.
1942   The fauna of Papago Springs Cave, Arizona. Bull. Amer. Mus. Nat. Hist. 80 (6):143-222 (fossil land snails:152).

Smith, A. G.
1958   Some land and fresh-water shells from Montezuma Castle National Monument. Minutes Conch. Club South. California 126:7-9.
1971   Note on *Micrarionta harperi* (Bryant, 1900). The Veliger 13 (2):202-203.

Smith, H. V.
1956   The climate of Arizona. Agr. Exp. Sta. Univ. Arizona, Tucson, Bull. 278:1-99.

Spence, G. C.
1928   Note on a "double-mouthed" *Holospira cockerelli* Dall. Nautilus 41 (3):94.

Springer, A.
1902   On some living and fossil snails of the genus *Physa* found at Las Vegas, New Mexico. Proc. Acad. Nat. Sci. Philadelphia 54:513-516; Pl. 26 (*P. virgata* in Salt Riv at Tempe).

Stearns, R. E. C.
1877   On the vitality of certain mollusks. American Naturalist, 11:100-102.

1879   Remarks on fossil shells from the Colorado Desert. American Naturalist, 15:141-154.

1882   On the history and distribution of the fresh water mussels and the identity of certain alleged species. Proc. California Acad. Sci. (Nov. 20):1-21; Pls. 1-6.

1883a   On the shells of the Colorado Desert and the region farther east. American Naturalist 17 (10):1014-1020.

1883b   Description of a new hydrobiinoid gastropod from the mountain lakes of the Sierra Nevada, with remarks on allied species. Proc. Acad. Nat. Sci. Philadelphia 35:171-176.

1889   Notes and comments on the distribution of *Planorbis* (*Helisoma*) *bicarinatus*. West American Scientist 6 (47): 110-112 (record from Sonora, about mouth of Yaqui Riv near Guaymas:110).

1890 (Sept.)   Scientific results of explorations by the U.S. Fish Commission's steamer Albatross. No. XVII. Descriptions of new West American land, fresh-water, and marine shells, with notes and comments. Proc. U.S. Nat. Mus. 13 (for 1890) (813):205-225; Pls. 15-17.

1891 (July)   List of North American land and fresh-water shells received from the U.S. Department of Agriculture, with notes and comments thereon. Proc. U.S. Nat. Mus. 14 (for 1891) (844):95-106.

1892   *Patula strigosa* in Arizona. Nautilus 6 (1):1-2.

1893   Report on the land and fresh-water shells collected in California and Nevada by the Death Valley Expedition, including a few additional species obtained by Dr. C. Hart Merriam and assistants in parts of the southwestern United States. North American Fauna (U.S. Dept. Agric.) 7:269-283.

1894a (Feb.)   Notes on recent collections of North American land, fresh water, and marine shells received from the U.S. Department of Agriculture. Proc. U.S. Nat. Mus. 16 (for 1893) (971):743-755.

1894b (July)   The shells of the Tres Marias and other localities along the shores of Lower California and the Gulf of California. Proc. U.S. Nat. Mus. 17 (for 1894) (996):139-204.

1894c   *Triodopsis Mesodon.*—Distribution, etc. Nautilus 8 (1): 6-8.

1901 (Dec.)   The fossil fresh-water shells of the Colorado Desert, their distribution, environment and variation. Proc. U.S. Nat. Mus. 24 (for 1902) (1256):271-299; Pls. 19-24.

Sterki, V.

1891   On *Pupa rupicola* Say and related forms. Nautilus 4 (12):139-143.

1892   Preliminary list of North American Pupidae (North of Mexico). Nautilus 6 (1):2-8.

1893 (July)   Genus *Vallonia* Risso. *In* H. A. Pilsbry, Manual of Conchology (2) 8, (Philadelphia):247-261; Pls. 32-33.

1896   Small land Mollusca from New Mexico. Nautilus 9 (10): 116.

1898a   *Bifidaria ashmuni,* a new species of Pupidae. Nautilus 12 (4):49-50.
1898b   New species of *Bifidaria.* Nautilus 12 (8):90-92.
1899   New Pupidae. Nautilus 12 (11):127-129.
1903   New North American *Pisidia.* Nautilus 17 (4):42-43.
1906   New species of *Pisidium.* Nautilus 20 (2):17-20.
1916   A preliminary catalogue of the North American Sphaeriidae. Ann. Carnegie Mus. 10:429-474.

Sterki, V., and G. H. Clapp
1909   *Bifidaria bilamellata* Sterki and Clapp, n. sp. Nautilus 22 (12):126-127; Pl. 8.

Stimpson, W.
1865a   Researches upon the Hydrobiinae and allied forms; chiefly made upon materials in the Museum of the Smithsonian Institution. Smithson. Misc. Coll. 201:1-59.
1865b   Diagnoses of newly discovered genera of gastropods, belonging to the subfamily Hydrobiinae, of the family Rissoidae. Amer. Journ. Conch. 1 (1):52-54; Pl. 8.

Taylor, D. W.
1950   Three new *Pyrgulopsis* from the Colorado Desert, California. Leaflets in Malacology 1 (7):27-33.
1957   Pliocene fresh-water mollusks from Navajo Co., Arizona. Journ. of Paleontology 31 Pt. 1 (3):654-661.
1960   Late Cenozoic molluscan faunas from the High Plains. U.S. Geol. Surv. Profess. Paper 337:i-iv + 1-94; Pls. 1-4.
1965   The study of Pleistocene nonmarine mollusks in North America. *In* H. E. Wright and D. G. Frey, "The Quaternary of the United States":597-611.
1966a   An eastern American freshwater mussel, *Anodonta,* introduced into Arizona. The Veliger 8 (3):197-198; Pl. 28.
1966b   Summary of North American Blancan nonmarine mollusks. Malacologia 4 (1):1-172; 8 Pls.
1966c   A remarkable snail fauna from Coahuila, Mexico. The Veliger 9 (2):152-228; Pls. 8-19.
1967a   Freshwater clam, *Sphaerium transversum* (Say) in Arizona. Southwestern Natur. 12 (2):202-203.
1967b   Freshwater mollusks collected by the United States Boundary Surveys. The Veliger 10 (2):152-158.
1970   West American freshwater Mollusca, 1. Bibliography of Pleistocene and Recent species. San Diego Soc. Nat. Hist. Memoir 4:1-73.

Taylor, D. W., and W. L. Minckley
1966   New world for biologists. Pacific Discovery 19 (5):18-22.

Thiele, J.
1931-   Handbuch der systematischen Weichtierkunde. (Jena):
1935   1-1154 (I, Pt. 1, 1931:1-376; Pt. 2, 1931:277-778 + i-vi. II, Pt. 3, 1934:779-1022; Pt. 4, 1935:1023-1154 + i-v).

Thompson, D. G.
1929   The Mohave Desert region, California; a geographic, geologic and hydrologic reconnaissance. U.S. Geol. Surv. Water Supply Paper 578:1-759 (freshwater mollusks of Death Valley:59).

Thompson, F. G.
1964   Systematic studies on Mexican landsnails of the genus *Holospira,* subgenus *Bostrichocentrum.* Malacologia 2 (1):131-143; 3 Pls.

Tryon, G. W.
1863   Descriptions of new species of fresh water Mollusca, be-
(June)   longing to the families Amnicolidae, Valvatidae and Limnaeidae inhabiting California. Proc. Acad. Nat. Sci. Philadelphia 15 (3) (published in late June or early July): 147-150; Pl. 1.
1866a   Descriptions of new fluviatile Mollusca. Amer. Journ. Conch. 2 (2):111-113; Pl. 10.
1866b   Monograph of the terrestrial Mollusca of the United States. Amer. Journ. Conch. 2 (4):306-327.
1867-   Monograph of the terrestrial Mollusca of the United
1868   States. Amer. Journ. Conch. 3 (1) (1867):34-80; Pls. 2-5; (1867):155-188; Pl. 7-10; (4) (1868):298-324; Pl. 21.
1887   Manual of Conchology (2) 3, Helicidae. (Philadelphia):1-313.

Twenter, F. R.
1962   New fossil localities in the Verde Formation, Verde Valley, Arizona. New Mexico Geol. Soc., 13th Field Conference, Mogollon Rim 1962 Guidebook:109-114 (34 Sites of allegedly Late Pliocene mollusks; unnamed sample in Fig. 3).

Twenter, F. R., and D. C. Metzger
1963   Geology and ground water in Verde Valley. — The Mogollon Rim Region, Arizona. U.S. Geol. Surv. Bull. 1177:i-v + 1-132; 1 Pl. (figs. of allegedly Late Pliocene mollusks from Sites listed by F. R. Twenter, 1962).

Vanatta, E. G.
1902   List of land shells collected in the Sacramento Mts, New Mexico. Nautilus 16 (5):57-58.
1915   *Pupoides inornatus* n. sp. Nautilus 29 (8):95.

Vanatta, E. G., and H. A. Pilsbry
1906   On *Bifidaria pentodon* and its allies. Nautilus 19 (1): 121-128; Pls. 6-7.

Vignal, L.
1923   De la durée de la vie chez l'*Helix spiriplana* Oliv. Journ. de Conchyliologie 67:262.

Waldén, H. W.
1961   On the variation, nomenclature, distribution and taxonomical position of *Limax (Lehmannia) valentianus* Férussac. Arkiv för Zoologi (2) 15 Pt. 1 (for 1962) (3):77-96; 1 Pl. (records from Tucson and Phoenix:90).
1963   Historical and taxonomic aspects of the land Gastropoda in the North Atlantic Region. In "North Atlantic Biota and their History" (Oxford, London, New York, and Paris):153-171.

Walker, B.
1909   Notes on *Planorbis.* II. *P. bicarinatus.* Nautilus 23 (1):1-10; (2):21-32; Pls. 1-3.

1915   A list of shells collected in Arizona, New Mexico, Texas
            and Oklahoma by Dr. E. C. Case. Occas. Pap. Mus. Zool.
            Univ. Michigan 15:1-11.
1917   Revision of the classification of the North American
            patelliform Ancylidae. Nautilus 31 (1):1-10; Pls. 1-3.
1918   A synopsis of the classification of the fresh-water Mol-
            lusca of North America, north of Mexico, and a cata-
            logue of the more recently described species, with notes.
            Misc. Publ. Mus. Zool. Univ. Michigan 6:1-213; 1 Pl.

Walton, M. L.
1963   Length of life in West American land snails. Nautilus
            76 (4):127-131.
1970   Longevity in *Ashmunella, Monadenia* and *Sonorella.*
            Nautilus 83 (3):109-112.

Webb, G. R.
1954   The life-history and sexual anatomy data on *Ashmunella,*
            with a revision of the triodopsin snails. Gastropodia 1
            (2):13-18; Pls. 7-11.

Weber, W. A.
1965   Theodore Dru Allison Cockerell, 1866-1948. Univ. Colo-
            rado Studies Bibliography 1:1-124; portrait.

Westerfelt, C. A., Jr.
1961   The biology and systematic status of *Limax (Lehman-
            nia) valentianus* Férussac 1823. [MS Master Sci. Disser-
            tation, Dept. Zoology, Univ. Arizona, Tucson:1-42].

White, C. A.
1886   On the relation of the Laramie molluscan fauna to that
            of the succeeding fresh-water Eocene and other groups.
            Bull. U.S. Geol. Surv. 34:389-442; Pl. 1-5.

Wiggins, I. L.
1969   Observations on the Vizcaino Desert and its biota. Proc.
            California Acad. Sci. (4) 36 (11):317-346; 3 Pls. (mollusks:
            322-330).

Willett, G.
1929   Description of two new land shells from southern Cali-
            fornia. Bull. South. California Acad. Sci. 28 (2):17-19.
1930a   Notes on *Micrarionta rixfordi* Pilsbry. Bull. South. Cali-
            fornia Acad. Sci. 29 (1):15-16.
1930b   The *Micrariontas* of the *indioensis* Group, with the de-
            scription of a new subspecies. Nautilus 43 (4):115-116.
1930c   Desert helicoids of the *Micrarionta hutsoni* group. Nau-
            tilus 44 (1):4-7.
1931   Two new helicoids from the Mohave Desert, California.
            Nautilus 44 (4):123-125.
1935a   Further notes on the desert snails of Riverside Co., Cali-
            fornia. Bull. South. California Acad. Sci. 34 (1):1-2.
1935b   Three new *Micrariontas* from the central Colorado Des-
            ert. Nautilus 49 (1):14-16.
1937a   A new land shell from the Riverside Mountains, Colorado
            Desert. Bull. South. California Acad. Sci. 36 (1):6-7.

1937b   *Micrariontas* of the southwest Colorado Desert. Nautilus 50 (4):122-125.

1939   *Micrariontas* of desert ranges bordering the east side of Coachella Valley and Salton Sink, California. Bull. South. California Acad. Sci. 38 (1):14-16.

1940   A new land shell from Lower California. Bull. South. California Acad. Sci. 39 (1):80-82; Pl. 12.

1941   California desert snails. Minutes Conchol. Club South. California for November, 1941 (unnumbered issue, without pagination).

Williams, H.
1936   Pliocene volcanoes of the Navajo-Hopi country. Bull. Geol. Soc. America 47:111-171; 4 Pls. (fossil freshwater mollusks of White Cone sediments:130).

Wilson, E. D.
1931   Marine Tertiary in Arizona. Science 74:567-568 (brackish water fossil mollusks in S Yuma Co., Arizona).

Woolstenhulme, J. P.
1942   New records of Mollusca. Bull. Univ. Utah 32 (11):1-14 (Arizona records of *Helisoma t. trivolvis*:12, and *Physa ampullacea*:13, need verification).

Wright, H. E., and D. G. Frey (Editors)
1965   The Quaternary of the United States. A review volume for the VII Congress of the International Association for Quaternary Research. (Princeton, New Jersey):i-x + 1-922.

Yarrow, H. C.
1875   Report upon the collections of terrestrial and fluviatile Mollusca made in portions of Colorado, Utah, New Mexico and Arizona during the years 1872, 1873 and 1874. U.S. Geogr. Geol. Surv. West of 100th Meridian, Final Reports 5:923-954.

Yates, L. G.
1890a   A new variety of *Helix carpenteri* from southern California. Nautilus 4 (5):51-52.

1890b   A new variety of *Helix*. Nautilus 4 (6):63 (completes 1890a).

Yom-Tov, Y., and M. Galun
1971   Note on feeding habits of the desert snails *Sphincterochila boissieri* Charpentier and *Trochoidea* (*Xerocrassa*) *seetzeni* Charpentier. The Veliger 14 (1):86-88; 1 Pl.

Zilch, A.
1959-
1960   Handbuch der Paläozoologie. Band 6, Gastropoda; Teil 2, Euthyneura. (Berlin):i-xii + 1-835 (1959:1-400; 1960:i-xii + 401-835).

INDEX

Main references in **bold face** type.